Lecture Notes in Physics

New Series m: Monographs

Springer-Verlag Berlin Heidelberg GmbH

The Editorial Policy for Monographs

The series Lecture Notes in Physics reports new developments in physical research and teaching - quickly, informally, and at a high level. The type of material considered for publication in the New Series m includes monographs presenting original research or new angles in a classical field. The timeliness of a manuscript is more important than its form, which may be preliminary or tentative. Manuscripts should be reasonably self-contained. They will often present not only results of the author(s) but also related work by other people and will provide sufficient motivation, examples, and applications.

The manuscripts or a detailed description thereof should be submitted either to one of the series editors or to the managing editor. The proposal is then carefully refereed. A final decision concerning publication can often only be made on the basis of the complete manuscript, but otherwise the editors will try to make a preliminary decision as definite as they can on the basis of the available information.

Manuscripts should be no less than 100 and preferably no more than 400 pages in length. Final manuscripts should preferably be in English, or possibly in French or German. They should include a table of contents and an informative introduction accessible also to readers not particularly familiar with the topic treated. Authors are free to use the material in other publications. However, if extensive use is made elsewhere, the publisher should be informed. Authors receive jointly 50 complimentary copies of their book. They are entitled to purchase further copies of their book at a reduced rate. As a rule no reprints of individual contributions can be supplied. No royalty is paid on Lecture Notes in Physics volumes. Commitment to publish is made by letter of interest rather than by signing a formal contract. Springer-Verlag secures the copyright for each volume.

The Production Process

The books are hardbound, and quality paper appropriate to the needs of the author(s) is used. Publication time is about ten weeks. More than twenty years of experience guarantee authors the best possible service. To reach the goal of rapid publication at a low price the technique of photographic reproduction from a camera-ready manuscript was chosen. This process shifts the main responsibility for the technical quality considerably from the publisher to the author. We therefore urge all authors to observe very carefully our guidelines for the preparation of camera-ready manuscripts, which we will supply on request. This applies especially to the quality of figures and halftones submitted for publication. Figures should be submitted as originals or glossy prints, as very often Xerox copies are not suitable for reproduction. For the same reason, any writing within figures should not be smaller than 2.5 mm. It might be useful to look at some of the volumes already published or, especially if some atypical text is planned, to write to the Physics Editorial Department of Springer-Verlag direct. This avoids mistakes and time-consuming correspondence during the production period.

As a special service, we offer free of charge LaTeX and TeX macro packages to format the text according to Springer-Verlag's quality requirements. We strongly recommend authors to make use of this offer, as the result will be a book of considerably improved technical quality.

Manuscripts not meeting the technical standard of the series will have to be returned for improvement.

For further information please contact Springer-Verlag, Physics Editorial Department II, Tiergartenstrasse 17, D-69121 Heidelberg, Germany.

José González   Miguel A. Martín-Delgado
Germán Sierra   Angeles H. Vozmediano

Quantum Electron Liquids and High-T_c Superconductivity

 Springer

Authors

José González
Instituto de Estructura de la Materia
CSIC, Serrano 123
E-28006 Madrid, Spain

Miguel A. Martín-Delgado
Departamento de Física Teórica I
Facultad de Ciencias Físicas
Universidad Complutense de Madrid
E-28040 Madrid, Spain

Germán Sierra
Instituto de Matemáticas y Física Fundamental
CSIC, Serrano 123
E-28006 Madrid, Spain

Angeles H. Vozmediano
Departamento de Matemáticas
Universidad Carlos III de Madrid
E-28913 Leganés (Madrid), Spain

Cataloging-in-Publication data applied for.

Die Deutsche Bibliothek - CIP-Einheitsaufnahme

Quantum electron liquids and high-t superconductivity / J.
González ...

(Lecture notes in physics : N.s. M, Monographs ; Vol. 38)
ISBN 978-3-662-14012-3   ISBN 978-3-540-47678-8 (eBook)
DOI 10.1007/978-3-540-47678-8
NE: Gonzalez, José; Lecture notes in physics / M

ISBN 978-3-662-14012-3

Originally published by Springer-Verlag Berlin Heidelberg New York in 1995
Softcover reprint of the hardcover 1st edition 1995

Typesetting: Camera-ready by the authors
SPIN: 10481151   55/3142-543210 - Printed on acid-free paper

Preface

This book originated from a course given at the Universidad Autónoma of Madrid in the Spring of 1994 and in the Universidad Complutense of Madrid in 1995. The goal of these courses is to give the non-specialist an introduction to some old and new ideas in the field of strongly correlated systems, in particular the problems posed by the high-T_c superconducting materials. As theoretical physicists, our starting viewpoint to address the problem of strongly correlated fermion systems and related issues of modern condensed matter physics is the renormalization group approach applied both to quantum field theory and statistical physics. In recent years this has become not only a powerful tool for retrieving the essential physics of interacting systems but also a link between theoretical physics and modern condensed matter physics. Furthermore, once we have this common background for dealing with apparently different problems, we discuss more specific topics and even phenomenological aspects of the field. In doing so we have tried to make the exposition clear and simple, without entering into technical details but focusing in the fundamental physics of the phenomena under study. Therefore we expect that our experience may have some value to other people entering this fascinating field.

We have divided these notes into three parts and each part into chapters, which correspond roughly to one or two lectures.

Part I, Chaps. 1–2 (A.H.V.), reviews the essentials of the Landau Fermi liquid theory and the modern approach of the renormalization group methods as applied to fermionic systems.

Part II, Chaps. 3–5 (J.G.), discusses the 1d electron systems and the Luttinger liquid concept using different techniques: the renormalization group approach, bosonization, and the correspondence between exactly solvable lattice models and continuum field theory.

Part III, Chaps. 6–11 (M.A.M.-D. and G.S.), introduces the basic phenomenology of the high-T_c compounds and the different theoretical models to explain their behaviour: Hubbard, t-J, Heisenberg. A modern review of the real-space renormalization group method is also given.

We would like to express our gratitude to all the people who have helped us through discussions but especially to J.L. Alonso, L. Brey, J.G. Esteve, J. Ferrer, G. Gómez-Santos, F. Guinea, F. Jiménez, and C. Tejedor.

M.A.M.-D. wishes to thank Artemio González-López for many computer hints in preparing part III of the manuscript and for sharing with us his access to the Alpha machine *Ciruelo* with which some of the computations in this book were carried out.

One of us (MAHV) wants to thank Xenia de la Ossa for her help with the figures and the hospitality of the Institute for Advanced Study of Princeton where her part of the manuscript was completed.

J. González
M.A. Martín-Delgado
G. Sierra
A.H. Vozmediano

Madrid
August 1995

Contents

I

III

Part I

Part 1

1. Fermi Liquid in $D \geq 2$

1.1 Introduction

One of the purposes of these lectures is to provide the non-specialized students with the basic ideas that will enable them to read and understand the papers concerning the controversial *normal* state of high-T_c superconductors. A great deal of them are concerned with the so-called Fermi liquid behavior of electronic systems (FL), together with the marginal Fermi liquid (MFL), or non-Fermi liquid (NFL). The first two chapters of the book are devoted to the description of the normal state of ordinary metals in spatial dimension $D \geq 2$. The basis for the understanding of this state is the Landau Fermi liquid theory (LFLT) settled as a phenomenological description by Landau in 1956. That will be the subject of the first paragraph. Next we will go over the quantum field theory description characterizing the Fermi liquid state by means of the fermion correlators. The second chapter is an introduction to the renormalization group (RG) both from the field theory as well as from the statistical physics point of view. In the last part of this chapter we will identify the Landau Fermi liquid theory as the low-energy effective action for an interacting Fermi system whose infrared behavior is controlled by the free Fermi gas fixed point.

The contents of the first chapter can be found in the reference books [1], and the last topic have been mostly taken from [2] and [3].

1.2 The Landau Fermi Liquid Theory

Let us start by describing the physical system of interest. The real thing is a solid seen as a set of electrons moving in the periodic lattice formed by the ions. The metallic materials are characterized by the existence of partially filled bands or, equivalently, a well defined Fermi surface. The phenomena we are interested in are low energy processes involving only the electrons close to the Fermi surface. In many cases it is a good approximation to treat the conduction electrons as an almost free Fermi system with a well defined (either continuum or discrete) momentum. The reason why is that so, i.e., how can we neglect the strong Coulomb interactions among the electrons and of electrons and lattice vibrations is the main topic of this chapter.

We shall then be concerned with a continuum homogeneous system commenting on the discrete lattice only when needed .

The Landau theory of the Fermi liquid was originally devised by Landau to provide a description of the energy levels of the liquid ^{3}He that remains liquid at such low temperatures that the de Broglie wavelenth corresponding to the thermal motion of the atoms is comparable to the interatomic distances, hence the name of quantum liquid. It constitutes a macroscopic system whose properties are determined by quantum effects. The Landau theory was later extended to treat electrons in a metal and nuclear matter. The original problem is then that of a system of interacting electrons at very low temperatures and in three spatial dimensions. The conceptual framework to be used is that of many-body physics that combines the quantum mechanical description with ideas of statistical physics. The thermodynamic quantities of the macroscopic body are determined by its — discrete — energy level spectrum E_n through the partition function. In the macrocanonical ensemble it is given by

$$Z = \sum_{n,N} exp\{-(E_n - \mu N)/T\},$$

where μ is the chemical potential.

We will be interested in equilibrium properties such as the specific heat, compressibility and static susceptibilities, as well as transport properties : electrical and thermal conductivity, Hall coefficient, dynamic spin susceptibility, etc.

1.2.1 The Magnitudes of Interest

In the forecoming discussion, a fundamental role will be played by the following magnitudes (that are in many cases related to measurable quantities):

- The dispersion relation, i.e the dependence of the energy with the momentum, for the elementary excitations of the system over its ground state $\varepsilon(\mathbf{p})$ (called the energy spectrum in field theory). As said before, it is always assumed that the momentum is a good, conserved, quantum number.

- The density of states as a function of the energy (or momentum once the dispersion relation is known) $n(\varepsilon)$ defined as the number of one-particle states per unit energy range in the system at the given energy. It contains information on the equation of state as well as on the Fermi surface.

Of major interest are, first the existence of a Fermi surface in the interacting system. Its presence can be experimentally determined and characterizes the state as a metallic state. Second, the existence (or absence) of a gap in the energy spectrum of the excitations over the Fermi surface.

1.2.2 The Landau Hypothesis

The main hypothesis of Landau is that *the interacting system evolves adiabatically from a free Fermi gas*. This has the following implications:

1. The vacuum of the interacting system has the same symmetries as the free Fermi vacuum, i.e., it is homogeneous and isotropic.

   This assumption explicitly excludes the possibility of a phase transition as the formation of charge or spin density waves or the BCS transition. These states should be seen as instabilities of the vacuum, not as a breakdown of the FLT.

2. The low energy excitations of the interacting system can be labeled with the same quantum numbers as the energy levels of the free gas. In particular they will be excitations of charge e and spin $1/2$.

   This is the strongest consequence of the Landau hypothesis and can be verified in the field theoretical approach.

The macroscopic description that follows is sufficient to determine all the thermodynamical magnitudes.

1.2.3 Review of the Fermi Gas

In order to set the notation and establish the definition of the magnitudes that will play a role in this discussion we shall here make a brief review of the relevant properties of a free Fermi gas. We start by characterizing the vacuum state and we describe afterwards the energy spectrum of its elementary excitations.

The *dispersion relation* is that of non relativistic free electrons:

$$\varepsilon(\mathbf{p}) = p^2/2m.$$

This relation allows us to introduce the *Fermi surface* in momentum space which *represents the limit of occupation of the different single-particle momentum states in the ground state of the system*. All states with momentum contained within this surface (a sphere in this free case) are occupied, all those with momentum outside this surface are unoccupied. The Fermi surface playing a major role in the discussion that follows, we shall advance here that the former definition ceases to make sense when the interaction among the fermions is taken into account.

The Fermi momentum p_F is the radius of the Fermi sphere which can be computed as a function of the density of the gas to be

$$p_F = (3\pi^2 \frac{N}{V})^{1/3} = (3\pi^n)^{1/3} \ , \tag{1.1}$$

where N is the total number of electrons in the volume V.

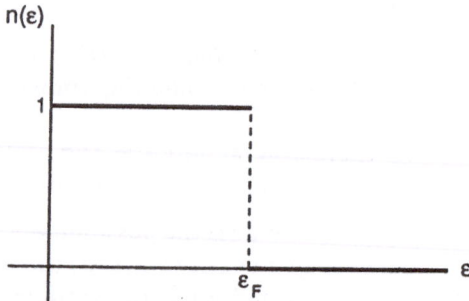

Fig. 1.1. The density of states of the free Fermi gas as a function of the energy.

exercise 1.1 *Prove the relation (1.1).*

The Fermi energy is defined by

$$\varepsilon_F = \frac{p_F^2}{2m} \quad ;$$

its order of magnitude is of a typical atomic binding energy, some electron volts[1].

The *distribution function* of the free Fermi gas is the known Fermi-Dirac distribution

$$n(\varepsilon) = \frac{1}{e^{(\varepsilon - \mu)/T} + 1} \quad , \tag{1.2}$$

where μ is the chemical potential whose value, at zero temperature, equals the Fermi energy ε_F. From (1.2) and from the dispersion relation, we can obtain the density of states of the free gas,

$$N = \frac{V}{3\pi^2} \left[\frac{2m\varepsilon}{\hbar^2} \right]^{3/2} \quad ,$$

$$D(\epsilon) = \frac{dN}{d\epsilon} = \frac{V}{2\pi^2} \left(\frac{2m}{\hbar^2} \right) \epsilon^{1/2} \quad .$$

exercise 1.2 *Show that the density of states at the Fermi level is a constant in spatial dimension $D = 2$.*

In the limit of very low temperatures the distribution function goes to a step function with a jump at the Fermi energy (**Fig. 1.1**). This is one of the characteristic features that will survive after the interaction is switched on in the LFLT.

We then characterize the ground state of the free gas by a $n(\varepsilon)$ with a discontinuity at ε_F with $\Delta n = 1$; $p_F \sim n^{1/3}$.

---

[1]The orders of magnitude and phenomenological details of most of this section can be seen in [4].

The low energy excitations of the free Fermi gas are pairs of spin 1/2 particles with momentum $p > p_F$ and "holes" with $p < p_F$ with p arbitrarily close to p_F. Any small amount of energy is sufficient to excite a pair; *the spectrum of excitations is gapless*. The dispersion relation of an excited particle with energy $\varepsilon(\mathbf{p})$ above the Fermi surface can be written in the form

$$\varepsilon(\mathbf{p}) - \mu = \frac{p^2}{2m} - \frac{p_F^2}{2m} \simeq \left.\frac{\partial E}{\partial \mathbf{p}}\right|_{p_F}(\mathbf{p} - \mathbf{p}_F) \equiv \mathbf{v}_F(\mathbf{p} - \mathbf{p}_F) , \qquad (1.3)$$

where the Fermi velocity $\mathbf{v}_F$ is defined by $\mathbf{v}_F = \left.\frac{\partial E}{\partial \mathbf{p}}\right|_{p_F}$. (The order of magnitude of v_F is of about one percent of the velocity of light).

The total energy per unit volume of the system is given by

$$E = \int n(\mathbf{p})\varepsilon(\mathbf{p})\frac{d^3\mathbf{p}}{(2\pi\hbar)^3}.$$

1.2.4 The Concept of Quasiparticles. The Phonon Analogy

We shall now derive quantitative consequences of the Landau hypothesis. The central concept in the LFLT is the idea of quasiparticle. In general for an interacting system with a large N it is not possible to determine the energy levels E_n. Nevertheless, at very low temperatures, only weakly excited states with energies very close to the ground state energy are present. In that case the energy spectrum can be determined from general grounds independently of the details of the interaction (provided that it is weak enough). The best example are phonons in a lattice. In the harmonic approximation (keeping from the interaction uniquely the terms quadratic on the deviation from the equilibrium position), we get a system of coupled oscillators with frequencies w_i such that the total energy of the lattice is given by

$$E_{ph} = \sum_{i=1}^{3N} w_i(n_i + \frac{1}{2}) . \qquad (1.4)$$

The lattice vibrations can be described as a superposition of plane waves with a wave vector $\mathbf{k}$, frequency w and an index s that labels the type of wave. The dispersion relation $w(\mathbf{p})$ has several branches (acoustical, optical, etc). In quantum mechanics each plane wave describes a set of particles, phonons, of momentum $\mathbf{p}$ and energy $w(\mathbf{p})$. An excited state of the lattice can be described as a set of independent particles, phonons, moving freely around the crystal. This leads to a set of energy levels similar to those of the ideal gas. If the anharmonic terms are taken into account, (1.4) of the energy is no longer exact and transitions between the different modes are possible. In the phonon language this describes scattering among the phonons with the possibility of creation and annihilation processes. Regarding the phonons as free particles is only an approximation that remains valid as far as the induced interactions are weak.

The role of the anharmonic terms grows as the amplitude (energy) of the oscillations increases until the phonon description ceases to make sense. We could then replace (1.4) by

$$E = \sum_{i=1}^{3N} w_i (n_i + \frac{1}{2}) + E_{incoh}.$$

The phonon description of the collective excitations of the lattice remains valid as soon as $E_{incoh} << Eph$.

We will see that the Landau quasiparticles constitute a very similar description for the collective excitations of a system of weakly interacting electrons in the conduction band of a metal. We will make frequent use of the phonon analogy in what follows.

Let us now derive quantitative consequences of the Landau hypothesis. As mentioned before, the adiabatic evolution from the free gas implies that the spectrum of elementary excitations of the liquid can be classified with the same quantum numbers as the free gas excitations, i.e., momentum $\mathbf{p}$, charge one, and spin 1/2. These elementary excitations are called quasiparticles and are to be interpreted as the quanta of the collective excitations of the Fermi liquid. *They are not the electrons of the original system.*

The Landau hypothesis has the following implications:

1. The entropy per unit volume of the system is that of a free Fermi gas.

2. By charge conservation the total number of quasiparticles is equal to the number N of electrons of the free system. This, together with the previous assumption, implies that

$$n(\epsilon) = \frac{1}{e^{(\epsilon - \mu)/T} + 1} \quad , \tag{1.5}$$

   where now the energy ϵ is a complicated functional of n to be discussed later.

3. *The ground state of the FL has a Fermi surface that encloses the same volume in momentum space than the unperturbed system.* In the isotropic case it is a sphere whose radious is given as a function of the density by the same relation (1.1) than that of the gas. The justification of this last statement is in a theorem due to Luttinger [5] that makes use of the second quantization formalism and will be discussed in the next section.

The most delicate issue in the phenomenological approach of the LFLT is to define the energy of the quasiparticles. This definition introduces a set of phenomenological parameters, the Landau parameters, that encode our ignorance of the dynamics but that can be determined by the experiments. Each of the quasiparticles has a definite momentum $\mathbf{p}$. The momentum distribution of the quasiparticles is $n(\mathbf{p})$ given by (1.5), normalized by the condition

$$\int n(\mathbf{p})d^3p/(2\pi\hbar)^3 = N/V \quad .$$

As happened in the phonon analogy, the total energy E of the liquid is not simply the sum of the energies of the quasiparticles; due to the "anharmonic" part of the interaction, E is a functional of the distribution function that can be determined as a response function. The change of E due to an infinitesimal change in the distribution function can be written as

$$\delta E = 2V \int \epsilon(\mathbf{p})\delta n(p)\frac{d^3\mathbf{p}}{2\pi\hbar^3} \quad , \tag{1.6}$$

where $\varepsilon(\mathbf{p})$ is

$$\epsilon(\mathbf{p}) = \frac{\delta E}{\delta n(\mathbf{p})} \quad , \tag{1.7}$$

and corresponds to the change in energy of the system when a single quasiparticle with momentum $\mathbf{p}$ is added. It plays the role of the Hamiltonian function of a quasiparticle in the field of the other particles. The form of (1.7) depends on the distribution of all the particles in the liquid.

We have so far ignored the spin of the quasiparticles as most of the results that we are getting are independent of it. Taking spin into account, (1.7) is written as

$$\delta(\frac{E}{V}) = \sum_{\alpha,\beta} \int \epsilon_{\alpha,\beta}(\mathbf{p})\delta n_{\alpha,\beta}(\mathbf{p})d\mathbf{p} \quad , \tag{1.8}$$

where α, β are spinorial indices. Making use of the phonon analogy, the quasiparticle energy can be written as

$$\epsilon(\mathbf{p},\sigma) = \epsilon^{(0)}(\mathbf{p},\sigma) + tr_{\sigma'} \int f(\mathbf{p},\sigma,\mathbf{p}',\sigma')\delta n(\mathbf{p}',\sigma')d\mathbf{p}' \quad , \tag{1.9}$$

where the first term $\epsilon^0(\mathbf{p})$ is the equivalent of E_{ph} in the phonon analogy and represents the equilibrium energy of the quasiparticles neglecting the interaction. The second term, the equivalent to the "anharmonic" contribution to the quasiparticles energy, depends on the interaction and is parametrized in terms the *interaction function* $f(\mathbf{p},\sigma,\mathbf{p}',\sigma')$. As defined by (1.9), it is the second variational derivative of E with respect to δn and constitutes a phenomenological characteristics of each particular Fermi liquid. It is related with the so-called Landau parameters and can be determined from the thermodynamical magnitudes (specific heat). As we will see in the next chapter, the interaction function is related with the four point function of the quantum field theory.

1.2.5 The Theory of the Effective Mass

Neglecting the interactions among the quasiparticles (not among the original electrons), the energy excitations of the FL are a set of free fermions for which $n(\mathbf{p}) = \theta(\mathbf{p} - p_F)$ (this distribution function should not be confused with the one corresponding to the original interacting electrons that will be discussed in the next chapter). Near the Fermi surface we have

$$\epsilon - \epsilon_F = \epsilon^{(0)}(\mathbf{p}) - \mu(T = 0) \sim \mathbf{v}_F(\mathbf{p} - \mathbf{p}_F),$$

where now

$$\mathbf{v}_F = [\frac{\partial \epsilon}{\partial \mathbf{p}}]_{p_F}$$

is a parameter which represents the velocity of the quasiparticles at the Fermi surface. In analogy with the free gas case, the *effective mass* of the quasiparticles is defined by

$$m^* = \frac{p_F}{v_F}.$$

This is the main parameter of a Fermi liquid and it determines the entropy and specific heat of the liquid which are given by the same relations as for the free gas substituting m by m^*:

$$S = C = V\gamma T$$

with

$$\gamma = \frac{m^* p_F}{3\hbar^2} = \frac{m^*}{3\hbar^2}(3\pi^n)^{1/3}.$$

In practice one determines m^* by measuring the specific heat of the particular liquid while p_F is determined by the liquid density. As an example, the effective mass of ^3He is of about 3 times the mass of the original He particles and $p_F = 0.76 \cdot 10^8 cm^{-1}$. For ordinary metals $m^* \sim m_e$ and $p_F \sim 1 - 2 \cdot 10^8 cm/s$.

exercise 1.3 *Prove that the specific heat of a free Fermi gas grows linearly with T at very low temperatures.*

The effective mass is related to the original mass of the fermions that constitute the liquid and with the interaction function by

$$\frac{1}{m^*} = \frac{1}{m} - \frac{p_F}{2(2\pi)^3}tr_{\sigma\sigma'}\int f(\xi)cos\xi d\Omega \quad . \tag{1.10}$$

exercise 1.4 *Look for the result (1.10) in the Abrikosov book [1].*

It is important to note that, although the Fermi liquid can be for most phenomenological purposes regarded as a free gas of quasiparticles with effective mass m^*, there is an essential difference. The (often negligible) quasiparticle interactions allow the liquid to support the BCS instability and to develop the superconductivity gap. Nothing like that can ever happen in a truly free gas.

We shall end this paragraph defining the quasiparticles as "an exact low energy elementary excitation of the Fermi liquid that has all the properties of a real free electron but can have a modified velocity, mass, and so forth and contains all of the high-energy effects of the interactions" [6].

1.3 Method of Second Quantization and Green's Functions

The Landau Fermi liquid theory can be provided a microscopic justification by using field theory techniques based on the second quantization formalism. This developments were started by Galitski and Migdal [7] as early as in 1958 and followed by Luttinger, Noziéres and many others in the early 60's.

The main idea of this approach is to trade the — too vast — knowledge of the wave function of the quantum interacting many-body system by the correlation functions which do contain less information but enough for most physical purposes. Besides, this formalism allows direct application of computational devices like ressummation of infinite series that have been developed for the Green's functions of Quantum Field Theories and can be adapted easily to the non relativistic many body physics. Finally, the latest developments that make use of Renormalization Group ideas not only provide a physical understanding of what is going on but allow to explore unknown kinematical regions as described in the introduction of [2].

The two main quantities to compute will be, first, the one-particle Green's function (propagator) that contains all the information on the single-particle operators, in particular on the particle distribution function, and on the dispersion relation of fermionic excitations (energy spectrum). And, second, the two-particle Green's function (four-point amplitude) that contains the information on the bosonic excitations, Landau interaction function and other response functions.

1.3.1 Similarities to and Differences from Quantum Field Theory. The Dirac Versus the Fermi Sea

Before entering into the details of this section, we shall make a few comments that can be useful for both the condensed matter and QFT-based students.

First it is worth mentioning that, although the second quantization formalism was mostly developed to account for relativistic processes where creation and annihilation of particles takes place, there is nothing specifically relativistic in the formalism which can be fully used for quantum mechanical problems being specially adequate for describing interactions which do not conserve the number of particles. It is the usual technique in many-body Physics and is also very adequate for studying quasiparticles whose interactions lead to decay processes very similar to those occurring in quantum field theory.

Despite the previous remark, we will see that the non-relativistic character of the electronic problem that we are dealing with, has important implications. Besides the lack of Lorentz covariance of the formalism, there are two main differences between the QFT and the many-body problem. The first is the absence, in the case of the many-body problem, of any of the infinities the QFT is

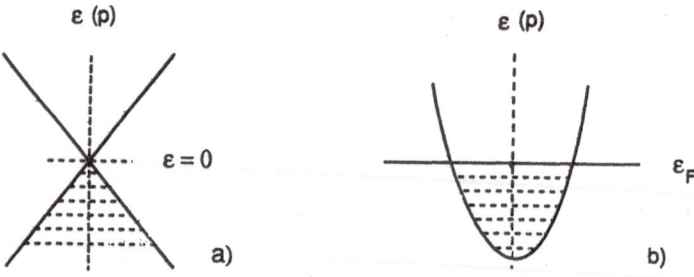

Fig. 1.2. The Dirac versus the Fermi sea.

plagued with. This is a good feature that will be discussed at length when we come to the renormalization group chapter.

The second and main difference lies in the character of the physical vacuum. In this case the difference goes against us. To be precise, for the case of a Fermionic system we have to deal with the difference between the Dirac and the Fermi seas. The vacuum of the QFT problem is the relativistic Dirac sea which extends to (negative) infinity in energy (this is one of the origins of the QFT infinities), but is such that the frontier of occupied versus unoccupied states is always a point (see **Fig. 1.2**). Whenever excitations are created from the vacuum, we are perturbing about *a single point in momentum space*. In the case of the Fermi sea on the contrary, we have to deal with a whole Fermi surface in momentum space from which excitations are created. We also have to deal with the peculiar fact that the vacuum depends on the number of particles in the original system so normal-ordering with respect to the many-body vacuum can be tricky. In both cases the dispersion relation for low energy excitations about the vacuum is linear in the momentum.

1.3.2 Second Quantization and the Field Operators

In order to settle notation and define the field operators, we shall start by writing down the Schrödinger equation in the second quantization formalism. This description can be found in full detail in [8] although students with a condensed matter basis may prefer [9], and [10].

The Hamiltonian of a system of interacting particles can be written as a sum of terms in the form:

$$H = \sum_a H_a^{(1)} + \sum_{a,b} U^{(2)}(\mathbf{r}_a, \mathbf{r}_b) + \sum_{a,b,c} U^{(3)}(\mathbf{r}_a, \mathbf{r}_b, \mathbf{r}_c) + \ldots$$

where $H_a^{(1)}$ is the one-particle Hamiltonian

$$H_a^{(1)} = -\frac{\nabla_a^2}{2m} + U(\mathbf{r}_a) \quad ,$$

and $U^{(2)}(\mathbf{r}_a, \mathbf{r}_b), U^{(3)}(\mathbf{r}_a, \mathbf{r}_b, \mathbf{r}_c)$ represent the two, three etc. - particle interactions. In most physical cases (and certainly in the only tractable ones) only two-particle interactions will be considered.

The wave function of the N-particle system can usually be expanded in a complete set of one-particle wave functions:

$$\psi(r_1, ..., r_N) \sim \varphi(r_1)...\varphi(r_N).$$

The one-particle functions $\varphi(r_i)$ are adequate to the symmetries of the problem, plane waves in the free case, Bloch waves for crystals, etc.

In the occupation number basis the state is given by specifying the number n_i of particles that are found in the state φ_i:

$$\psi(r_1, ..., r_N) \leftrightarrow \psi(n_1, n_2, .., n_N)$$

meaning that there are n_1 particles in the state φ_1, n_2 in φ_2, etc. For fermions, $n = 0$ or 1.

Example 1.5 .

$$\psi(n_1 = 1, n_2 = 0, ..., n_N = 0) \sim \varphi_1(r_1);$$

$$\psi(n_1 = 1, n_2 = 1, n_3 = 0, .., n_N = 0) \sim \frac{1}{\sqrt{2}} \left(\varphi_1(1)\varphi_2(2) \pm \varphi_1(2)\varphi_2(1) \right).$$

The creation and annihilation operators act on the occupation number wave functions as:

$$a_j^\dagger \psi(n_1, .., n_j, .., n_N) = \psi(n_1, .., n_j + 1, .. n_N),$$
$$a_j \psi(n_1, .., n_j, .., n_N) = \psi(n_1, .., n_j - 1, .. n_N),$$

and obey the usual anticommutation relations:

$$\{a_i, a_k^\dagger\} = \delta_{ik},$$

$$\{a_i, a_k\} = \{a_i^\dagger, a_k^\dagger\} = 0 \quad . \tag{1.11}$$

The number operator is $a_n^\dagger a_n = n_n$, and the full state can be created from the Hilbert state vacuum $\psi_{vac} = |0, ...0>$ by

$$\psi(n_1, .., n_N) = (a_1^\dagger)^{n_1} .. (a_N^\dagger)^{n_N} \psi_{vac}.$$

In terms of these operators, H can be written as

$$H^{(1)} = \sum_i \epsilon_i a_i^\dagger a_i \equiv \sum \epsilon_i N_i$$

$$H^{(2)} = \sum_{i,k,l,m} U_{lm}^{(2)ik} a_i^\dagger a_k^\dagger a_l a_m \quad ,$$

where ϵ is the energy of the single-particle state i, N_i is the number operator, and

$$U_{lm}^{(2)ik} = \int d\mathbf{r} d\mathbf{r}' \varphi_l^\dagger(\mathbf{r}) \varphi_m^\dagger(\mathbf{r}') U^{(2)}(\mathbf{r}, \mathbf{r}') \varphi_i(\mathbf{r}) \varphi_k(\mathbf{r}') \quad .$$

Now we can define one of the basic concepts of this section: the *field operators* are defined as the linear combination of creation and annihilation operators:

$$\psi_\alpha(\mathbf{r}) = \sum_i \varphi_i(\mathbf{r}, \alpha) a_i$$

$$\psi_\beta^\dagger(\mathbf{r}) = \sum_i \varphi_i^*(\mathbf{r}_i, \beta) a_i^\dagger \quad ,$$

where the coefficients are the single-particle wave functions and the sum is over the complete set of single-particle quantum numbers. In particular, for spin 1/2 fermions, $i = \{\mathbf{p}, \sigma\}$ and the wave functions φ are two-component spinors.

The field operator $\psi_\alpha^\dagger(\mathbf{r})$ acting on the Hilbert space vacuum, creates a full quantum state at $\mathbf{r}$ with spin α. From (1.11) it can be seen that the field operators obey the commutation relations

$$\{\psi_\alpha(\mathbf{r}), \psi_\beta^\dagger(\mathbf{r}')\} = \delta_{\alpha\beta}\delta(\mathbf{r} - \mathbf{r}'),$$

all other combinations giving zero.

The Hamiltonian and all the single-particle operators can be written in terms of the fields $\psi, \psi^\dagger$ as

$$H = \int [-\frac{1}{2m}\nabla\psi_\alpha^\dagger(\mathbf{r})\nabla\psi_\alpha(\mathbf{r}) + \psi_\alpha^\dagger(\mathbf{r})U^{(1)}(\mathbf{r})\psi_\alpha(\mathbf{r})]d^3\mathbf{r}$$

$$+ \frac{1}{2}\int\int d\mathbf{r}d\mathbf{r}'\psi_\alpha^\dagger(\mathbf{r})\psi_\beta^\dagger(\mathbf{r}')U^{(2)}(\mathbf{r}, \mathbf{r}')\psi_\beta(\mathbf{r}')\psi_\alpha(\mathbf{r}) + ... \qquad (1.12)$$

where summation over spin projections is understood. The density of particles at a given point is written as

$$n(\mathbf{r}) = \sum_a \delta(\mathbf{r} - \mathbf{r}_a) = \int \psi_\alpha^\dagger(\mathbf{r}_a)\delta(\mathbf{r} - \mathbf{r}_a)\psi_\alpha(\mathbf{r}_a)d^3r_a = \psi^\dagger(\mathbf{r})\psi(\mathbf{r}),$$

and the total number of particles has the form

$$N = \int n(\mathbf{r})d\mathbf{r}.$$

1.3.3 The One-Particle Green's Function

The most important magnitude of this formalism is the two-point Green's function, called the one-particle Green's function in condensed matter physics, the correlator in statistical mechanics, and the field propagator in QFT. In any case, its formal definition is

$$G_{\alpha\beta}(x, x') \equiv -i < 0|T(\psi_\alpha(x)\psi_\beta^\dagger(x')|0 > \quad , \quad x \equiv \mathbf{r}, t \quad , \qquad (1.13)$$

where $\psi_\alpha(x)$ is the field operator in the Heisenberg representation, the vacuum expectation value is taken with respect to the physical vacuum of the full theory and T is the temporal ordering operator. For the case of fermions, this ordering will be important later; it means:

$$G_{\alpha,\beta}(x, x') = \begin{cases} -i < \psi_\alpha(x)\psi_\beta^\dagger(x') > & \text{if } t > t', \\ i < \psi_\beta^\dagger(x')\psi_\alpha(x) > & \text{if } t < t' \end{cases} \quad . \qquad (1.14)$$

Apart from the different vacua mentioned before, this function is the same than that of the QFT. In statistical mechanics, the vacuum expectation value is to be read as an statistical average with respect to the density matrix:

$$< O > \equiv \frac{tre^{-\beta(H-\mu N)}O}{tre^{-\beta(H-\mu N)}}.$$

In QFT, the propagator describes the process where a positive-energy electron is created out of the vacuum at x, transported to x', and reabsorbed into the vacuum at x' (this will only make sense if $t' > t$ since we can't destroy an electron before creating it. For $t' < t$, the propagator does not vanish but describes the equivalent process of the propagation of a positive-energy positron (hole) created at x' forward in time, to the point x where it is destroyed. Hence the time ordering in the definition).

The one-particle Green's function contains all the information about:

1. The spectrum of *fermionic* excitations of the problem, i.e. , the dispersion relation.

2. The energy as a function of volume, and, hence, the equation of state of the system at zero temperature.

3. The average value of any one-particle operator $F^{(1)}$ in the vacuum (called vacuum expectation value in QFT) and, hence, of the thermodynamical magnitudes:

$$F^{(1)} = \int \psi_\alpha^\dagger(x) f_{\alpha\beta}^{(1)} \psi_\beta(x) \quad , \quad f_{\alpha\beta}^{(1)} = < \varphi_\alpha \mid F \mid \varphi_\beta >$$

$$< F^{(1)} > = -i \int \lim_{\mathbf{r}'\to\mathbf{r}, t'\to t+} f_{\alpha\beta}^{(1)}(x) G_{\alpha\beta}(x, x') d\mathbf{r} \quad .$$

In particular the particle density can be computed as

$$N(x) = -i \lim_{\mathbf{r}'\to\mathbf{r}, t'\to t+} G_{\alpha\alpha}(x, x') \quad . \qquad (1.15)$$

(We use the symbol $N(x)$ for the electron particle distribution to distinguish it from the quasiparticle distribution $n(x)$).

exercise 1.6 *Prove that the density matrix of a statistical fermionic system is:*

$$\rho(\mathbf{r}, \mathbf{r}') = - \lim_{\tau \to 0^+} G(\mathbf{r}, t; \mathbf{r}', t + \tau).$$

exercise 1.7 *Derive (1.15) from the definition (1.13) of the Green's function.*

1.3.4 General Properties of the Green's Function

Irrespective of the particular form of the interactions among the particles which constitute the original system (electrons in our case), the Green's functions has general properties coming from the symmetries of the problem and from very general quantum mechanical principles.

In the absence of external magnetic fields, the spin structure of the matrix $G_{\alpha\beta}$ is diagonal:

$$G_{\alpha\beta} = G\delta_{\alpha\beta},$$

so, from now on we will drop the spin indices. For a homogeneous and isotropic system we also have

$$G(x, x') = G(x - x').$$

In that case (translational invariance), we can define the Fourier transform

$$G(x - x') = \int G(\mathbf{p}, \omega) e^{i[\mathbf{p} \cdot (\mathbf{r} - \mathbf{r}') - \omega(t - t')]} \frac{d\mathbf{p} \, d\omega}{(2\pi)^4}.$$

exercise 1.8 *Derive the following important expression*

$$N(\mathbf{p}) = -i \lim_{t \to 0^-} \int_{-\infty}^{\infty} G(\omega, \mathbf{p}) e^{-i\omega t} \frac{d\omega}{(2\pi)} \tag{1.16}$$

for the density of particles in momentum space $N(\mathbf{p})$ and show that it has the correct normalization:

$$2 \int N(\mathbf{p}) \frac{d^3 p}{(2\pi)^3} = \frac{N}{V}.$$

The function $G(\mathbf{p}, \omega)$ can be easily computed for the case of a free Fermi gas to be

$$G^{(0)}(\mathbf{p}, \omega) = \frac{1}{\omega - \epsilon_0(\mathbf{p}) + i\delta \, \mathrm{sgn}(|\mathbf{p}| - p_0)}, \tag{1.17}$$

where δ is a small quantity used to regularized the Fourier integral and $\epsilon_0(\mathbf{p})$ is the dispersion relation of the free gas. This function will play a fundamental role in the diagram technique.

exercise 1.9 *Derive (1.17) from the free propagator in coordinate space:*

$$G^0(x) = -\frac{i}{V} \sum_{\mathbf{p}} e^{i[\mathbf{p} \cdot \mathbf{r} - \epsilon_0(\mathbf{p})t]} \begin{cases} 1 - n_{\mathbf{p}}, & t > 0 \\ -n_{\mathbf{p}}, & t < 0. \end{cases}$$

1.3.5   Analytic Properties of the Green's Function

From (1.17) we can see that $G(\mathbf{p}, \omega)$ is a meromorphic (analytic except for a finite set of isolated singularities) function of ω with simple poles of residue 1 (the 1 appearing in the numerator) at the energies of the elementary excitations $\omega = \epsilon(\mathbf{p})$. The dependence on $\mathbf{p}$ occurs only through the dispersion relation. We will see that these features survive in the interacting theory and that we can derive them without knowing the precise form of the interaction. The following spectral representation was originally derived for QED by Källén in 1952 and by Lehmann in 1954 for a scalar field. Field theory students can see it in [11]; a very appealing exposition of its application to our actual problem can be found in chapter 5 of ref. [12]. We shall here review the main line of arguments and its physical consequences.

The main idea is to expand the wave functions that appear in $G(x, x')$ in a complete set of *stationary* eigenstates of the full Hamiltonian (so the only property that the interaction must have is the existence of this complete set). Then use is made of the translational invariance of the system to deal with the time and space dependences. Last the Fourier transform is done by explicitly performing the space-time integrals in the complex plane.

Let $|n>$ be a complete set of eigenstates of the full Hamiltonian operator that we are using: $H - \mu N$.

$$< n|\psi_\alpha(t, \mathbf{r})|m >= e^{i\omega_{nm}t} < n|\psi_\alpha(0, \mathbf{r})|m >,$$

with

$$\omega_{nm} = E_n - E_m - \mu(N_n - N_m).$$

As the operator ψ decreases (and $\psi^\dagger$ increases) by one the number of particles in the system, we must have $N_n = N_m - 1$ so that

$$\omega_{nm} = E_n(N) - E_m(N+1) + \mu.$$

The space dependence can be computed in the same way making use of the total momentum operators:

$$< n|\psi_\alpha(0)|m >= e^{i\mathbf{k}_{nm}\mathbf{r}} < n|\psi_\alpha(\mathbf{r})|m >,$$

We then have

$$< n|\psi_\alpha(t, \mathbf{r})|m >= e^{i(\omega_{nm}t - \mathbf{k}_{nm}\mathbf{r})} < n|\psi_\alpha(0)|m > ,$$

$$< n|\psi_\alpha^\dagger(t, \mathbf{r})|m >=< m|\psi_\alpha(t, \mathbf{r})|n >^* ,$$

where $\mathbf{k}_{nm} = \mathbf{P}_n - \mathbf{P}_m$.

The next delicate issue is to notice that the function $G(t, \mathbf{r})$ is discontinuous due to the time ordering (1.14). This discontinuity oblies us to divide the t-integration of the Fourier transform into two parts, from $-\infty$ to 0, and from 0 to ∞. The corresponding integrals must be regularized by giving a small imaginary part to ω with the appropriate sign.

exercise 1.10 *Deduce the sign of the imaginary part of ω.*

Inserting the complete set of states $|n>$ in the part with $t = t - t' > 0$, we get:

$$G(t, \mathbf{r}) = -\frac{1}{2}i \sum_m < 0|\psi_\alpha(x)|m >< m|\psi_\alpha(x')|0 >$$

$$= -\frac{1}{2}i \sum_m |< 0|\psi_\alpha(0)|m >|^2 e^{i(\omega_0 m t + \mathbf{P}_m \cdot \mathbf{r})},$$

where now $\omega_{0m} = E_0(N) - E_m(N+1) + \mu$.

Performing the space-time integrals and doing the same for the $t < 0$ part of the propagator, we obtain the final result:

$$G(\omega, \mathbf{p}) = \frac{1}{2}(2\pi)^3 \sum_m \{ \frac{A_m \delta(\mathbf{p} - \mathbf{P}_m)}{\omega + \mu + E_0(N) - E_m(N+1) + i\delta}$$

$$+ \frac{B_m \delta(\mathbf{p} + \mathbf{P}_m)}{\omega + \mu + E_m(N-1) - E_0(N) - i\delta} \} , \qquad (1.18)$$

where A_m and B_m are real and positive with values between 0 and 1:

$$A_m = |< 0 | \psi_\alpha(0) | m >|^2 , \quad B_m = |< m | \psi_\alpha(0) | 0 >|^2 ,$$

$E_0(N) - E_m(N+1)$ is the change in ground state energy as one extra particle is added to the system, and $(\mathbf{P}_m, E_m)$ are the momentum and energy of the state m. The numbers A_m, B_m represent the probability for the field operator to create from the vacuum the state of energy m.

All this has been deduced for the discrete system with a finite number of particles N. In order to study the real macroscopic system we have to go to the thermodynamic limit $N \to \infty, V \to \infty$, N/V fixed. In that case we can define the functions

$$A(\mathbf{p}, E)dE = (2\pi)^3 \sum_s |(a_\mathbf{p}^\dagger)_{s0}|^2 \delta(\mathbf{p} - \mathbf{p}_s) , \quad E \leq \epsilon_s \leq E + dE ,$$

$$B(\mathbf{p}, E)dE = (2\pi)^3 \sum_s |(a_\mathbf{p})_{s0}|^2 \delta(\mathbf{p} + \mathbf{p}_s) , \quad E \leq \epsilon_s \leq E + dE ,$$

in terms of which, we have

$$G(\mathbf{p}, \omega) = \int_0^\infty \{ \frac{A(\mathbf{p}, E)}{\omega - E - \mu + i\delta} - \frac{B(\mathbf{p}, E)}{\omega + E - \mu - i\delta} \}dE . \qquad (1.19)$$

This is the spectral decomposition of the the Green's function. As an inmediate consequence of it, we can read the asymptotic behavior:

$$\lim_{\omega \to \infty} G(\omega, \mathbf{p}) \sim \frac{1}{\omega} ,$$

over which we shall come back later.

exercise 1.11 *(for QFT students): Compare the spectral representation (1.19) with the one appearing in QFT:*

$$G(p) = \int_0^\infty d\sigma^2 \rho(\sigma^2) G_{free}(p, \sigma) \qquad (1.20)$$

for a massive theory. (Bjorken and Drell op.cit. (16.33), page 139). Try to rewrite (1.20) for the case of a massless spinor.

1.3.6 Physical Meaning of $G(p,\omega)$. The Spectrum

Let us now review what we can learn from the spectral decomposition (1.19).

The chief property of the Green's function in momentum space is that *its poles can only be at the points $\omega = \epsilon_m - \mu$ where ϵ_m are the discrete excitations energies of the system.* Each of these energies corresponds to a definite value of the momentum $\mathbf{P}_m$ of the system, as is explicit from the presence of the corresponding δ-function in each pole term in (1.19). All that picture holds in the absence of interaction among the excitations (quasiparticles), i.e., when the complete set of energy eigenvalues is a stable set.

The effect of the interaction (anharmonic part in the phonon analogy), is to shift the Green's function's pole into the complex plane. The energies $\epsilon(\mathbf{p})$ acquire an imaginary part $\gamma(\mathbf{p})$ meaning that the elementary excitations are no longer stable but have a finite lifetime proportional to $\gamma(\mathbf{p})$. The quasiparticle picture will remain valid as soon as $\gamma(\mathbf{p}) << Re\epsilon(\mathbf{p})$. This condition is satisfied for weakly excited states of the system for whose $p - p_F << 1$.

The lifetime of the quasiparticles is one of the characteristic values of a Fermi Liquid. It is defined by $\tau^{-1}(\epsilon) = Im\Sigma(\epsilon)$, and can be computed by the standard techniques of the next paragraph. A very simple argument based on the exclusion principle (see for instance page 346 of [4]) shows that

$$\frac{1}{\tau} = a(\epsilon - \epsilon_F)^2 + b(k_B T)^2 , \qquad (1.21)$$

where a and b are constant. The life time is related with the electric resistivity of the Fermi Liquid by the formula

$$\rho \equiv (v_F \tau)^{-1} ,$$

1.21 implies that the electronic contribution to ρ is proportional to the squared of the temperature. One of the most puzzling features of the "normal" state of high-T_c superconductors is the linear dependence of the resistivity with the temperature [6] a clear departure from the Fermi liquid behavior.

Near a pole, the Green's function has the form:

$$G(\mathbf{p}, \omega) \approx \frac{Z}{\omega + \mu - \epsilon(\mathbf{p})},$$

From the spectral representation it can be seen that the *renormalization constant Z* (the residue of G at the pole) is a number with $0 < Z < 1$ whose physical meaning will become clear in what follows. *A constant non-zero value of Z is another definig characteristic of a Fermi Liquid.* This quantity can be computed directly from the Green's function and is also anomalous in the high-T_c materials.

Let us now study the physical meaning of the Green's function in the condensed matter context. Suppose that at the initial state we prepare the system to be in the state $\Phi(0) = a_{\mathbf{p}}\Phi_0$, where Φ_0 is the ground state of the N-particle system (the physical vacuum). At time $\tau > 0$, the wave function of the system is

$$\Phi(\tau) = e^{-iH\tau} a_{\mathbf{p}}^{\dagger}\Phi_0.$$

The function $G(\mathbf{p}, \tau)$ is then the probability amplitude to find the system in the state $\Phi(0)$ at time τ:

$$(\Phi(0), \Phi(\tau)) = (\Phi_0 a_{\mathbf{p}} e^{-iH\tau} a_{\mathbf{p}}^{\dagger}\Phi_0) = G(\mathbf{p}, \tau).$$

To obtain the function $G(\mathbf{p}, t)$ we have to evaluate the integral

$$G(\mathbf{p}, t) = \int_{-\infty}^{\infty} G(\mathbf{p}, \omega) e^{-i\omega t} \frac{d\omega}{2\pi}$$

in the complex ω-plane. With the usual techniques of contour integration we get the result that the main contribution to it comes from the poles. When we study the behavior of $G(\mathbf{p}, \omega)$ for large, positive times, we get

$$G(\mathbf{p}, \tau) \approx Z e^{-i\epsilon_R(\mathbf{p})\tau - \gamma\tau} + O[(\frac{\gamma}{\epsilon(\mathbf{p})})^2].$$

This result means that the state prepared at initial time as $a_{\mathbf{p}}^{\dagger}\Phi(0)$, evolves in time to a wave packet which behaves like a quasiparticle with energy $\epsilon_R(\mathbf{p})$ and decays in time according to the law $e^{-\gamma t}$. The renormalization constant can be interpreted as follows: in the state $\Phi(0)$ there is present, with amplitude Z, a wave packet representing a quasiparticle with energy $\epsilon_R(\mathbf{p})$ and damping γ. *The existence of a finite wave function renormalization Z is one of the distinctive characteristics of a Fermi liquid.* As we shall see in the next chapters, this function vanishes in the Luttinger liquid which does not support fermionic excitations. It also goes to zero in the marginal Fermi liquid of ref. [13].

Had we considered the case $\tau < 0$, we would have obtained a similar relation for a quasihole. Thus, $G(\mathbf{p}, \omega)$ may be written in the form:

$$G(\mathbf{p}, \omega) = Z \left[\frac{1 - n_{\mathbf{p}}}{\omega - \epsilon(\mathbf{p}) + i\gamma} + \frac{n_{\mathbf{p}}}{\omega - \epsilon(\mathbf{p}) - i\gamma} \right] + G_{\text{incoh}} \equiv$$

$$\equiv Z G_{QP} + G_{\text{incoh}},$$

where $\bar{G}_{QP}$ is the Green's function of the quasiparticle having the same form as the free Green's function with poles at the quasiparticles energies but with residue $0 < Z < 1$, and a finite imaginary part. The piece G_{incoh} has no pole structure and is to be seen as the incoherent part coming from the "anharmonic" interaction.

The values of $\epsilon(\mathbf{p})$ and γ are determined by the position of the poles of $G(\mathbf{p}, \omega)$. The dispersion relation is obtained in general from the equation:

$$G^{-1}(\epsilon - \mu, \mathbf{p}) = 0.$$

exercise 1.12 *(for QFT students): Explain why the pole of the propagator seems to be always 1 en QFT.*

1.3.7 The Momentum Particle Distribution

We shall now analyze the form of the density of states that can be inferred from the pole structure of the Green's function. To exemplify the line of thoughts, we start with the case of the free gas. Its Green's function is

$$G^0(\mathbf{p}, \omega) = \left[\omega - \frac{p^2}{2m} + \mu + i0 \cdot \text{sgn}\omega\right]^{-1}.$$

Taking into account that $\mu = p_F^2/2m$ and for weakly excited states $p \sim p_F$, we can make the substitution $p^2/2m \approx \mu + v_F(p - p_F)$, and write the Green's function *for such states* as:

$$G^0(\mathbf{p}, \omega) = \left[\omega - v_F(p - p_F) + i0 \cdot \text{sgn}\omega\right]^{-1}.$$

In all integrations, the imaginary part of ω is only important near the pole were we can write $\omega \approx v_F(p - p_F)$ and hence the sign of ω can be replaced by $\text{sgn}(p - p_F)$. The interest of this replacement is that now we have the Green's function analytic in the entire ω-plane:

$$G^0(\mathbf{p}, \omega) = \left[\omega - \frac{p^2}{2m} + \mu + i0 \cdot \text{sgn}(p - p_F)\right]^{-1}.$$

From this we can readily compute the density of states of the free gas as the integral (1.16):

$$N(\mathbf{p}) = -\frac{i}{2\pi} \int \frac{e^{-i\omega t} d\omega}{\omega - \frac{p^2}{2m} + \mu + i0 \cdot \text{sgn}(p - p_F)}.$$

For $t < 0$ we close the contour in the upper half plane (and then take the limit $t = 0$). What we found is that for $p > p_F$ there is no pole and $N(\mathbf{p}) = 0$, while for $p < p_F$ the pole at $\omega = p^2/2m - \mu$ contribution gives $N(\mathbf{p}) = 1$.

What we learn from the free case is that the discontinuity of the density of states comes from the pole structure of the Green's function.

Let us now study the Fermi liquid. We do not know the detailed structure of the Green's function in that case but we know that it has a pole at:

$$\omega = \epsilon(\mathbf{p}) - \mu \approx v_F(p - p_F) \quad , \quad v_F = p_F/m^* .$$

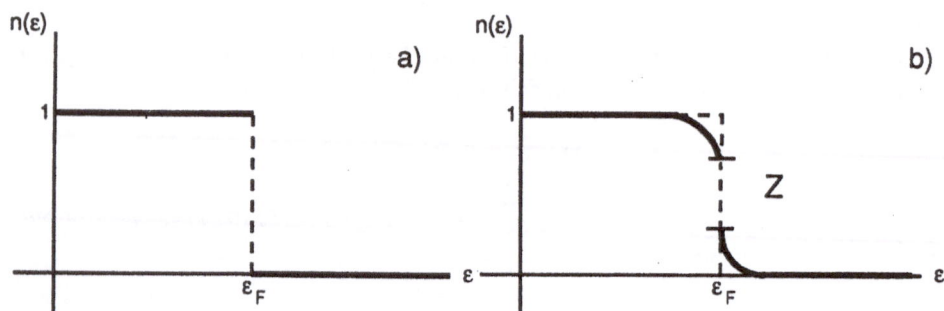

Fig. 1.3. The momentum particle distribution of a) the quasiparticles, b) the electrons, in a Fermi Liquid.

It can therefore be written as:

$$G(\mathbf{p}, \omega) = \frac{Z}{\omega - v_F(p - p_F) + i0 \cdot \mathrm{sgn}\omega} + G_{\mathrm{incoh}}(\mathbf{p}, \omega),$$

where the function G_{incoh} is finite at the pole. Now we can verify that *the density of states of the liquid also has a jump at the Fermi momentum, of magnitude Z*. In order to see that, we compute the difference $N(p_F - q) - N(p_F + q)$ in the limit $q \to 0^+$. It is clear that only the pole part contributes to this difference:

$$N(p_F - q) - N(p_F + q) = -i \int_{-\infty}^{\infty} \left\{ \frac{Z}{\omega + v_F q - i0} - \frac{Z}{\omega - v_F q + i0} \right\} \frac{d\omega}{2\pi};$$

closing the contour in either half plane we obtain

$$N(p_F - 0) - N(p_F + 0) = Z \quad , \quad 0 < Z < 1 \ .$$

This is one of the defining characteristics of the Fermi liquid. The particle distributions corresponding to quasiparticles and real electrons in a Fermi liquid are shown in **Fig. 1.3**.

1.3.8 Computing the Green's Function. Wick Theorem and Feynman Diagrams

We now come to the issue of how $G(\mathbf{p}, \omega)$ can be computed by using QFT techniques. We will be very schematic in this section as the results can be found in almost any book.

The main idea is to develop a systematic way to obtain any term in the perturbative series by using as building blocks the Green's function of the free system and the particular interaction.

The starting point is the *interaction representation* for the field operators:

$$|\psi_I(t)> = e^{i/\hbar H_0 t}|\psi_S(t)>,$$

that obeys the Schrödinger equation

$$\partial_t \psi_I = H_I \psi_I,$$

where H_I is the interaction part of the Hamiltonian.

In terms of ψ_I, the Green's function can be written as:

$$G(x,x') = \frac{-i<T\psi(x)\psi^\dagger(x')S(\infty)>}{<S(\infty)>},$$

where $S(\infty)$ is the operator:

$$S(\infty) = T\exp\{-i\int_{-\infty}^{\infty} H_I(t)dt\} \ . \tag{1.22}$$

This operator can be expanded in powers of the interacting Hamiltonian:

$$S(\infty) = 1 - i\int_{-\infty}^{\infty} H_{int}(t)dt + \frac{(-i)^2}{2}\int_{-\infty}^{\infty} dt_1 dt_2 T(H_{int}(t_1)H_{int}(t_2)) + \dots$$

Then we get:

$$G(x,x') = -\frac{i}{<S(\infty)>}\sum_{n=0}^{\infty}\frac{(-i)^n}{n!}$$

$$\int_{-\infty}^{\infty} dt_1...dt_n < T\big(\psi(x)\psi^\dagger(x')H_{it}(t_1)...H_{it}(t_n)\big)> \ .$$

Now we notice that all the terms inside the T-product are sums of products of ψ, $\psi^\dagger$ fields mixed with functions $U^{(2)}(x,x')$ (as H_{int} is made out of those (1.12)). The main result now at use is the *Wick theorem* which assures that the vacuum expectation value of any product of field operators can be decomposed in pairs as follows:

$$< T(ABC...XYZ)> = < T(AB)><T(CD)> \dots <T(YZ)> \pm$$

$$\pm <T(AC)><T(BD)> \dots <T(YZ)> \pm \dots$$

In our case, only the normally ordered pairs $<T(\psi\psi^\dagger)>$ will be different from zero and those are precisely free Green's functions.

A very good way of ordering the different terms in the previous sum is given by the *Feynman diagrams* technique: Use as building blocks the free propagators shown in **Fig. 1.4**:

To obtain the full Green's function at a certain given order in the perturbation expansion, construct all possible diagrams that contribute to that order. We will not insist on the Feynman rules that can be found in any book. We shall instead compute one of the diagrams.

Example 1.13 . Computation of $G(p)$ at one-loop order.

Fig. 1.4. The Feynman blocks.

Fig. 1.5. Diagrams contributing to $G(p)$ at one loop.

We shall not specify the form of the interaction. The diagrams that contribute at this order are shown in **Fig. 1.5.**
From them we get:

$$G^{(1)}(p) = G_A(p) + G_B(p),$$

The diagram in A represents the interaction of a particle with the medium to first order in U. The closed loop is nothing but the density of states in the vacuum.

$$G_A(p) = -i(G^0(p))^2 U(0) \int \frac{d^4 k}{(2\pi)^4} G^0(k) = i(G^0(p))^n(\mu)U(0).$$

$$G_B(p) = -\int \frac{d^4 k}{(2\pi)^4} G^0(p)G^0(k)U(p-k)G^0(p) =$$

$$= G^0(p)[-\int \frac{d^4 k}{(2\pi)^4} G^0(k)U(p-k)]G^0(p).$$

We have used here the four-vector notation of QFT where $k = (\mathbf{k}, \omega)$ to shorten the writing. In general, in QFT calculations, $U(p)$ represents the propagator of the field that carries the interaction and has a real dependence on all four components. In non-relativistic computations as the one we want to describe here, it should be read:

$$U(x) = U(\mathbf{r})\delta(t)$$

as the interaction is not retarded. Similarly we have for the Fourier transform:

$$U(q) = U(\mathbf{q}) = \int U(\mathbf{r})e^{-i\mathbf{q}\mathbf{r}}d^3 r.$$

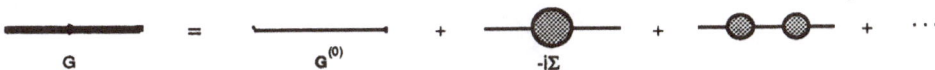

Fig. 1.6. The electron self-energy.

In any case, the schematic computation shown above serves to exemplify some general features. Notice in first place that at any loop order there will appear, while playing no role, the two free propagators corresponding to the external legs of the two-point diagram. Cutting them up, we define the *electron self-energy* $\Sigma(\mathbf{p}, \omega)$ as the sum of graphs which do not contain parts joined by a single line (the 1PI graphs of QFT). In terms of it, the full propagator obeys the integral equation shown graphically in **Fig. 1.6** (*Dyson equation*). It can be summed to get $G = G^0 + G\Sigma G^0$. Or, equivalently,

$$\frac{1}{G(p)} = \frac{1}{G^0(p)} - \Sigma(p) \ .$$

The evaluation of the self-energy involves a number of diagrams substantially smaller than that of the full Green's function. We shall come back to this when discussing the renormalization.

1.3.9 The Four Point Function. Bosonic excitations

To complete the microscopic description of the LFLT, we shall say a few words about the four-point function (called the two-particle Green's function in condensed matter books). Its definition (with spin indices suppressed) is

$$G_2(x_1, x_2, x_3, x_4) = < T\psi(x_3)\psi(x_4)\psi^\dagger(x_1)\psi^\dagger(x_2) > \ .$$

It gives the transition amplitude for the case in which the initial and final states correspond to two particles (or holes, or a particle and a hole) while the remaining $(N - 2)$ particles are in the ground state at both initial and final times. The two-particle Green's function contains all the information on the density-density correlation function whose poles (singularities of the compressibility), determine the bosonic spectrum of the theory and some of its transport properties. It will play a major role in the description of the Luttinger liquid later.

In the diagrammatic approach, the four-point function is represented by anything having four external legs. As happened with the two-point function, all the information is encoded in the *vertex function* $\Gamma(1, 2; 3, 4)$ where the external legs and all parts connected by a single fermionic line, or disconnected, are suppressed. For the study of the Fermi liquid, the only relevant diagrams are those for which the momentum transfer, i.e., the available momentum of the excitations, is very small (this corresponds to treating only excitations close to

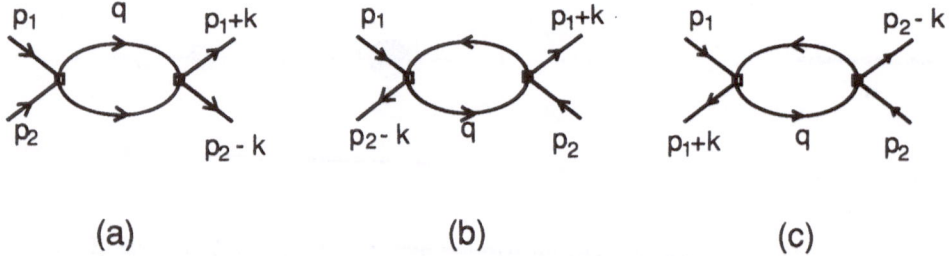

Fig. 1.7. Diagrams contributing to the vertex function of the LFLT.

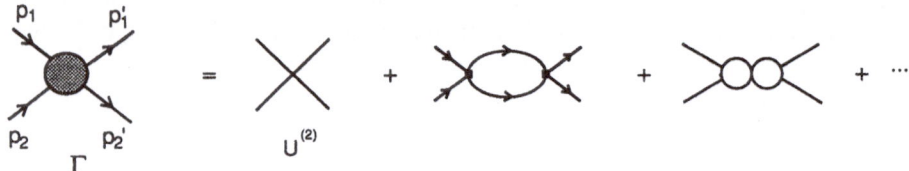

Fig. 1.8. The two-particle scattering amplitude.

the Fermi surface). The simplest diagrams contributing to this case are shown in **Fig. 1.7.** We will be interested in the limit $k \to 0$.

Diagrams of the type (a) represent the motion, between the collision events, of two quasiparticles (or quasiholes) while the types (b) and (c) describe the motion of a particle and a hole.

The scattering amplitude involves graphs of type (a) only and can be represented by the sum of **Fig. 1.8**:

This sum can be written symbolically as:

$$\Gamma = U + UGGU + UGGUGGU + \dots$$

whose sum gives rise to the equation:

$$\Gamma = U + UGG\Gamma,$$

where the function G is the free-particle Green's function and U is the two-particle interaction potential.

The bosonic excitations of the Fermi liquid are determined by the singularities (poles) of the complete Γ function. They are of two different types. One coming from diagrams where the energy-momentum transfer is small, and another one arising when the energies ω_1, ω_2 and *the sum* $p_1 + p_2 = s$ of the incoming particles is small.

The first set is related to the forward scattering amplitude of two quasiparticles with $|p_1| = |p_2| = p_F$. It corresponds to the singular limit $k \to 0$ in diagrams of the type (c) in **Fig. 1.7**, the singularity arising as the poles of the two Green's functions merge. They obey Bose statistics, as they correspond to operators that are bilinear in the Fermi operator, and correspond to various branches of the zero-sound spectrum. Such an acoustic excitation can then be seen as a bound state of a quasiparticle and a hole with similar values of the momentum. The spin-wave excitations are also of this type.

The second set of singularities of the two-point function comes from the diagrams (a) of **Fig. 1.7** that, for the particular kinematics **s** → 0, have the poles of the two Green's function lying close together. They could be regarded as bound states of two particles or two holes. The singularity arises only for negative bare coupling (attractive interaction), and gives rise to the BCS instability.

The reader is encouraged to look at the detailed results of this paragraph in chapter 4 of the Abrikosov book [1]. We will come back to that in the next chapter.

A summary of the phenomenological Fermi liquid characteristics follows. Violation of any of these, signals the departure from Fermi liquid behavior.

SUMMARY OF FERMI LIQUID CHARACTERISTICS.

1. The ground state has a sharp Fermi surface visualized as a discontinuity of the momentum density of states at the Fermi momentum of magnitude Z, $0 < Z < 1$. The volume within the Fermi surface is an invariant of FLT, and is the same as that of the free gas.

2. The low energy elementary fermionic excitations are quasiparticles of charge one and spin one half. They are poles of the fermion propagator, which can be written as:

$$G(\mathbf{p}, \omega) = \frac{Z}{\omega - \mu - v_F(p - p_F) + i\gamma \cdot \mathrm{sgn}(p - p_F)} + G_{\mathrm{incoh}}(\mathbf{p}, \omega) \ .$$

3. The electron has a finite wave function renormalization Z.

4. The lifetime of the quasiparticles is

$$\gamma(\omega) = a(\omega - \omega_F)^2 + bT^2 \ .$$

5. There is no gap in the energy spectrum of the quasiparticles. The system is at a critical point.

6. It also supports bosonic excitations of zero energy visualized as singularities of the four point function.

7. The transport and response functions go as follows:

$$\frac{C}{V} \sim \alpha T \ , \quad \rho \sim \rho_{res} + AT^2 + \rho_{ph} \ ,$$

where the residual resistivity ρ_{res} is a constant that does not get renormalized by Z, and ρ_{ph} is the phonon contribution.

1.3.10 Marginal-Fermi Liquids

We shall end this chapter by saying a few words on the early attempts to describe non-Fermi liquid behavior existing in the literature. Let us first advance that there is nothing like Fermi liquid behavior in spatial dimension $D = 1$. A system of correlated electrons in this dimensionality adopts systematically a Luttinger liquid behavior which will be studied in detail later. It is characterized by the absence of quasiparticle excitations manifested as a vanishing of the renormalization constant Z, and a separation of the charge and spin degrees of freedom. It only supports bosonic excitations with either charge (charge-density waves) or spin (spin-density waves).

As will become clear in the next chapters, the general pattern is that correlated electrons organize themselves as a Fermi liquid in $D \geq 2$ and as a Luttinger liquid in $D = 1$. When talking about non-Fermi liquid behavior it is always understood in $D > 1$.

The Luttinger type of behavior was left as a low dimensional curiosity until very recently when interest on it has exploded for two main reasons. First, the experimental availability of one-dimensional devices as quantum wires where the theoretical ideas can be tested, and second and more important, the observed behavior of the so-called normal state of the high T_c superconductors. These are certainly two — or may be three — dimensional electronic systems that do possess a Fermi surface and whose behavior departs strongly from the Fermi liquid described before. They will be discussed at length in the last part of these lectures (for a list of the experimental anomalies, see ref. [6]).

One of the first attempts of describing a non-Fermi liquid behavior was that of ref. [13]. By making the hypothesis that the electron self-energy is of the form:

$$\Sigma(\mathbf{k}, \omega) \sim g^2 N^2(0) \left[\omega \log \frac{x}{\omega_c} - i \frac{\pi}{2} x \right] \ , \tag{1.23}$$

where $x = \max(|\omega|, T)$, ω_c is an ultraviolet cutoff, and g is a coupling constant. Then, the wave function renormalization constant is

$$Z_k^{-1} = [1 - \frac{\partial Re \Sigma}{\partial \omega}]_{\omega = E_k} \sim \log |\omega_c / E_k|.$$

Near the Fermi surface, $E_k \to 0$, and Z vanishes logarithmically. This is the so-called *marginal Fermi liquid*. What kind of interaction would produce the given behavior of Σ is not discussed in the original paper although has been the subject of many subsequent papers. These have been produced following the renormalization group techniques and will be commented in the next chapter.

P. W. Anderson's passionate defense of the non-Fermi liquid idea in $D = 2$ was first settled in [14] and again will be described in detail when talking about the Hubbard model.

References

[1] A. A. Abrikosov, L. P. Gorkov and I. E. Dzyaloshinski, "Methods of Quantum Field Theory in Statistical Physics", Dover, New York, (1975); E. M. Lifshitz and L. P. Pitaevskii, Statistical Physics, Vol. 2, Pergamon Press, Oxford (1980).

[2] R. Shankar, Physica **A177** (1991) 530. For an extense review see R. Shankar, "Renormalization Group Approach to Interacting Fermions", Rev. Mod. Phys., **66** (1994) 129.

[3] For a theoretical physicist's vision of the problem see J. Polchinski, "Effective Field Theory and the Fermi Surface", Proceedings of the 1992 Theoretical Advanced Institute in Elementary Particle Physics, eds. J. Harvey and J. Polchinski, World Scientific, Singapore (1993).

[4] N. W. Ashcroft and N. D. Mermin, "Solid State Physics", CBS pub. (1976).

[5] J. M. Luttinger, "Fermi Surface and Some Simple Equilibrium Properties of a System of Interacting Fermions", Phys. Rev. **119** (1960) 1153.

[6] P. W. Anderson, "Experimental Constraints on the Theory of High-T_c Superconductivity", Science **256** (1992) 1526.

[7] V. M. Galitski and A. B. Migdal, "Application of Quantum Field Theory Methods to the Many Body Problem", Sov. Phys. JEPT **7** (1958) 96; J. M. Luttinger and P. Noziéres, Phys. Rev. **127** (1962) 1423, 4131.

[8] A. L. Fetter and J. D. Walecka, "Quantum Theory of Many- Particle Systems", McGraw-Hill NY (1971).

[9] G. D. Mahan, "Many Particle Physics", Plenum, NY (1981).

[10] S. Doniach, and E. H. Sondheimer, "Green's Functions for Solid State Physicist", Addison-Wesley (1974).

[11] J. D. Bjorken and S. D. Drell, "Relativistic Quantum Fields", McGraw-Hill (1965).

[12] A. B. Migdal, "Qualitative Methods in Quantum Theory", W. A. Benjamin, Inc. (1977).

[13] C. M. Varma, P. B. Littlewood, S. Schmitt-Rink, E. Abrahams, and A. E. Ruckenstein, " Phenomenology of the Normal State of Cu-O High-Temperature Superconductors", Phys. Rev. Lett. **63** (1989) 1996.

[14] P. W. Anderson, "Luttinger-Liquid Behavior of the Normal Metallic State of the 2D Hubbard Model", Phys. Rev. Lett **64** (1990) 1839.

2. Effective Actions and the Renormalization Group

2.1 Introduction

In this chapter we will identify the Fermi liquid as a fixed point of the renormalization group where the transformation is generated by systematically eliminating the high energy modes of the system and keeping only the modes close to the Fermi surface.

The idea behind this approach is taken from the critical phenomena in statistical physics. What is observed there is that many different physical systems behave in the same way in the proximity of a critical point and this behavior can be described by a universal lagrangian (Ginzburg–Landau). The physical reason underlying this universality is that near a critical point, the correlation length becomes much bigger than the typical dimensions of the system whose behavior is then insensitive to local perturbations.

The universal behavior of correlated electronic systems subject to different interactions, together with the gapless nature of the effective description (Fermi liquid), suggests a similar physical principle acting in this case. The Fermi liquid must be an effective theory for the low–energy excitations of a system near criticality.

The organization of this chapter is as follows. First we review the idea of renormalization as originally arises in quantum field theory with the aim of introducing a little formalism of the path integral, the cutoff, and the freedom to move it around while keeping the physical observables.

Then we briefly recall the uses of the RG in the study of critical phenomena in statistical physics mostly to introduce the Wilsonian effective action idea and the classification of operators. A good reference to compare the field theory and statistical physics RG points of view is the review article by Brézin in [1].

Finally we come to the implementation of the former ideas to the case of correlated electrons. We closely follow refs. [2] and [3] in this part.

2.2 The Renormalization Group in Quantum Field Theory

The aim of this section is to introduce the notion of renormalization as used in quantum field theory with the double purpose of setting the ideas as well as some computational tools that will be used later in the study of correlated electrons. We will describe the origin of the infinities in QFT, the renormalization procedure, and the original use of the renormalization group in this context. Namely, as a way of giving a meaning to renormalized quantum field theories.

We will also introduce the path integral formalism which is universally used in both QFT and statistical physics and is the simplest form to show the scaling phenomena we will be talking about.

Students with a field theoretical background can see most of the contents of this part clearly exposed in [4] and [5] A very detailed and quite clear account of renormalization of the scalar theory is also given in [6]. Condensed matter students may find useful the chapter 6 of the Migdal's book [7]. A very interesting historical review of the subject is given in [8].

2.2.1 The Physical Origin of the Divergences

Infinite quantities are by no means a privilege of QFT. They have often appeared in the physics of the last century usually leading to a deeper understanding of the phenomena where they occurred. A good example is the ultraviolet divergence in black–body radiation which brought the advent of the Planck quanta. Another one is the divergence found in the interaction of an electric charge with its radiation field. This example will help us to understand both the origin of the infinities and why and how one can still make sense of a theory having them. The main reason for that is, as we will see, that the infinite quantities are linked to unobservable parameters of the theory and can then be hidden on them.

Perhaps the oldest concept of renormalization comes from classical hydrodynamics. When studying the motion of a material body in the interior of an incompressible fluid – or any continuum medium – it was worked out in the middle of the 19th century that the net effect of the medium was to modify the mass of the material body. In particular, the total kinetic energy of a material object of mass M_0 moving with constant velocity V in the interior of a fluid is $1/2MV^2$ where $M = M_0 + M'$. The added mass M' depends on the geometry of the object and on the density of a fluid. Now if an object can only live in the interior of a fluid, there is no way to directly meassure its "naked" mass M_0. It could be estimated by determining the experimental value of M for different types of motion but it will never be directly observable.

Working by analogy, J. J. Thompson introduced in 1881 the idea of the electromagnetic mass of a charge conceived as a material – extended – object moving in the continuum medium of the radiation field (the aether on that time). The observable mass of an electron should be $m = m_0 + m'$ where m_0 is its mechanical mass and the electromagnetic mass m' depends on the – unknown – internal structure of the electron. Now as a moving charge creates its own magnetic field, the mechanical mass of the electron is unobservable, and the physical mass m has to be determined by experiments becoming an empirical constant of Maxwell theory. The electromagnetic mass of a spherical charge distribution of radius a being proportional to $e^2/2a$, it diverges for a point particle. This is one of the infinites that appear in a purely classical theory.

The infinite electromagnetic charge of the point electron is the classical analog of the ultraviolet divergences that we will address in QFT. The divergence occurring in black–body radiation is the classical counterpart of another infinity of QED occurring at low photon energy. It arises through the emission of an infinite number of photons during any change in the electron's motion. This was called the "infrared catastrophe" and is one of the infrared divergences that are not directly related to the subject of this chapter although they are very important being specially relevant in low–dimensional systems.

2.2.2 The Expression of the Divergences in QFT

The infinities in QFT appear in the perturbation series when computing processes involving virtual intermediate states. They were present in the first stages of Quantum Electrodynamics (QED) where it was clear that the only finite results were related with the noninteracting EM field or with simple processes involving absortion or emission of real photons (tree level processes). In any other case, the sums over the – infinite – set of intermediate states would give a divergent contribution to the process. The actual form of the infinities is mathematically related with the singularities of the local operators.

In the foundation of Quantum Field Theory (QFT) is the assumption of locality. In real space all fields are local (operators and derivatives defined in a single point of space–time) and the interactions take place also at a single point. The locality assumption is essential for both quantum and relativistic theories and it substantiates physical causality. Infinities arise whenever we must compute the product (convolution) of two – singular – Green's functions defined at the same point.

In Fourier space the short–distance singularities appear as ultraviolet divergences in the upper limit of the integral over the momentum running on an internal loop (as in the diagram of **Fig. 1.5 B**). Such a diagram represents the emission and absorption of a virtual particle (photon in QED) whose momentum can be as large as possible if the interaction takes place at a single point. Small–distance singularities correspond to large momentum singularities as a simple consequence of the uncertainty principle.

These are the UV singularities object of the renormalization program. Notice that in condensed matter systems defined on a lattice such singularities do not arise.

2.2.3 Renormalization. Substracting Infinities

Renormalization is, in its origins, the name given in QFT to the process of getting rid of the ultraviolet divergences of the theory. The basic idea of how this can be done and still be left with some sensible physics is that the parameters that appear in the "bare" action (masses and coupling constants) do not correspond to the physical masses and couplings of the interacting theory. We can redefine the bare parameters to absorb the infinity quantities that will be "substracted out".

To see in a toy model how infinities can be substracted out, consider the integral [8]

$$f(x) = \int_1^\infty \frac{dy}{x+y} \; .$$

Due to the logarithmic divergence of the integral for large y, the function $f(x)$ is undefined for every x. We can however look at the difference of $f(x)$ for two different values of x:

$$f(x) - f(0) = \int_1^\infty [\frac{1}{x+y} - \frac{1}{y}] dy$$

$$= -x \int_1^\infty \frac{dy}{y(x+y)} = \bar{f}(x) \; .$$

By making the – illegal – step of taking the substraction inside the integral, we see that, apart from an infinite additive constant, $f(x)$ is a well defined function:

$$f(x) = A + \bar{f}(x) \; .$$

Obviously not all divergent integrals can be controlled in that way but most of the ones we will be interested on, are.

In order to fix the ideas and to show concrete examples of how the renormalization works we will choose the simplest field theory, that of a self–interacting real scalar field. Although non realistic as the description of elementary particles, it is throughly used in the context of effective field theory and will be the basis of our condensed–matter application. The lagrangian is written in terms of only two parameters, the mass and the coupling (the freedom of redefining (rescaling) the field will introduce another constant).

The scalar field theory is described by the action

$$S[\Phi] = \int d^4x \mathcal{L}(\Phi, \partial_\mu \Phi)$$

$$= \int d^4x [\frac{1}{2} \partial_\mu \Phi \partial^\mu \Phi - \frac{1}{2} m^2 \Phi^2 - \frac{\lambda}{4!} \Phi^4] \; . \tag{2.1}$$

The construction and computation of the Green's function or any other quantity follows the rules described in Sect. 1.3.8. The building blocks for the Feynman rules in momentum space are the free propagator

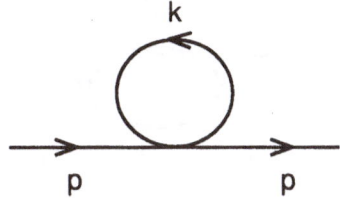

Fig. 2.1. Diagram contributing to the mass correction at one loop order.

$$G(p) = \frac{1}{p^2 - m_0^2} ,$$

and the coupling constant g. Each closed loop will carry an integration over the momentum running on the loop.

The complete propagator is given in perturbation theory by the sum of the corresponding Dyson series

$$G_F(p) = \frac{1}{p^2 - m^2 - \Sigma(p)} ,$$

and will have a pole at the value of the physical mass $p^2 = m_R^2$. As we discused earlier, the interaction causes a shift of the pole of $G_F(p)$ from m_0^2 to m_R^2. What we want to remark here is that, even if there would be no infinities, the value of m_0 is unknown; it has to be determined by fitting m_R to an observed value. As we will see, the self–energy $\Sigma(p)$ is one of the places where there are divergences and hence the bare mass m_0 will be infinity. The second origin of the divergences will be in the vertex function for the four–point Green's function what will renormalize the coupling constant.

We shall demonstrate the above ideas by working out two simple examples.

Example 2.1 Let us start by computing the correction to the mass arising from the diagram of **Fig. 2.1**. That would be the first diagram in the Dyson series of the self–energy. In order to identify the divergences we introduce a cutoff Λ in the momentum integral.

The computation of the field self–energy at this order gives

$$\Sigma_{(1)} = g \int^\Lambda \frac{d^4 k}{(2\pi)^4} \frac{1}{k^2 - m^2} \sim g\Lambda^2 . \tag{2.2}$$

We see that the one–loop contribution to the field self–energy coming from the given diagram diverges quadratically in the UV limit. We could try to redefine

$$m_0^2(\Lambda, g) = m^2 - gA_1\Lambda^2 m^2 = m^2(1 - gA_1\Lambda^2) ,$$

where A_1 is a constant containing the numerical factors that we have obviated. This is equivalent to adding a new term to the lagrangian with the exact form of the mass term but with a Λ–dependent coefficient:

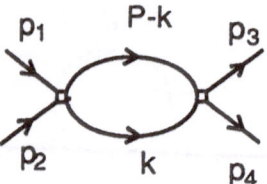

Fig. 2.2. First order diagram contributing to the vertex function.

$$\mathcal{L}_{ct}^{(m)} = \frac{1}{2}A_1^2(\Lambda)\Phi^2 \quad , \quad A_1(\Lambda) = gA_1\Lambda^2 \ ,$$

(observe also the wrong sign of the new "mass term"). We associate to the new term a new Feynman rule which contributes to the determination of the field self–energy with a tree–level value that exactly cancels the one–level divergence of **Fig. 2.1**.

All together, with the redefined action and at the given order in perturbation theory, the computation of the self–energy gives a result which is finite and independent of Λ. To that order,

$$G(p) = \frac{1}{p^2 + m^2 + \Sigma_1(p)} \ .$$

Let us remark that to the next order in g, there appears a new divergence associated to the so–called rising sun diagram, that can be absorbed in a wave function renormalization constant by the rescaling

$$\Phi_0 = Z_\Phi^{1/2}\Phi \ .$$

Example 2.2 We will now work out the contribution to the coupling constant g coming from the diagram of **Fig. 2.2**.

The contribution to the vertex function coming from this diagram is

$$\Gamma_{(1)} \sim g^2 \int^\Lambda \frac{d^4k}{(2\pi)^4} \frac{1}{k^2 - m^2} \frac{1}{(P-k)^2 - m^2} \sim g^2 \log \Lambda \ , \tag{2.3}$$

where $P = p_1 + p_2$. We find in this case a logarithmic divergence that can be cured by the redefinition:

$$g_0(\Lambda, g) = g - B_1 g^2 \log \Lambda = g(1 - gB_1 \log \Lambda) \ ,$$

where B_1 is a numerical constant. This redefinition is again equivalent to adding to the lagrangian a new –fictitious – Φ^4 interaction with a Λ–dependent coefficient whose tree–level contribution to the vertex function cancels the one–loop infinity of the given diagram,

$$\mathcal{L}_{ct}^{(g)} = \frac{1}{4!}(g^2 B_1 \log \Lambda)\Phi^4 \ .$$

Now the point is, it would be possible, in principle, to repeat the former procedure ad infinitum. Every divergent diagram that we find at a given order in the perturbative series can be cancelled out by adding a proper counterterm to the action that will act at the previous order in the series. As there is an infinite number of divergent diagrams, we would end up with a finite theory with an infinite number of interactions. (This is in fact the case for non–renormalizable theories as, for instance, Quantum Gravity).

What can be proven in the case of the scalar Φ^4 theory – as well as in QED – is that all the infinities of the perturbative series can be taken care of by redefining a finite set of parameters, in our case, the mass and the coupling constant, and performing a proper rescaling of the fields. Each infinite appearing at higher order in perturbation theory will give rise to a new term in the definition of the parameters of the theory that end up being defined by a perturbative series of the form:

$$m_0^2(\Lambda, g) = m^2(1 + g A_1(\Lambda) + .. + g^n A_n(\Lambda) + ..) \ ,$$

$$g_0(\Lambda, g) = g(1 + g B_1(\Lambda) + .. + g^n B_n(\Lambda) + ..) \ , \tag{2.4}$$

and with a Lagrangian

$$\mathcal{L}_{ren} = \frac{1}{2}(\partial\Phi)^2 - \frac{1}{2}m^2\Phi^2 - \frac{g}{4!}\Phi^4$$

$$+ \frac{1}{2}\delta Z(\partial\Phi)^2 - \frac{1}{2}\delta m^2\Phi^2 - \frac{\delta g}{4!}\Phi^4 \tag{2.5}$$

$$\equiv \mathcal{L} + \mathcal{L}_{ct},$$

where the counterterm lagrangian includes the series defining the parameters to the given order.

As a summary, a QFT is said *renormalizable* when all the possible divergences arising at any order in perturbation theory can be absorbed in the bare parameters of the action in the way described above. This implies that at any order in perturbation theory, the result obtained in the computation of any given graph will have the same general structure as a term already existing in the action. As we will see later the requierement of renormalizability is a very restrictive one that leaves us with a very short list of theories.

The previous examples serve to raise two main questions. First, how is the substraction to be done? (as the physical parameters end up being the finite remnant of it, will they depend on the substraction point?). And, second, when will a given theory be renormalizable?.

Let us answer the last question first. There are in the QFT books precise prescriptions as to when a given theory is renormalizable and we encourage the students to go through them [9]. What we want to enlight here is the relation between the space–time dimensions and the renormalizability. In the previous example we saw a diagram with a quadratic dependence on the cutoff because we were working in four dimensions. Should we have asked the same question in two dimensions, we would get exactly the same expression for the self–energy correction but with the integral to be done in d^2k. In that case we would have encountered a logarithmic divergence. The diagram of **Fig. 2.2** would then have been finite. This shows trivially how renormalizability is linked with space–time dimensionality. In general, for our purposes we can say that a given theory is renormalizable if all the couplings — except the masses — are dimensionless where the dimension of a coupling is understood as everybody's naive scaling dimension. We shall come back to that later. As an example, it can be shown that in d=3+1 self–interacting scalar theories Φ^n with $n > 4$ are not renormalizable, while in d=1+1 there are no UV divergences (apart from those coming from normal ordering). All scalar interactions are renormalizable. As a curiosity we mention that no scalar theory is renormalizable in dimension $d \geq 7$.

2.2.4 Renormalization Prescriptions. The Renormalization Group

We will now answer the first of the previous questions, namely whether the physical observables of a QFT will depend on the way in which we redefine our bare parameters, i.e., on the renormalization procedure.

This procedure occurs in two steps. We must first devise a *regularization scheme*, namely, we must parametrize the infinities in a certain way by introducing an adequate cutoff. The two most popular schemes are the cutoff shown before, and dimensional regularization where one redefines the space–time dimension to be $d = 4 - \epsilon$ and the infinities occur in the physical limit $\epsilon \to 0$. It can be shown — although it is far from obvious — that a change in the regularization procedure amounts to a finite renormalization, i.e. , shifts the bare parameters in a finite amount.

The next step is to decide the "amount" of counterterm which we will add. As we have seen, the only requirement of a counterterm for a graph is that it cancels the divergence of that particular graph but it may contain any amount of finite part. Finite parts can be freely moved from $\mathcal{L}$ to $\mathcal{L}_{ct}$ (this translates in a reorganization of the perturbative series, a procedure which can be very useful even in the absence of infinities) and remain largely arbitrary. *A rule for choosing the value of the counterterm is called a renormalization prescription.*

The most common way of fixing the input parameters is to define

$$-\Sigma(p, m_R) \equiv p^2 - m_R^2 \quad at \quad p^2 = 0 \quad ,$$

$$\Gamma(p_1, p_2, p_3, p_4) \equiv g_R \quad at \quad p_i = 0 \quad .$$

In the absence of infrared divergences, this prescription is well defined.

Another possibility physically more appealing is to change the substraction point to a physical value:

$$-\Sigma(p, m_R) \equiv p^2 - m_R^2 \quad at \quad p^2 = \mu^2 \quad ,$$

$$\Gamma(p_1, p_2, p_3, p_4) \equiv g_R \quad at \quad p_i p_j = \mu^2 \left(\delta_{ij} - \frac{1}{4}\right) \ .$$

the later point is chosen so that the Mandelstam variables take the value $s = t = u = M^2$. As an example, we can identify the vertex function of QED with the physical charge of the electron in the Thompson limit.

In any case we end up with an action depending not only on the UV cutoff Λ but also on an arbitrary mass or momentum μ coming from the substraction point.

Let us notice that to find a physical point to define the parameters of the theory is usually not possible and the renormalization point remains arbitrary. There are nevertheless some common practices, for instance, a rescaling of the fields is usually done so that each propagator has a pole of unit residue (recall exercise 1.11). That fixes the wave function renormalization Z.

The essential feature provided by the renormalization group invariance is that, when computing an observable quantity, no matter how we choose the renormalization prescription, the final result is independent of the renormalization point.

One way of seing it is to notice that starting from the lagrangian (2.5) we obtain *physical* one–particle irreducible Green's functions $\Gamma_0^{(N)}(p_1, .., p_N; m_0, g_0, \Lambda)$ that depend on the bare parameters m_0, g_0 (2.4) —and, through them, on the cutoff Λ — and are finite in the limit $\Lambda \to \infty$. Since they are written in terms of the bare parameters, they know nothing about the renormalization point μ and we must have

$$\mu \frac{d}{d\mu} \Gamma_0^{(N)}(p_1, .., p_N; m_0, g_0, \Lambda) = 0 \ .$$

Expressing the bare parameters in terms of the physical parameters m, g, we have

$$\Gamma_0^{(N)}(p_1, .., p_N; m_0, g_0, \Lambda) = Z_\Phi^{-n/2} \Gamma^{(N)}(p_1, .., p_N; m, g, \mu, \Lambda) \ , \tag{2.6}$$

and using the chain rule to differenciate with respect to μ, we obtain the differential equation

$$\left[\mu \frac{\partial}{\partial \mu} + \mu \frac{\partial m}{\partial \mu} \frac{\partial}{\partial m} + \mu \frac{\partial g}{\partial \mu} \frac{\partial}{\partial g} - \frac{n}{2} \mu \frac{\partial \ln Z_\Phi}{\partial \mu}\right] \Gamma^{(N)} = 0 \ . \tag{2.7}$$

This is (a form of) the renormalization group equation. As mentioned, it only involves the renormalized Green's function $\Gamma^{(N)}$ which is finite in the limit $\Lambda \to \infty$. It is accoustumed to define the coefficients

$$\beta\left(g, \frac{m}{\mu}, \Lambda\right) \equiv \mu \frac{\partial g}{\partial \mu} \ . \tag{2.8}$$

$$\gamma_\Phi\left(g, \frac{m}{\mu}, \Lambda\right) \equiv \frac{1}{2} \mu \frac{\partial \ln Z_\Phi}{\partial \mu}$$

$$\gamma_m\left(g, \frac{m}{\mu}, \Lambda\right) \equiv \frac{1}{2}\mu\frac{\partial m^2}{\partial\mu}$$

They are analytic in the limit $\Lambda \to \infty$ and dimensionless. In terms of this *beta functions* (2.2.4), (2.7) is written

$$\left[\mu\frac{\partial}{\partial\mu} + \beta(g)\frac{\partial}{\partial g} + \gamma_m\frac{\partial}{\partial m} - \frac{N}{2}\gamma_\phi\right]\Gamma^{(N)} = 0 \ . \tag{2.9}$$

This is the usual form of the renormalization group equation. Its physical meaning is that *any change in the momentum scale where we are measuring (or defining) the theory, is absorbed in a redefinition of the parameters so that the physical result is unaffected.* The most representative quantity of the RG is the beta function of the coupling $\beta(g)$ that tells how the coupling constant of the theory changes when we change the energy at which the theory is defined (or the parameters are measured). This idea of a coupling constant changing with the energy is the source of all the later developments of the RG idea.

The renormalization group equation is an expression of the invariance of the physical magnitudes under the continuos group of rescalings $\Lambda(t) = \Lambda_0 e^{-t}$.

$$\Lambda(t_1 + t_2) = \Lambda(t_1)\cdot\Lambda(t_2) \ , \quad \beta(g) = \frac{dg}{dt} \ .$$

2.2.5 Uses of the Renormalization Group in QFT. Fixed Points

One of the most important practical implications of the previous result is the possiblity of moving the momentum scale at which a certain quantity is computed so that, starting from a strongly interacting theory, we can go to a kinematical region where g is small and perturbation theory is reliable. The way in which this is done is by computing the β function of the coupling and then integrating (2.2.4). We will discuss qualitatively examples of different behaviors that will be useful in the forecoming discussion.

exercise 2.3 *Compute the β–function of the Φ^4 theory at first order in the coupling constant g. Show that (up to factors of 2π) it is*

$$\mu\frac{\partial g}{\partial\mu} = \frac{3g^2}{16\pi^2} + O(g^3) \ . \tag{2.10}$$

Taking as an example our scalar theory, we see that (2.10) can be integrated to give

$$g = g_s\frac{1}{1 - \frac{3}{16\pi^2}g_s\ln\frac{\mu}{\mu_s}} \ , \tag{2.11}$$

where g_s is the value of g at some scale μ_s. From the solution we see that g increases with μ. If we start with a small coupling at some scale μ_s and increase the scale, we will eventually reach a regime in which perturbation theory is no longer applicable.

exercise 2.4 *Find the scale at which the solution (2.10) ceases to be reliable as a function of g_s, μ_s, i.e. , $g = \infty$ (Landau point).*

The scalar Φ^4 theory is an example of a theory in which perturbation theory becomes more reliable at large distances. The perturbative approach to define asymptotic states is to be trusted. QED is another such theory. (That there was a point at which g_{QED} became infinity was discovered by Landau in 1955 and led people to suspect QFT for a long time).

It is clear from this analysis that the behavior of a theory is ruled by the *sign* of its (complete) beta function. Would it have been negative in our previous example, we had encounter the opposite behavior: perturbation theory is only reliable at very short distances and we can not define the asymptotic states of the field. A theory in which the beta function starts at the origin and is always negative is called asymptotically free. Examples are provided by non abelian gauge theories as QCD.

A very interesting situation occurs when there is a point at which $\beta(g_F) = 0$. This is called a *fixed point* because if the coupling was originally at g_F it would remain there. The behavior near a fixed point can be analyzed by expanding β about it:

$$\mu\frac{\partial g}{\partial \mu} = (g - g_F)\beta'(g_F) + ...$$

The sign of $\beta'(g_F)$ is crucial. If $\beta'(g_F) < 0$, then g will be driven towards g_F as μ increases. The fixed point is *called ultraviolet stable*. The origin $g = 0$ of asymptotically free theories is this type of fixed point. The origin is *infrared stable* when $\beta'(0) > 0$.

QFT students are encouraged to read the complete discussion of these behaviors in section 4.6 of the Ramond book [4]. The statistical physics point of view can be seen in chapter 9 of [10].

2.2.6 Effective field theory. The Classification of Operators

The modern approach to the renormalization group as used in statistical physics differs in espirit from what has been described here. It is based on the crucial observation that cutting off the high–energy oscillations (virtual particles) of a theory can by compensated by the addition of new interactions (the counterterm Lagrangian) in the low–energy theory plus a modification of the parameters of the theory. One possibility then is to keep the cutoff finite and treat the modified Lagrangian as an effective action which will produce sensible results for the physics of processess involving energies much lower than the cutoff. The idea is that in any case virtual processes of arbitraryly large energies will always be unaccesible experimentally, the only manifestation of them will be throug their influence on the low–energy physics. What renormalization teaches us is that, at least for renormalizable theories, this influence produces only tiny effects.

With this new point of view, the cutoff is promoted to a physical quantity which sets the frontier between the low and the high–energy physics. Mathematically it is a way of controlling the badness of the divergences, physically it is equivalent to introducing an auxiliary particle to cancel the infinite contribution of real high–energy particles. The failure of QFT at very high energies indicated by the ultraviolet divergences implies that the region in which the existing framework of of QFT is valid should be separated from the region in which it is invalid. In this sense, all QFT's are effective. In the invalid region new physics and interactions should enter the scene. This point of view is described at length in [11] and [12].

To see how that can work in practice and to make the connection with the statistical physics applications clearer, let us introduce the basic notation for the path integral formulation of QFT [13] and show how the infinities arise in this context. We will stay with our real scalar field and use units $\hbar = c = 1$. The action for the scalar field is given, as mentioned in the former section by

$$S[\varPhi] = \int d^4x \mathcal{L}(\varPhi, \partial_\mu \varPhi) = \int d^4x \left[\frac{1}{2}\partial_\mu \varPhi \partial^\mu \varPhi - \frac{1}{2}m^2\varPhi^2 - \frac{\lambda}{4!}\varPhi^4\right] .$$

Following the same techniques as in the Feynman path integral formulation of quantum mechanics [14], the N–point Green's function of the theory, can be computed as

$$G(x_1, .., x_N) = N \int \mathcal{D}\varPhi e^{iS[\varPhi]}\varPhi(x_1)...\varPhi(x_N),$$

where the integral stands for the sum over all possible field configurations weighted with the corresponding classical action. N is an (infinite) normalization constant that ensures that $< 0|0 >= 1$:

$$N = Z^{-1} \quad , \quad Z = \int \mathcal{D}\varPhi e^{iS[\varPhi]} .$$

(Notice that this form of the Green's function is very similar to the one found in statistical mechanics. We shall insist on that later.) Formally, the Green's functions can be obtained by functional differentiation from the generating functional

$$W[J] = N \int \mathcal{D}\varPhi \exp\{i[S(\varPhi) + J\varPhi]\}.$$

where J is an external source. In particular, the two–point function (1.13) is obtained as

$$< 0|T\varPhi(x)\varPhi(y)|0 >= \frac{1}{W[0]} \frac{\delta^2}{\delta J(x)\delta J(y)} W[J]\Big|_{J=0}.$$

The only — great — difference with the quantum mechanical formulation is in the (un)–definition of the integration measure $\mathcal{D}\varPhi$. To make sense of the former definitions we must:

1. Perform a Wick rotation $t \rightarrow -i\tau$ to go to the Euclidean action:

$$-S_E[\Phi] = -\int d\tau d^3x \left[-\frac{1}{2}(\partial\Phi)^2 + P(\Phi) \right].$$

2. Discretize the spacetime introducing an ultraviolet cutoff a such that $x^\mu = n^\mu a$, and enclose the system in a box of volume V with $\tau \in [-T, T]$. Now we have an ordinary integral over a finite number of variables.

3. Take the limits $a \rightarrow 0$, $V \rightarrow \infty$, $T \rightarrow \infty$.

4. Go back to Minkowski space by analytical continuation.

The divergences are found in the third step. They are of two kinds: the infrared divergences $V, T \rightarrow \infty$ are common to all physical problems and one deals with them in no special way; the ones associated to the limit $a \rightarrow 0$ are the ultraviolet divergences characteristic of QFT.

Let us notice the similarity of the euclidean path integral defined in step 2 with the one corresponding to a statistical system *in spatial dimensions D=4*. Taking the action (2.1) as an example we can write:

$$S_E[\Phi] = g^{-1} \int d^4x \left[\frac{1}{2}(\partial\hat\Phi)^2 - \frac{1}{2}m^2\hat\Phi^2 - \frac{\hat\Phi^4}{4!} \right],$$

where $\hat\Phi = g^{1/2}\Phi$.

$$Z = \int \mathcal{D}\Phi e^{-S_E[\Phi]} , \tag{2.12}$$

is the partition function of a *classical* statistical system in $D = 4$ and at temperature $T = 1/g$.

The advantage of the path integral formulation in what follows is that we can separate high from low–energy modes at once in a very neat way. Readers uneasy with this slopy set of definitions can forget about everything and stay with the regulated euclidean partition function seing it in the statistical physics sense. QFT students should go to the text books cited and get acquinted with the subject.

In the Wilsonian effective field theory approach, the cutoff is used to separate different energy scales on a field theory as follows.

Say that we have a theory with a natural energy scale E_0, for instance the one associated to the lattice spacing a in a lattice system or to the width of the band in a solid, but we are only interested in the physics at very long distances (energies $E << E_0$). The way to proceed is, choose a cutoff Λ at an energy at or slightly below E_0 and divide the fields in the partition function (2.12) into high and low–energy frequency parts: $\Phi = \Phi_H + \Phi_L$, with

$$\begin{cases} \Phi_H = \Phi(w), & \text{for } 0 < w < \Lambda; \\ \Phi_L = \Phi(w), & \text{for } w > \Lambda. \end{cases}$$

(We will be more precise later in showing the concrete way in which this sepa-ration is done). By assumption we will only be interested in correlations of Φ_L. The first step in the RG program is to ask whether there is an effective action or Boltzmann weight $e^{S_\Lambda(\Phi_L)}$ such that when integrated over just the slow modes, will reproduce the low–frequency correlation functions. The answer is yes. By doing the path integral over the high–frequency part, we get

$$\int \mathcal{D}\Phi_L \mathcal{D}\Phi_H e^{S(\Phi_L, \Phi_H)} = \int \mathcal{D}\Phi_L e^{S_\Lambda(\Phi_L)},$$

where

$$e^{S_\Lambda(\Phi_L)} = \int \mathcal{D}\Phi_H e^{S(\Phi_L, \Phi_H)}.$$

We are left with an integral with an upper frequency cutoff Λ and an effective action $S_\Lambda(\Phi_L)$. This is known as the *low energy or Wilsonian effective action* to which we shall apply the renormalization group program.

The next step in the RG program is to lower the cutoff Λ to select only the very low energy modes of interest. We need to see how the parameters of the action evolve under this transformation (when this is done by choosing a layer of momenta – or energy – at the time and integrating out the modes on that layer, it can be seen as a discrete transformation). Now we use the basic idea of renormalization that all the effects of the very high-energy states on the Hibert space of the low-energy modes can be simulated by a set of new local interactions. In the case of renormalizable QFT, these local interactions (the counterterms) happenned to be of the same form as terms already existing in the action but this is not essential for our purposes[1]. In the actual context we will not insist on renormalizability and will consider in general that each layer of suppressed high-energy modes will produce local terms to which we shall only require that they preserve the symmetries of the action. Now instead of computing them for each particular process or diagram as we did in Sect. 2.2.3, we shall consider all possible terms as follows.

Expand the effective action S_Λ in terms of local operators $\mathcal{O}_i$,

$$S_\Lambda = \int d^D x \sum_i g_i \mathcal{O}_i \ . \tag{2.13}$$

The sum runs over all local operators (made of polynomials in the fields and derivatives) allowed by the symmetries of the problem.

Our interest is now to see which of those – infinitely many – terms will influence the low-energy physics. We then will follow their evolution as the energy is lowered. For that we need some dimensional analysis.

The scaling dimension δ_i of an operator $\mathcal{O}_i$ is estimated by assigning units to the field Φ so that the free action is dimensionless (in units $\hbar = c = 1$). It means that $\mathcal{O}_i$ scales whith the energy as E^{δ_i}. The accompanying coupling

---

[1]The vision that renormalizability is not an essential characteristics of useful field theories is strongly advocated in [11].

constant g_i has dimension units $E^{D-\delta_i}$ where D is the space-(time in the case of QFT) dimension. For instance, in the scalar field theory, the units of the field Φ are $E^{-1+D/2}$. An operator $\mathcal{O}_i$ with M Φ's and N derivatives has dimension

$$\delta_i = M(-1 + D/2) + N.$$

Now define *dimensionless* couplings $\lambda_i = \Lambda^{\delta_i - D} g_i$. For a physical process taking place at an energy scale E, the magnitude of a given term in the action (2.13) is estimated to be

$$\int d^D x \mathcal{O}_i \sim E^{\delta_i - D},$$

so that the given i'th term is of the order

$$\lambda_i \left(\frac{E}{\Lambda}\right)^{\delta_i - D} . \tag{2.14}$$

This is the key result of the analysis. From (2.14) we see that if $\delta_i > D$, this term becomes less and less important at low energies and is termed *irrelevant*. In a QFT analysis such an interaction is non renormalizable. Similarly, if $\delta_i < D$ the operator becomes more important as the energy is lowered and is called *relevant*. The corresponding QFT interaction is superrenormalizable. An operator with $\delta_i = D$ is equally important at all scales and is called *marginal* (strictly renormalizable).

What this analysis shows is that *the low-energy physics depends on the microscopic Hamiltonian only through the relevant and marginal couplings* (irrelevant couplings would become important if we measure small enough effects). That explains in the context of critical phenomena why very different microscopic systems share the same long-range physics.

As a summary what we have done is:

1. Start with an action that is a fixed point of the renormalization group transformations consisting of integrating out slices of high-energy fields. (The choice of the original action should be dictated by experience or physically-grounded prejudices as in the case of the Fermi liquid discussed in Sect. 2.4.2).

2. Perturb this critical action with all possible terms compatible with the symmetries of the problem.

3. Compute the scaling dimensions of the couplings and keep only those that are relevant or marginal. Analyze where the original action is driven by the new couplings.

2.3 The Renormalization Group in Statistical Physics

The field theory point of view about the renormalization group was radically changed after the work of Kadanoff (1965) [15] and Wilson (1971) [16].

Trying to understand the phase transition phenomena in spin lattice systems, Wilson succeeded in transferring the RG philosophy from the quantum relativistic field to the new arena.

The basic argument originates from the insight of Kadanoff that a diverging correlation length implies that there is a relationship between the coupling constants of an effective Hamiltonian and the length scale over which the order parameter is defined. He designed his trick of spin blocking with simultaneous finite renormalization of the coupling constant to obtain the scaling relations that were known from experiments. His argument was correct in spirit but not quite right in detail mostly due to the basic – wrong – assumption that the block spin Hamiltonian is of the same form as the original Hamiltonian.

The conceptual importance of Kadanoff's argument lies in the recognition that the coupling constants vary with a change in length scales. The question of precisely how they vary under repeated elimination of short length scales was addressed in the work of Wilson and is at the heart of the modern applications of the RG.

In this section we will remind briefly the use of the renormalization group in statistical mechanics, more concretely in the analysis of critical phenomena. The ideas and notation settled here can be directly translated to our problem. We will be very schematic in this part but students not familiar with it are strongly advised to study it either in some of the classical reviews on the subject [17], [18] or in the text books [19, 20, 21, 22].

2.3.1 Renormalization Group Analysis of Critical Phenomena

We come now to the implementation of the former ideas in the description of critical phenomena.

Consider a cubic lattice in D dimensions with a real scalar field $\Phi(\mathbf{n})$ defined at each site. The classical statistical mechanics of this system is described by the partition function

$$Z = \int \prod_{\mathbf{n}} d\Phi(\mathbf{n}) e^{S[\Phi(\mathbf{n})]} \quad , \tag{2.15}$$

that, as mentioned before, can be read either as a statistical physics partition function or as the regularized Feynman path integral of a $(D-1)+1$ QFT. We have again chosen for concreteness a real scalar field defined this time at the sites of a cubic lattice.

The relevant information for our problem is contained in the two–point function

$$G(\mathbf{n}_1 - \mathbf{n}_2) = \frac{\int \prod_{\mathbf{n}} d\Phi(\mathbf{n}) \Phi(\mathbf{n}_1) \Phi(\mathbf{n}_2) e^{S[\Phi(\mathbf{n})]}}{\int \prod_{\mathbf{n}} d\Phi(\mathbf{n}) e^{S[\Phi(\mathbf{n})]}} \quad .$$

For long separations this correlation function typically falls off exponentially as

$$G(\mathbf{n}_1 - \mathbf{n}_2) \sim e^{-|\mathbf{n}_1 - \mathbf{n}_2|/\xi},$$

where ξ is the correlation length. It is related to the mass gap m (the lowest excitation energy above the ground state) by the equation $\xi = 1/m$. When the system is at a critical point, the two–point function falls off like a power:

$$G(\mathbf{n}_1 - \mathbf{n}_2) \sim \frac{1}{|\mathbf{n}_1 - \mathbf{n}_2|^x},$$

where x is a critical exponent characteristic of the critical point. The critical point of QFT occurs when the mass gap is $m = 0$ and the correlation length diverges. *The starting point of the RG analysis is the remark that different systems with microscopically distinct Hamiltonians have the same critical exponents.* This leads to the hypothesis of universality according to which the physics of the critical point is governed by an effective action independent of the local interactions.

We will follow here the analysis sketched in Sect. 2.2.6. The scalar field sitting on the cubic lattice is described in Fourier space by

$$Z = \int \prod_{|k| \leq \pi/a} d\Phi(\mathbf{k}) e^{S(\Phi(\mathbf{k}))} \ ,$$

the allowed momenta $\mathbf{k}$ lie within a cube of side $2\pi/a$.

Let us start by analyzing more closely what the mode elimination does to the action. We will be interested only on modes within tiny ball of size Λ/s (with s very large) centered at the origin so we define the slow and fast modes as

$$\Phi_L = \Phi(k) \quad \text{for } 0 < k < \Lambda/s \quad \text{(slow modes)} \ ,$$

$$\Phi_H = \Phi(k) \quad \text{for } \Lambda/s \leq k \leq \Lambda \quad \text{(fast modes)} \ .$$

The construction of the Wilsonian effective action proceeds by integrating out the fast modes. To do that we notice that the action can be written as

$$S(\Phi_L, \Phi_H) = S_0(\Phi_L, \Phi_H) + S_I(\Phi_L, \Phi_H) \ ,$$

where the free part is quadratic and separates into the slow and fast pieces while the interaction S_I mixes the two. Integration over the fast modes gives:

$$Z = \int \prod_{0 \leq k \leq \Lambda/s} d\Phi_L(k)$$

$$\times \prod_{\Lambda/s \leq k \leq \Lambda} d\Phi_H(k) e^{S_0(\Phi_L)} e^{S_0(\Phi_H)} e^{S_I(\Phi_L, \Phi_H)}$$

$$\equiv \int [d\Phi_L] \int [d\Phi_H] e^{S_0(\Phi_L)} e^{S_0(\Phi_H)} e^{S_I(\Phi_L, \Phi_H)}$$

$$= \int [d\Phi_L] e^{S_0(\Phi_L)} \int [d\Phi_H] e^{S_0(\Phi_H)} e^{S_I(\Phi_L, \Phi_H)}$$

$$\equiv \int [d\Phi_L] e^{S'(\Phi_L)} \ .$$

This formula defines the effective action. It can also be written as

$$e^{S'(\Phi_L)} = e^{S_0(\Phi_L)} \langle e^{S_I(\Phi_L, \Phi_H)} \rangle_{0H} \ , \tag{2.16}$$

where $\langle \ \rangle_{0H}$ denotes averages with respect to the fast modes with action S_0. This formula can be expanded in a Wick expansion similar to the one applied to the S operator (1.22) in Sect. 1.3.8.

In order to see the influence of the modes that we have eliminated on the parameters of the low-energy theory, we will move the cutoff changing s in a way that is equivalent to integrate a single slice of high-energy phase space at a time. (Ultimately, the region of interest will be $s \to \infty$). Let us start by moving the cutoff as $\Lambda \to \Lambda/s$.

Suppose that before the mode elimination we had

$$S(\Phi) = r\Phi^2 + u\Phi^4 + \dots \ ,$$

and after,

$$S'(\Phi_L) = r'\Phi_L^2 + u'\Phi_L^4 + \dots \ .$$

What we would like is to compare r to r', u to u', etc. The problem with this is that the new and old theories are defined in two different kinematical regions as in this approach the couplings are also functions of the momenta; for instance, the $u\Phi^4$ could be the shorthand for

$$\int dk_1 dk_2 dk_3 dk_4 \delta(k_4 + k_3 + k_2 + k_1) \cdot u(k_4, k_3, k_2, k_1) \Phi(k_4)\Phi(k_3)\Phi(k_2)\Phi(k_1) \ .$$

What we do is to define new momenta after mode elimination: $k' = sk$ which run over the the same range as k did before the mode elimination. This transformation is the renormalization group transformation. The last step is to rescale the fields so as to absorb any trivial change of parameters and write

$$\Phi'(k') = Z^{-1}\Phi_L(k'/s) \ ,$$

choosing Z so as to obtain a canonical kinetic term, for instance. The final action S' will then be expressed in terms of this new field.

Now we have completed the three steps of the RG transformation in this context that follow closely but not exactly what we did in QFT. As a summary, we quote the steps together with their QFT counterparts.

1. Start with an action with two different energy scales on it, parametrized by certain couplings $S(u, r)$. Introduce a cutoff of the order of the (big) energy scale of the system and eliminate the fast modes by reducing the cutoff from Λ to Λ/s. Get a (different) effective action parametrized in terms of new couplings $S'(r', u')$.

   Introduce a cutoff in the momentum integrals and absorb the influence of the eliminated modes on a change of parameters. Go from $\mathcal{L}(g_0)$ to $\mathcal{L}(g')$.

2. Introduce rescaled momenta $k' = sk$ that now go all the way to Λ, and rescaled fields $\Phi'(k') = Z^{-1}\Phi_L(k/s)$.

   Redefine the parameters and fields so as to absorb the influence of the eliminated modes – the cutoff – into a constant fixed at a certain scale. Send the cutoff to infinite. Changes in the scale produce changes in the coupling constants of the theory so that the solutions of it (physical Green's functions) remain invariant.

3. Express the new action S' in terms of the new fields and couplings.

   Use the *same action* parametrized with the renormalized couplings $\mathcal{L}_{\text{ren}}$ to compute finite, cutoff-independent results.

Before proceeding further let us notice that in QFT the RG transformations really form a group of – continuum – symmetry transformations of the theory (or at least of its solutions). Under the RG transformations, the action of the system remains the same action parametrized by a different set of couplings. The transformations defined here are of a very different nature. The procedure of mode elimination (a mimic of the Kadanoff blocking) diminishes the number of degrees of freedom and simplifies the physical problem, while preserving its long-range properties. However, at each step of the RG transformation, we create a *different* model for the same physical phenomena, based on preserving the long-range properties of a big statistical system. *What the RG do is to map actions defined in a certain phase space to actions in the same phase space.* Thus if the initial action is represented as a point in the coupling constant space, it will flow under the RG to another point in the same parameter space. We can obviously compose two RG transformations R_1, R_2 to produce a third $R_1 \cdot R_2 = R_{12}$ (what amounts to move down the cutoff in two succesive steps and eliminate the corresponding modes) but we certainly cannot take the "inverse" transformation and restore the eliminated degrees of freedom. Althoug retainig the name for historical reasons, in this case the renormalization group transformations only form a semigroup and do not correspond to any symmetry of the action.

The concept of a fixed point is now operating on a space of different theories (actions). *A fixed point S^* is an action that does not flow under the RG transformations.* To see the connection with the critical phenomena, we observe that if a system has a correlation length ξ before the RG scaling, it will flow under the RG to a system with with $\xi' = \xi/s$. This implies that *if a system is a fixed point of the RG, its correlation length must be either zero or infinite.*

The RG explains the universality of critical phenomena as follows. Let S_A and S_B be two different critical actions defined in the full k space. Each is parametrized by a set of coupling constants and they will be represented as two different points in the space of all possible actions. They do have different observables. Now we want to study their very long distance behavior, in particular the decay of the two–point function. We eliminate the high–energy modes and study the RG flow. What we see is that both actions end up at the same fixed point action S^* where the flow stops. This explains why they share the same-long distance physics and, in particular, the same critical exponents.

The previous classification of operators is described in this language as follows. In the space of all actions (whose dimension is given by the number of coupling constants parametrizing the theories), we sit at the fixed point S^* and study deviations $S - S^*$. Any deviation lying at the critical hyperplane is termed irrelevant as it renormalizes to zero and preseves the long distance physics. In particular a gapless system will stay gapless if an irrelevant interaction is added. A deviation that takes the system off the critical hyperplane is relevant and will drive the system to a different fixed point with entirely different physics. Marginal perturbations are the most important and must be studied closely as they may become relevant (usually when they are associed with diagrams that develop infrared divergences). They should not open a gap if they are really marginal.

The study of the phase diagram of a given theory, i.e. , the localization of the fixed points and the different flows around them is crucial in order to know what is the real physics that the microscopic action describes. This analysis is usually very complicated as it involves the knowledge of the complete β–function of the theory and can only be done in very special cases. Concrete examples of such an analysis will be provided in the chapters devoted to the Luttinger liquid behavior.

2.4 Renormalization Group Analysis of the Fermi Liquid

The concepts of renormalization and RG ideas in condensed matter go back to the P. Anderson's "poor man scaling" [23] that led to his solution of the Kondo problem [24], and have been a constant presence in the field ever since[2]. However the problems treated with the RG method in condensed matter were essentially one-dimensional.

The new approach that we will discuss here started in [26], and [27] where RG techniques were used to study the stability of gapless Fermi systems under perturbations in spatial dimensions $D \geq 2$. This new vision has gained increasing popularity due not only to the beauty of the formulation but to the expectations of obtaining new results in the vast and complicated field of correlated electrons. Although it is not as yet clear that such new results will be attainable at all in the higher-dimensional case (the RG approach has certainly produced new results in the D=1 case as will be discussed in the next chapter), this approach gives us a new insight on the problem and has opened a way of studying possible departures from Fermi liquid behavior as will be described at the end of this section.

The organization of the section is as follows. First we introduce the notation and give a presentation to the precise technique of the RG by solving the Gaussian model. Next we give a general description of how the Fermi liquid problem

---

[2]See specially the RG diagram in the P. W. Anderson's Conference Summary of the 1973 Nobel Symposium quoted in [25].

fits into the previous scheme following the presentation of [3]. The aim here is to catch on the concepts and the line of reasoning, hence we ignore in this part a set of delicate issues that are inessential for the understanding of the subject. We comment on these in the next paragrph.

We end up with an overview of recent ideas on non-Fermi liquid fixed points that have arisen as direct applications of the former and whose permanence as time goes on is unclear. In this sense the last paragraph differs very much from the previous ones.

2.4.1 The Gaussian Model

We will start by demonstrating the details of how the RG is used in the actual context by solving the Gaussian model in which all the steps of the RG including the path-integral can be done exactly. The discussion of the Fermi liquid will be done following the same steps and using the same notation. The Gaussian model can be obtained from many different physical systems; consider for instance a system of spins in a hypercubic lattice in $D = 4$ whose partition function is

$$Z = \int_{|k|<\Lambda} [d\Phi d\Phi^*] e^{S(\Phi,\Phi^*)} ,$$

where Φ is a complex scalar field; the integration measure is

$$[d\Phi d\Phi^*] = \prod_{|k|<\Lambda} \frac{1}{\pi} dRe\Phi \, dIm\Phi ,$$

and the free action is

$$S(\Phi,\Phi^*) = \int_{|k|<\Lambda} \Phi^*(k) J(k) \Phi(k) \frac{d^4k}{(2\pi)^4} ,$$

$$J(k) = 2[(\cos k_x - 1) + (\cos k_y - 1)(\cos k_z - 1) + (\cos k_t - 1)] .$$

This action is obtained by Fourier transformation of the nearest-neighbors action

$$S(\Phi,\Phi^*) = -\frac{1}{2} \sum_{n,i} |\Phi(n) - \Phi(n+i)|^2 .$$

Since we are interested in small k, we approximate $J(k)$ by its leading term in the Taylor series and have

$$S_0(\Phi,\Phi^*) = -\int_{|k|<\Lambda} \Phi^*(k) k^2 \Phi(k) \frac{d^4k}{(2\pi)^4} , \qquad (2.17)$$

This is the Gaussian model. The functional integrals are products of ordinary Gaussian integrals, one for each **k**. All its correlations can be computed by using Gaussian integrals only. Charge conservation implies that the only non-vanishing averages to compute are these containing an even number of fields in the combination $\Phi\Phi^*$.

The two-point function of the Gaussian model is

$$\langle \Phi^*(\mathbf{k_1})\Phi(\mathbf{k_2})\rangle = \frac{(2\pi)^4 \delta^4(\mathbf{k_1} - \mathbf{k_2})}{k_1^2} \equiv (2\pi)^4 \delta^4(\mathbf{k_1} - \mathbf{k_2})G(\mathbf{k_1}) \equiv <\bar{2}1> \ ,$$

and the four-point function is written in the same notation as

$$\langle \Phi^*(\mathbf{k_4})\Phi^*(\mathbf{k_3})\Phi(\mathbf{k_2})\Phi(\mathbf{k_1})\rangle = <\bar{4}2><\bar{3}1> + <\bar{4}1><\bar{3}2> \ .$$

The power-law decay of the two-point function ($1/k^2$ decay in Fourier space is equivalent in D=4 to $1/r^2$) means that (2.17) is a critical point. We will now demonstrate that it is also a fixed point of the RG.

The first step in the RG is to integrate over the modes in the stripe

$$\Lambda > k > \Lambda/s \ .$$

As there is no interaction term present, (2.17) splits into fast and low pieces and we are left with an effective action

$$S'(\Phi_L) = -\int_{|k|<\Lambda/s} \Phi_L^*(\mathbf{k})k^2\Phi_L(\mathbf{k})\frac{d^4 k}{(2\pi)^4} \ . \tag{2.18}$$

Now we rescale the fields and find their appropriate scaling dimension:

$$\Phi'(\mathbf{k'}) = \zeta^{-1}\Phi_L(\mathbf{k'}/s) \ , \tag{2.19}$$

By counting powers of momenta in (2.18) we see that (2.18) remains invariant under the scaling $k \to k' = sk$ if the field Φ scales with $\zeta = s^3$, i.e.,

$$\Phi'(s\mathbf{k}) = s^{-3}\Phi_L(\mathbf{k}) \ . \tag{2.20}$$

With this choice we have

$$S'(\Phi_L') = S'(\Phi') = S(\Phi) = S^* \ .$$

The next step is to study the stability of the fixed point under perturbations and classify these as relevant, irrelevant or marginal. The only perturbations that we have to consider involve the even combinations of fields discussed before.

The first are quadratic. They are of the form

$$\delta S = -\int_{|k|<\Lambda} \Phi^*(\mathbf{k})r(k)\Phi(\mathbf{k})\frac{d^4 k}{(2\pi)^4} \ . $$

Assuming that the perturbations are short-ranged, the coupling function r can be expanded as

$$r(k) = r_0 + r_2 k^2 + \dots \ .$$

The interactions induced by the quadratic perturbation scale as

$$r'(k') = s^2 r(k'/s) \ ;$$

using (2.20) we find the scaling dimension of the quadratic operators to be

$$r_0' = s^2 r_0 \ , \quad r_2' = r_2 \ , \quad r_4' = s^{-2} r_4 \quad \dots \ .$$

The first term is a mass term $r_0 = m_0^2$ which is relevant, r_2 is marginal, and the rest are irrelevant.

Quartic perturbations

They can be discussed in the same way as the quadratic perturbations but now we will encounter all the complications of true interactions. Quadratic perturbations are of the form

$$\delta S = -\frac{1}{4!} \int_{|k|<\Lambda} \prod_{i=1}^{4} \frac{d^4 k_i}{(2\pi)^4} (2\pi)^4 \delta(k_4 + k_3 - k_2 - k_1)$$

$$\cdot u(k_1 k_2 k_3 k_4) \Phi^*(k_4) \Phi^*(k_3) \Phi(k_2) \Phi(k_1)$$

$$\equiv -\int_{\Lambda} u(4321) \Phi^*(4) \Phi^*(3) \Phi(2) \Phi(1) ,$$

where the coupling function is invariant under the exchange of the first two (or last two) arguments. The new feature is that quadratic perturbations induce true interactions that mix the fast and slow modes. We use (2.16)

$$e^{S'(\Phi_L)} = e^{S_0(\Phi_L)} \langle e^{\delta S(\Phi_L, \Phi_H)} \rangle_{0H} \equiv e^{S_0 + \delta S'} ,$$

and to it we apply the cummulant expansion

$$\langle e^{\Omega} \rangle = e^{\langle \Omega \rangle + \frac{1}{2}(\langle \Omega^2 \rangle - \langle \Omega \rangle^2) + \cdots}$$

to get

$$\delta S' = \langle \delta S \rangle + \frac{1}{2}(\langle \delta S^2 \rangle - \langle \delta S \rangle^2) + \cdots . \tag{2.21}$$

The next steps follow exactly the expansion of (1.22). The Wick theorem and the Feymann rules can be translated to this problem with only minor changes.

Using the symmetries of the problem and the Feymann rules adequated to this case we find that the first term in the expansion, $\langle \delta S \rangle$ produces, after averaging over the fast modes, two types of terms to be kept in the effective action. A tree-level term comes from terms in the expansion with only slow modes, plus there are mixed terms with two slow and two fast modes.

The *tree-level* term can be obtained by simple substitution of Φ by Φ_L in δS. Rewriting it in terms of rescaled momenta and fields it gives rise to the interaction

$$\delta S'_{4,tree} = -\frac{1}{4!} \int_{|k|<\Lambda} \prod_{i=1}^{4} \frac{d^4 k_i}{(2\pi)^4} (2\pi)^4 \delta(k_4 + k_3 - k_2 - k_1)$$

$$u(k_4'/s, k_3'/s, k_2'/s, k_1'/s) \Phi'^*(k_4') \Phi'^*(k_3') \Phi'(k_2') \Phi'(k_1') .$$

The renormalized four-point function

$$u'(k_1', k_2', k_3', k_4') = u(k_4'/s, k_3'/s, k_2'/s, k_1'/s)$$

can be Taylor expanded as $u = u_0 + O(k)$ where u_0 is marginal and the rest of the terms are irrelevant. We then see why the scalar Φ^4 interaction in four dimensions has only a coupling constant instead of a coupling function. The only fine point of this analysis is to notice that the delta function scales oppositely to the momenta.

Now we come to the issue of the mixed terms. *What they do is to renormalize the quadratic interaction.*

Note 2.5 (For QFT students) This is why masses are generated by radiative corrections in QFT even if we don't put them in the action, unless there is a symmetry – like chiral symmetry – preventing them from appear.

The mixed terms are of the form

$$\delta S_2'(\Phi_L) = \frac{1}{4!} u_0 \langle \int_{|k|<\Lambda} [\Phi_H^*(4)\Phi_L^*(3) + \Phi_L^*(4)\Phi_H^*(3)]$$

$$[\Phi_H(2)\Phi_L(1) + \Phi_L(2)\Phi_H(1)]\rangle_{0H} \ .$$

Performing the averages over the fast modes it can be seen that all the products give the same contribution to the action that ends up being

$$\delta S_2'(\Phi_L) = -u_0 \int_{|k|<\Lambda/s} \frac{d^4k}{(2\pi)^4} \Phi_L^*(k)\Phi_L(k) \int_{\Lambda/s}^{\Lambda} \frac{d^4k}{(2\pi)^4} \frac{1}{k^2} \ ,$$

after performing the last integral we get

$$\delta S_2'(\Phi'(k')) = -u_0 s^2 \int_{|k|<\Lambda} \frac{d^4k}{(2\pi)^4} \Phi'^*(k')\Phi'(k')\Lambda^2 \frac{1}{2}[1 - \frac{1}{s^2}]\frac{2\pi^2}{(2\pi)^4} \ . \qquad (2.22)$$

Diagrammatically, the terms coming from the cummulant expansion (2.21) inducing corrections to the quadratic coupling can be understood as a tree-level four-point function in which the two external legs corresponding to fast modes close on each other to form the diagram of **Fig. 2.1** with slow modes in the external legs and with the momentum k running on the loop being of order Λ. (Diagrams with four slow modes are included in the tree-level action while those with four fast modes degenerate in a disconnected two-loop diagram).

From (2.22) we can read the change induced by the quartic interaction on the mass term:

$$\delta r_0 = \frac{u_0 \Lambda^2}{16\pi^2}(s^2 - 1) \ .$$

This analysis teaches us that we must consider both quadratic and quartic terms to start with. To the order at which we are working, the flow of the (dimensionless) couplings is

$$r_0' = s^2[r_0 + \frac{u_0}{16\pi^2}(1 - \frac{1}{s^2})] \qquad (2.23)$$

$$u_o' = u_o \ .$$

Notice that (2.23) is nothing but the mass correction (2.2). If we consider an infinitesimal change $s = 1 + t$, the former equations can be written as the differential equations

$$\frac{dr_0}{dt} = 2r_0 + \frac{u_0}{8\pi^2}$$

$$\frac{du_0}{dt} = 0 .$$

One-loop corrections

The analysis done so far has been a tree-level analysis. To see the flow of the quartic coupling we must go to the next order.

Note 2.6 Observe that the expansion we are refering to is a loop expansion corresponding to a power series in $\hbar$ where each particular term may contain different powers of the coupling constant. Graphs like those in **Fig. 2.1** (of order g) and **Fig. 2.2** (order g^2) are both one-loop graphs.

The flow of u_0 at one loop comes from the next term in the cummulant expansion

$$\frac{1}{2}[\langle(\delta S)^2\rangle - \langle\delta S\rangle^2] . \tag{2.24}$$

Working out all the terms coming from (2.24) we end up with the diagrams of **Fig. 1.7** where the graphical representation of the cummulant expansion is again such that only fast modes propagate in the internal lines. We note that in [2] these diagrams are named as:

$$(a) = \text{BCS} , \quad (b) = \text{ZS}' , \quad (c) = \text{ZS} .$$

We shall come back to them later.

The computation to be done is the one in Example 2.2 of Sect. 2.2.3 with the only proviso that the integral in the loop is limited to the stripe $\Lambda < k < \Lambda/s$. We can read off the result from (2.3) and write down

$$u_0' = u_0 - bu_0^2 ,$$

where b is now proportional to log Λ. Writing

$$\frac{|d\Lambda|}{\Lambda} = dt \quad \text{i.e.} \quad \Lambda(t) = \Lambda_0 e^{-t} ,$$

we obtain the final result at one loop:

$$\frac{dr_0}{dt} = 2r_0 + au_0$$

$$\frac{du_0}{dt} = -bu_0^2 , \tag{2.25}$$

where a and b are positive numbers. The equation for u can be integrated to obtain the known result (2.11) that in our notation is

$$u_0(t) = \frac{u_0(0)}{1 + bu_0(0)t} \ .$$

The conclusion is the one stated on the example of Sect. 2.2.5 (notice that the energy increased with μ there as we were interested in the limit of very high energies). A positive coupling increases as the energy is increased, but decreases to zero at low energies. Repulsive interactions are *marginally irrelevant*. If we start with a negative coupling (what in QFT does not make sense), we find that it grows larger at small energies. An attractive quartic coupling is then *marginally relevant*. (See the original ref. [2] for a diagram showing the RG flows of the Gaussian model).

We end this section by noting that the only fixed point of the RG equations (2.25) is the Gaussian point at the origin.

2.4.2 The Fermi Liquid as Fixed Point of the Renormalization Group

In this paragraph we will boldy apply the former procedure to the problem of interacting electrons in the conduction band of a metal in three spacial dimensions.

Generally speaking, the major difficulties that the RG treatment has in $D \geq 2$ come from the extended nature of the vacuum – the Fermi surface –. This translates first to the problem of properly define the scaling towards it and, second, it makes it hard to controle the low-energy processes as they do not necesarily carry small momenta. An example is the scattering process in which an electron – or quasiparticle – very close to the (circular) Fermi surface in D=2, gets scattered to a point on the Fermi surface in the oposed diameter. Such process has an almost zero energy and a momentum as large as $2k_F$.

In QFT as well as in critical phenomena the low-energy or long wavelengths modes are concentrated on a tiny ball around a single point, $k = 0$ in momentum space (see the comments of Sect. 1.2.1). The running couplings are functions of k that can be Taylor expanded around $k = 0$ to obtain a few constants (the more powers of momenta a coupling term has, the more irrelevant it is at low energy). Now the running couplings will be functions, not only of the momenta, but also of the coordinates of the Fermi surface that will survive the scaling. This leads often to having to consider the infinite set of coefficients of their expansions as is the case with the Landau parameters. We shall comment on that later.

As advertised in the introduction, for the time being we will ignore all these fine points and give a general understanding of the problem.

Let us first notice that our system fits in the effective action scheme. What we have is a metal with a natural energy scale associated to the width of its conduction band which is typically of order $1 - 10$ eV. We consider $c = M_{ions} =$

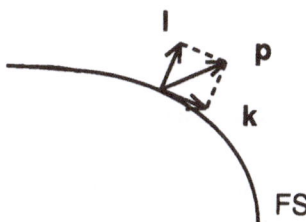

Fig. 2.3. The momentum decomposition.

∞. So we can set our high-energy cutoff at $E_0 \sim 1eV$. The physics we are interested on is that of the very low energy (about zero) excitations over the vacuum (Fermi surface) so we will try to integrate over the high-energy modes in the band and get an effective action for modes with $\epsilon(\mathbf{p}) - \epsilon_F << E_0$.

We will perform the analysis in various steps.

1. We start by writing down an effective QFT action for the low energy excitations of the form suggested by the Landau ideas, namely

$$S_0 = \int dt d^3\mathbf{p}\{i\Psi_\sigma^\dagger(\mathbf{p},t)\partial_t\Psi_\sigma(\mathbf{p},t) - [\epsilon(\mathbf{p}) - \epsilon_F]\Psi_\sigma^\dagger(\mathbf{p},t)\Psi_\sigma(\mathbf{p},t)\} , \qquad (2.26)$$

where we do not specify a particular form for the dispersion relation of the quasiparticles, (we do not even assume that it be linear in p), as it might get renormalized later. It will be treated as a parameter of the theory.

2. Our momenta live in the band $0 < p < E_0$ and we are interested in the behavior of the theory in the limit $E \to 0$ or, equivalently, $p \to p_F$. So we decompose the momenta as $\mathbf{p} = \mathbf{k} + \mathbf{l}$, where $\mathbf{l}$ is in the direction perpendicular to the Fermi surface and $\mathbf{k}$ is parallel to it (see **Fig. 2.3**).
Now we perform the rescaling $E \to sE$, $s < 1$. This in turn implies

$$\mathbf{k} \to \mathbf{k} \;\; ; \;\; \mathbf{l} \to s\mathbf{l} \;\; ; \;\; dt \to s^{-1}dt \;\; ; \;\; d\mathbf{l} \to sd\mathbf{l} \;\; .$$

Under this scale transformation, the action transforms as

$$S_0 \to \int dt d^2\mathbf{k}d\mathbf{l}\{i\Psi_\sigma^\dagger(\mathbf{p},t)\partial_t\Psi_\sigma(\mathbf{p},t) - lv_F(\mathbf{k})\Psi_\sigma^\dagger(\mathbf{p},t)\Psi_\sigma(\mathbf{p},t)\} ,$$

where we have made the substitution $[\epsilon(\mathbf{p}) - \epsilon_F] = lv_F(\mathbf{k}) + \mathcal{O}(l^2)$.

Now we see that if the scaling dimension of the field is $[\Psi] \sim s^{-1/2}$ the action S_0 remains invariant and hence is a fixed point of the renormalization group.

3. The next step is to perturb the fixed point action with all possible local interactions and see if there are relevant or marginal perturbations that could take the system out of the fixed point. Notice that this procedure is simpler but equivalent to finding the effective action and expanding it in local operators. We simply write down all possible terms compatible with the symmetries of the original system. Due to the charge conservation we only have to consider terms with pairs of fields $\Psi_\sigma^\dagger(\mathbf{p},t)\Psi_\sigma(\mathbf{p},t)$ and any number of momenta.

Quadratic perturbations

They are of the form

$$\int dt\, d^2k\, dl\, \mu(\mathbf{k})\Psi_\sigma^\dagger(\mathbf{p},t)\Psi_\sigma(\mathbf{p},t),$$

and a simple power counting shows that they scale as s^{-1} and are *relevant* perturbations in the limit $s \to 0$. They can be absorbed in a redefinition of the dispersion relation $\epsilon(\mathbf{p})$ exactly as we saw in the QFT discussion of the RG. What it means is that, should we have started out from a free Fermi gas, these terms would have changed the dispersion relation and hence the form of the Fermi surface.

Quartic perturbations

They are

$$\int dt\, d^2k_1\, dl_1\, d^2k_2\, dl_2\, d^2k_3\, dl_3\, d^2k_4\, dl_4\, [V(\mathbf{k}_1, \mathbf{k}_2, \mathbf{k}_3, \mathbf{k}_4)$$

$$\Psi_\sigma^\dagger(\mathbf{p}_1)\Psi_\sigma(\mathbf{p}_2)\Psi_\sigma'^\dagger(\mathbf{p}_3)\Psi_\sigma'(\mathbf{p}_4)\delta^3(\mathbf{p}_1 + \mathbf{p}_2 - \mathbf{p}_3 - \mathbf{p}_4)].$$

Assuming that the δ-function does not scale with s, which will be true for generic values of the momenta (as only the components in the normal direction scale), these perturbations scale as s and are *irrelevant* as $s \to 0$. This exhausts the possibilities. Any other term will contain higher power of momenta and be even more irrelevant as $s \to 0$.

exercise 2.7 *Check explicitly that terms with six or more fields are irrelevant. Think of a way to extend this analysis to electron–phonon interactions.*

The conclusion to extract from the former analysis is that if we start from a system of interacting electrons whose ground state has the symmetry of the free Fermi gas (or if we start with a Fermi gas system), its low energy behavior will always be that of the free gas with a modified dispersion relation (effective mass) irrespective of the nature of the interactions. This is what Landau assumed from intuition and can be seen as the effective-mass approach to the Fermi liquid.

But we know that this can not possibly be the complete story for various reasons. First, we know that the Fermi liquid supports bosonic excitations that can never be produced from the action (2.26). Moreover we also know that the Fermi liquid can develop a BCS instability. What was left out in the previous analysis is a more careful treatment of the δ–function scaling properties. In fact there are kinematical situations, i.e. , particular values of the incoming momenta for which the δ–function becomes degenerate and depends explicitly on l. It scales inversely to it $\delta(P) \sim s^{-1}$. Obviously that happens for the cases described in the previous chapter when discussing the bosonic branch of the

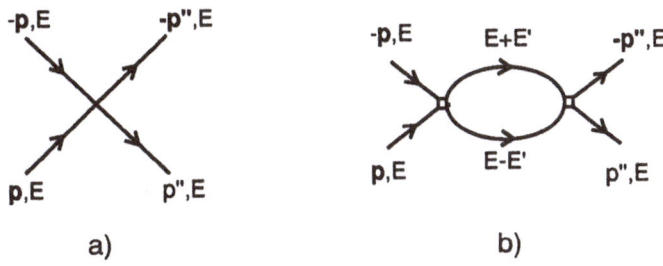

Fig. 2.4. Diagrams contributing to the beta function of the four–Fermi interaction at one loop order.

Fermi liquid spectrum. In particular if $p_1 = -p_2$, and the total incoming momentum is zero, (or if the total momentum P of any two lines in a four Fermi vertex is constrained to be zero) the quartic interactions become marginal. The graphs with no irrelevant interactions form a geometric series represented in **Fig. 1.8**. In order to take them into account we must sum over all graphs or, equivalently, perform the RG analysis which follows. The result is the LFL. Expectation values of currents are modified by the interactions whose contribution is parametrized by the Landau interaction function.

The final conclusion is then that *a system of free or interacting electrons behaves at very low energies as a Landau Fermi liquid with characteristic effective mass and interaction function.* (The free gas can be seen as a particular Fermi liquid with $m^* = m_e$ and $f = 0$.)

exercise 2.8 *Repeat the previous analysis for an electron system in (1+1) dimensions.*

The beta function of the marginal four Fermi interactions and the BCS instability

We shall end this paragraph with the explicit computation of the beta function of the marginal four–Fermi interaction generated for the particular kinematics discussed before. We shall choose the coupling *function* $V(k_1, k_2, k_3, k_4)$ to be a constant (any dependence on the momenta would render the interaction irrelevant). The diagrams to compute to first order in perturbation theory are shown in **Fig. 2.4**.

The contribution of diagram b) to the vertex function is

$$V^2 \int \frac{dE\, d^2k'\, dl'}{(2\pi)^4} \frac{1}{[(1 + i\epsilon)(E + E') - v_F(k')l'][(1 + i\epsilon)(E - E') - v_F(k')l']},$$

where the integral is cutoff in momenta at a scale E_0. In terms of the density of states at the Fermi energy

$$N = \int \frac{d^2k'}{(2\pi)^3} \frac{1}{v_F(k')},$$

the vertex function results to be

$$V(E) = V - V^2 N\{\ln(E_0/E) + \mathcal{O}(1)\} + \mathcal{O}(V^3).$$

The beta function of V is then

$$E\frac{\partial V(E)}{\partial E} = NV^2(E) + \mathcal{O}(V^3),$$

which can be integrated to give

$$V(E) = \frac{V}{1 + NV\ln(E_0/E)} \ . \tag{2.27}$$

This equation shows that a repulsive interaction $V > 0$ grows weaker at low energy – the coupling is marginally irrelevant – while an attractive interaction $V < 0$ as the one induced by phonons grows stronger. The coupling in this case is marginally relevant and drives the system away from the Fermi liquid fixed point to the BCS phase where a gap opens in the spectrum.

2.4.3 Comments on Fine Points

Some of the steps followed in the previous paragraph hide tricky points that we will now disentangle. The most noticeable, the slopy argument about the scaling of the δ-function making the quartic interactions irrelevant – or marginal –. This argument can be made rigorous as follows.

As mentioned in the opening of this section, most of the troubles with the electronic system come from the extended nature of the Fermi surface. To be precise we shall center the problem on the case of a circular Fermi surface in D=2. The free action for this case reads

$$S_0 = \int_{-\infty}^{\infty}\frac{d\omega}{2\pi}\int_0^{2\pi}\frac{d\theta}{2\pi}\int_{-\Lambda}^{\Lambda}\frac{dk}{2\pi}[\bar{\Psi}(\omega,\theta,k)(i\omega - vk)\Psi(\omega,\theta,k)] \ ,$$

and the mode elimination has to be done for all modes in the k stripe and for all values of θ and ω. There is no problem in showing that the free action is a fixed point, neither in the discussion of the quadratic interactions.

Tree-level analysis

The fine points arise in the discussion of the quartic interactions:

$$\delta S_4 = \frac{1}{4!}\int_{K,\omega,\theta}\bar{\Psi}(4)\bar{\Psi}(3)\Psi(2)\Psi(1)u(4,3,2,1) \ .$$

Let us analyze the integration measure

$$\int_{K,\omega,\theta} \equiv \left[\prod_{i=1}^3\int_0^{2\pi}\frac{d\theta_i}{2\pi}\int_{-\Lambda}^{\Lambda}\frac{dk_i}{2\pi}\int_{-\infty}^{\infty}\frac{d\omega_1}{2\pi}\right]\theta(\Lambda - |k_4|) \ ,$$

$$k_4 = |\mathbf{K_4}| - K_F .$$

The crucial role here is played by the θ-function. We have started with an interaction in Fourier space containing an integral over four ω's and over four momenta $\mathbf{K}_i$ subject to a momentum-conserving δ-function. Then we try to eliminate one of the four variables, say the fourth, in favor of the other three. There is no problem with the ω-integration; as all the ω's are allowed, the condition $\omega_4 = \omega_1 + \omega_2 - \omega_3$ is always satified for any choice of the first three frequencies. The trouble with the momenta is that they are constrained to live within the annulus of thickness 2Λ around the Fermi circle. Even if one fixes the first three momenta on the annulus, the fourth can have a length as large as $3K_F$. The role of the θ-function is to prevent that from happening. The problem is that its behavior under scaling is irregular as it depends not only on K but also on K_F which does not scale. To be precise, under $k \to k' = sk$, the θ-function scales as

$$\theta(\Lambda - |k_4(k_1, k_2, k_3, K_F)|) \to \theta(\Lambda - |k_4'(k_1', k_2', k_3', sK_F)|) .$$

This failure of the θ-function to rescale properly prevents us from knowing the scaling of the quartic couplings.

The only exception comes from the special kinematical situations that have been with us from the very beginning. This kinematics is the one allowed when we restrict all the momenta to lie on the Fermi line what amounts to take the limit $\Lambda \to 0$. The only freedom left in that situation is the choice of their directions or angles θ_i. Now the point is that we cannot choose three of these angles freely on the Fermi circle but only two. The momentum conservation leaves us with only two possibilities that will give rise to two marginal couplings:

$$u[\theta_4 = \theta_2; \theta_3 = \theta_3; \theta_1, \theta_2] \equiv F(\theta_{12}) ,$$

$$u[\theta_4 = -\theta_3; \theta_3, -\theta_1; \theta_1] \equiv V(\theta_{13}) .$$

There is no mistery on the fact that the F and V couplings correspond to the scattering procceses of **Fig. 1.7**. F corresponds to forward scattering while V is obtained when the two incoming particles have opposite directions. These are the two distinct possiblities for the argument of the delta-functions of the former paragraph to become degenerate. These couplings describe the only true fixed points of the RG for the free action at this level.

The flow of the couplings

The second fine point that we have overlooked in the previous exposition concerns the distinction between the F and V couplings. The main difference between them lies in their flows at the one-loop level. In particular the F coupling does not flow and defines the Fermi Liquid fixed point, while it is the V coupling the one that follows the analysis of the previous paragraph.

We will now give a hint on how things go at one loop for these two couplings. The one-loop computation of the beta functions for the couplings F and V is done by evaluating the diagrams of **Fig. 1.7.** with all the external legs at zero frequency and on the Fermi surface (k=ω=0) since the dependence on these variables is irrelevant.

A careful but straightforward analysis (we encourage the students to perform it) shows that none of the three diagrams renormalize the F coupling due to the fact that, for the particular kinematics of this coupling, in all cases the poles of the propagators are on the same side of the ω plane.

In the case of the V coupling, however, the BCS diagram is not trivial and the coupling flows fllowing the analysis of the previous paragraph.

Conclusion

The previous analysis, although being rather schematic, serves to exemplify the nature of the difficulties that arise when studying systems of correlated electrons. In the very simple cases of circular (spherical) Fermi surface in D=2 and D=3, the analysis can be carried on to the end with the results that we have obtained. *In D=2 there is a fixed point of the RG characterized, at tree level, by two coupling constants F and V. The flow of the couplings at one loop level is such the coupling F remains at the (free) fixed point while V renormalizes to zero if repulsive and drives the system to the BCS instability if attractive.* In D=3 the kinematics is more complicated but the physics remains very much the same. The coupling equivalent to F is a marginal coupling function (the Landau interaction function).

There are few results on more general situations. We end this section by strongly encouraging the reader to follow the details and the much richer discussion of the subject given in the original reference [2].

2.5 Non–Fermi Liquids

As was mentioned at the end of Chap. 1, finding a non-Fermi liquid behavior for electronic systems in spatial dimension greater than one has become a challenge for researchers in the area due to the anomalous properties measured in the high-T_c materials. The theoretical developments described in this chapter have given rise to a series of attempts to find fixed points of the renormalization group other than the Fermi liquid.

Before discussing some of the proposals encountered in the literature, we will start by quoting the results of ref. [2] on the subject: *No non–Fermi liquid behavior is found at D=2 if:*

1. *The Fermi surface is spherical (isotropic system).*

2. *The coupling is weak.*

3. The input interaction is short ranged.

4. We work at infinite volume from the start.

This quotation makes clear that any non-Fermi liquid behavior will be associated to breaking some of the conditions stated[3].

Long-range interactions

The first set of NFL fixed points that we will describe is indeed associated with the breakdown of the third condition and involves long-range interactions. The interest of them is twofold. On one side is the obvious fact that Coulomb interactions are, in principle, long-ranged – although screened in normal metals – [29, 30, 31, 32]. The second driving force of these works concerns the description of electrons in Hall states at particular filling fractions [33]-[34]. There is some evidence on the fact that the elementary excitations of these states have fractional statistics hence justifying the analysis of electrons in two dimensions coupled to magnetic or Chern-Simon fields which are both long-ranged [35].

What most of these works have in common is the magnitudes that are computed, differing mostly in the techniques for computing them (and, occasionally, on the interpretation of the results). Those are, first the electron self-energy which, as was discussed at length in Chap. 1, contains all the information on the equilibrium properties. The real part gives us the wave function renormalization – which is a constant between zero and one for Fermi liquids –, and the imaginary part provides the lifetime of the quasiparticles – which grows as ω^{-2} at low energies in the Fermi liquid.

As an example of how the ideas of the RG are being used, we will present the results of [30] which are somehow typical. They start with the action

$$H = \int d^D r \Psi^\dagger(\mathbf{r}) \left[\frac{1}{2m}(-i\boldsymbol{\nabla} - g\mathbf{A})^2 - \mu \right] \Psi(\mathbf{r}) + \frac{1}{2} \int d^D r \left[(\frac{\partial \mathbf{A}}{\partial t})^2 + (\boldsymbol{\nabla} \times \mathbf{A})^2 \right] \ .$$

(The scalar potential is not included). Computing the self-energy of the electron they find

$$Z_\Psi(k_F, \omega) \equiv \left[1 - \frac{\partial \Sigma(k_F, \omega)}{\partial \omega} \right]^{-1} \sim 1 - \frac{g^2 v_F}{4\pi^2} \ln(\frac{\Lambda}{\omega}) \ .$$

This gives for the spectral density the anomalous scaling behavior

$$G(k_F, \omega) \sim \frac{1}{\omega^{1-\eta}} \ ,$$

with

$$\eta = \frac{g^2 v_F}{4\pi^2} \ .$$

The model of [32] stands by itself in this general classification as it is, as stated, associated to a peculiar system, the half-filled hexagonal lattice. The

---

[3]A possible exception to this is the instability advocated by Anderson in two dimensions [28] associated to a possible breakdown of perturbation theory which we will not discuss here.

peculiar thing about this lattice is that its Fermi "line" at half filling reduces, as in the one dimensional case, to two discrete points. This lattice is realized in graphite sheets. The model which describes the system is a genuine QFT and the scaling is defined unambiguously. The special interest of this work is that it describes an anomalous (un)screened Coulomb interaction which is not suppressed by powers of v_F/c. The anomalous screening comes from the fact that the system has a density of states vanishing at the Fermi points. This is characteristic of semimetals. The effective action for the system about one of the Fermi points is

$$S = \int dt \int d^2r \bar{\Psi}(x)[-\gamma_0(\partial_0 - ieA_0) + v\gamma \cdot (\nabla - ie\mathbf{A})]\Psi(x) .$$

The computation of the wave-function renormalization of the electron using standard field-theory methods gives at two-loop order the result

$$Z_\Psi = 1 - \frac{c_1}{16\pi^2}\frac{e^4}{v^2}\log \Lambda ,$$

where e, v are the charge of the electron and the Fermi velocity respectively and e^2/v is the effective coupling constant of the system. This anomalous dimension signals a departure from Fermi liquid. The one-loop beta function for the effective coupling shows that it flows to zero in the infrared. The same behavior is found by computing the renormalization group factors to leading order in a $1/N$ expansion, which represents a nonperturbative solution of the Coulomb interaction in the effective coupling constant[36].

Van Hove Fixed Points

The second set of non-Fermi liquid behavior for electrons in higher dimensions concerns more directly the high-T_c systems and is related with the van-Hove singularities of two-dimensional lattice systems. These are points on the Fermi surface where the density of states has a logarithmic divergence. It is known that the two-dimensional square lattice at half-filling has van Hove singularities and their presence was invoked from the beginning as a possible explanation for the high critical temperature of the cuprates [37, 38]. Recently effective actions to describe the electrons close to a van Hove singularity have produced non-Fermi liquid results in two-dimensional systems (although a scaling argument was used as early as 1985 in [39]).

One of the most remarkable results obtained is that the computation of the electron wave function renormalization is exactly of the form (1.23) used qualitatively in [40] to explain the linear dependence of the resistivity with the temperature found in the cuprates. These were the first concrete models showing the marginal-Fermi liquid behavior.

The main virtue of the van Hove argument is that it is the only simple one brought forward to explain why the anomalous normal state coincides with optimal doping. On this respect it is worth mentioning the work of [41] which was directly inspired by the techniques discussed in this chapter. By using the RG technique applied to an effective action describing the electronic states near the van Hove singularities, it is found there that the model corresponds to a perturbatively renormalizable two-dimensional field theory. The starting point is an effective hamiltonian of the type

$$\mathcal{H} \sim -\frac{1}{2m} \int \Psi^\dagger(\mathbf{r})[\partial_x^2 - \partial_y^2]\Psi(\mathbf{r})dxdy$$

perturbed by standard quartic interactions. The computation of the particle-hole polarizability $\chi(\mathbf{q},\omega)$ gives the result

$$\mathrm{Re}\,\chi(\mathbf{q},\omega) = -\frac{1}{2\pi^2}\left[\log\left|\frac{4\varepsilon_c\varepsilon(\mathbf{q})}{\omega^2 - \varepsilon(\mathbf{q})^2}\right| - \frac{\omega}{\varepsilon(\mathbf{q})}\log\left|\frac{\omega + \varepsilon(\mathbf{q})}{\omega - \varepsilon(\mathbf{q})}\right| + 2\right]$$

$$\mathrm{Im}\,\chi(\mathbf{q},\omega) = \frac{2}{\pi}(|\omega + \varepsilon(\mathbf{q})| - |\omega - \varepsilon(\mathbf{q})|) \tag{2.28}$$

where $\varepsilon(\mathbf{q}) = (q_x^2 - q_y^2)/(2m)$ and ε_c is the cutoff in energy. The forward scattering channel is therefore renormalized in the infrared. More subtle, however, is the study of the renormalization of the BCS channel, as long as it is affected by $\log^2 \varepsilon_c$ divergences. This fact, together with the anisotropy of the effective interaction arising from (2.28), leaves open theoretical questions about the precise mechanism for superconductivity in the model.

Recently, the van Hove scenario for high-T_c superconductivity has attracted a lot of attention[42], as there seems to be compelling experimental evidence that in the case of the hole-doped high-T_c superconductors the Fermi level is pinned at a very flat region of the dispersion relation[43]. The stability of this situation is also predicted within the renormalization group approach of ref. [41].

References

[1]E. Brézin, "Applications of the renormalization Group to Critical Phenomena" in "Methods in Field Theory", R. Balian and J. Zinn-Justin eds., Les Houches (1975).

[2]R. Shankar, Physica A **177** (1991) 530. For an extense review see R. Shankar, "Renormalization Group Approach to Interacting Fermions", Rev. Mod. Phys. **66** (1994) 129.

[3]J. Polchinski, "Effective Field Theory and the Fermi Surface", Proceedings of the 1992 Theoretical Advanced Institute in Elementary Particle Physics, eds. J. Harvey and J. Polchinski, World Scientific, Singapore (1993).

[4]P. Ramond, "Field Theory: A Modern Primer", Addison-Wesley (1989).

[5]J. Collins, "Renormalization", Cambridge University Press (1984).

[6]L. S. Brown, "Quantum Field Theory", Cambridge University Press, 1992.

[7]A. B. Migdal, "Qualitative Methods in Quantum Theory", W. A. Benjamin, Inc. (1977).

[8]L. M. Brown, editor, "Renormalization: from Lorentz to Landau (and Beyond)". Springer-Verlag (1993).

[9]A very nice account of renormalizability can be found in the article "Renormalization: a Review for Non Specialists", of the book by S. Coleman, Aspects of Symmetry, Cambridge University Press (1988).

[10]N. Goldenfeld, "Lectures on Phase Transitions and the Renormalization Group", Addison-Wesley, 1992.

[11]G. P. Lepage in "What is Renormalization?" TASI'89 Summer School, Boulder, Colorado 1989.

[12]J. Polchinski, Nucl. Phys. B **231** (1984) 269.

[13]J. Glimm and A. Jaffe, "Quantum Physics: A Functional Integral Point of View", Springer–Verlag 1987; R. J. Rivers, "Path Integral Methods in Quantum Field Theory", Cambridge University Press 1987.

[14]R. P. Feynman, and A. R. Hibbs, "Quantum Mechanics and Path Integrals", McGraw-Hill, 1965.

[15]L. P. Kadanoff, " Scaling Laws for Ising Models near T_c, Physics **2** (1966) 263.

[16]K. G. Wilson, "Renormalization Group and Critical Phenomena", Phys. Rev. B **4** (1971) 3174.

[17]L. P. Kadanoff, Rev. Mod. Phys. **49** (1977) 267.

[18]K. G. Wilson, "Renormalization Group and Critical Phenomena", Rev. Mod. Phys. **55** (1983) 583; "Problems with Physics with Many Scales of Length", Sci. Am. **241** (1979) 158.

[19]C. Itzykson and J. M. Drouffe, "Statistical Field Theory", vol. I, Cambrigde University Press (1989).

[20]J. Zinn-Justin, "Quantum Field Theory and Critical Phenomena", Clarendon, Oxford, 1989.

[21]D. J. Amit, "Field Theory, the Renormalization Group and Critical Phenomena", World Scientific, 1984.

[22]S. K. Ma, "Modern Theory of Critical Phenomena", Benjamin, 1976.

[23]P. W. Anderson, J. Phys. C 3 (1970) 2346; see also "Basic Notions of Condensed Matter Physics", Benjamin, 1984.

[24]P. W. Anderson and G. Yuval, Phys. Rev. Lett. 23 (1970) 89; P. Noziéres, J. Low Temp. Phys. 17 (1974) 31.

[25]P. W. Anderson, "A Career in Theoretical Physics", World Scientific 1994.

[26]G. Benfatto and G. Gallavotti, Phys. Rev. B 42 (1990) 9967.

[27]R. Shankar, Physica A 177 (1991) 530.

[28]P. W. Anderson, Phys. Rev. Lett. 71 (1993) 1220.

[29]A. Houghton, H.-J. Kwon, J. B. Marston, and R. Shankar, "Coulomb Interactions and the Fermi-Liquid State: Solution by Bosonization", cond-mat/9312047 (1993).

[30]J. Gan and E. Wong, "Non-Fermi-Liquid Behavior in Quantum Critical Systems", Phys. Rev. Lett. 71 (1993) 4226.

[31]S. Chakravarty, R. E. Norton, and O. F. Syljuasen, "Transverse Gauge Interactions and the Vanquished Fermi Liquid", Phys. Rev. Lett. 74 (1995) 1423.

[32]J. González, F. Guinea, and M. A. H. Vozmediano, "Non-Fermi Liquid Behavior of Electrons in the Half-Filled Honeycomb Lattice. (A Renormalization Group Approach)", Nucl. Phys. B 424 (1994) 595.

[33]B. I. Halperin, P. Lee, and N. Read, Phys. Rev. B 47 (1993) 7312.

[34]C. Nayak and F. Wilczek, Nucl. Phys. B 417 (1994) 359; 430 (1994) 534; "Physical Properties of Metals from a Renormalization Group Standpoint", cond-mat/9507040 (1995).

[35]A recent review of this subject can be found in F. Wilczek, "Statistical Transmutation and Phases of Two-Dimensional Quantum Matter", Princeton preprint to appear (1995).

[36]J. González, F. Guinea, and M. A. H. Vozmediano, preprint in preparation.

[37]J. Labbé and J. Bok, Europhys. Lett. 3 (1987) 1225.

[38]D. M. Newns, H. R. Krishnamurty, P. C. Pattnaik, C. C. Tsuei, and C. L. Kane, Phys. Rev. Lett. 69 (1992) 1264, and references therein.

[39]H. J. Shulz, Europhys. Lett. 4 (1987) 609.

[40]C. M. Varma, P. B. Littlewood, S. Schmitt-Rink, E. Abrahams, and A. E. Ruckenstein, " Phenomenology of the Normal State of Cu-O High-Temperature Superconductors", Phys. Rev. Lett. 63 (1989) 1996.

[41]J. González, F. Guinea, and M. A. H. Vozmediano, "Renormalization group analysis of electrons near a van Hove singularity", cond-mat/9502095 (1995).

[42]A recent update of the situation of the van Hove singularity in relation with the high-T_c superconductors with a fairly complete list of references can be found in M. L. Horbach and H. Kajuter, Int. J. Mod. Phys. B 9 (1995) 1067.

[43]See Z.-H. Shen, W. E. Spicer, D. M. King, D. S. Dessau and B. O. Wells, Science 267 (1995) 343, and references therein.

Part II

3. Electronic Systems in $d = 1$

3.1 Introduction

In this chapter we undertake the general description of electrons in one dimension. There exist systems which are actually one-dimensional, as is the case of polyacetilene[1], and others which may be considered "quasi-one-dimensional", i. e. that show conduction properties in a prefered direction. This is the case of the well-known organic charge-transfer compounds[2]. The study of electronic properties in one dimension is not, therefore, purely academic. This is also supported by the relevance of these systems in the investigation of the so-called "quantum wires" [3] and by the increasing evidence that the edge excitations in the fractional quantum Hall effect are effectively that of a one-dimensional liquid[4].

When considering a physical problem of electrons in one dimension, one may either place the emphasis on the electron-phonon interaction or turn to consider as most relevant the electron-electron interaction. Following the scope of these notes we are going to focus our interest in the latter. Even in this perspective, one may adopt two different approaches depending on the problem. One of them is well-suited to problems where there are strong correlations among electrons and localization effects are important. These use to be properties of non-conducting systems, in which the physics is supposed to be well-described by the Hubbard model[5]. The other approach is appropriate to situations in which the interactions are not so strong, leading in general to metallic behavior. This is the point of view that we are going to adopt in what follows. It will allow us to consider perturbation theory as the starting point to investigate our systems and to make the map of the space of all the couplings constants. The main achievement of this approach is to show that, opposite to what happens in three spatial dimensions, the generic behavior in one dimension is not given by the Landau Fermi liquid picture, but by the so-called Luttinger liquid. This can be already understood from the breakdown of perturbation theory, which is spoiled by infrared divergences in one spatial dimension.

In the present chapter we undertake the analysis of perturbation theory and its failure in one dimension, which we fix with the machinery of the renormalization group. Chapter 4 is devoted to introduce the general techniques of bosonization for one-dimensional fermion systems and to develop the physical

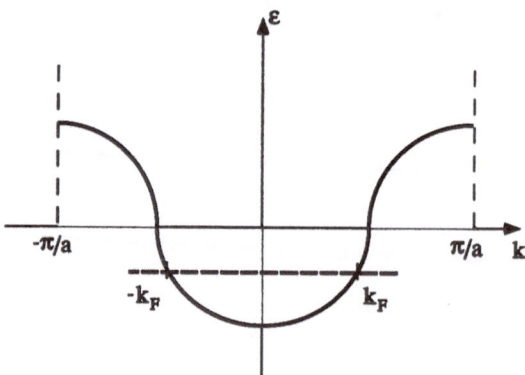

Fig. 3.1. Dispersion relation of electrons in a one-dimensional lattice.

picture of the Luttinger liquid. Finally, in Chap. 5 we discuss the correspondence of discrete models to the continuum field theory framework.

3.2 Perturbation Theory. Renormalization Group

Our one-dimensional systems have in general some kind of lattice support, given by the periodic disposition of atoms in a crystal. In tight-binding approximation, for instance, and considering only one conduction band, the electronic dispersion relation for Bloch states is given by

$$\varepsilon(k) = -t \cos ak \tag{3.1}$$

where a is the lattice spacing.

The one-particle hamiltonian in terms of the creation and annihilation operators $c_{k\sigma}^+, c_{k\sigma}$ of Bloch states is

$$H_0 = \sum_{k,\sigma} \varepsilon(k) c_{k\sigma}^+ c_{k\sigma} \tag{3.2}$$

The symbol σ denotes spin, $\uparrow$ or $\downarrow$, and the label k is restricted to the first Brillouin zone $(-\pi/a, \pi/a)$. The bandwith W is in general proportional to the hopping parameter t.

We will suppose that the Fermi level ε_F is placed somewhere in the middle of the band, not very close to its top nor to its bottom —we will see later that the effective strength of the interaction is inversely proportional to the slope at the Fermi momentum. The main feature of the electronic system in $d = 1$ is the existence of two Fermi points at $\pm k_F$, given by $\varepsilon(\pm k_F) = \varepsilon_F$. This makes possible to encode the physics of the low-energy excitations into a simple field theory, what is not feasible in general at higher dimensions. Let us consider, for instance, processes which involve an energy $\sim E_0$ much smaller than the bandwith W. Then it is justified to focus on the states in an interval $(-E_0, E_0)$ about the Fermi level. The dispersion relation may be linearized within this range, so that

$$\varepsilon(k) = \varepsilon_F + v_F(k - k_F) + \dots \quad \text{at the right Fermi point} \quad (3.3)$$
$$\varepsilon(k) = \varepsilon_F + v_F(-k - k_F) + \dots \quad \text{at the left Fermi point} \quad (3.4)$$

where

$$v_F = \frac{d\varepsilon}{dk}(k_F) \quad (3.5)$$

Higher orders in the expansion may be considered irrelevant, in the sense pointed out in the preceding chapter. This linear approximation, that we will use henceforth, is good in general if the interaction is not too strong and not long-ranged. The Coulomb interaction, for instance, may give rise to different effects than those considered here.

At this point we may distinguish between states attached to the right Fermi point, denoted by $a_{k\sigma}, a_{k\sigma}^+$, and states attached to the left Fermi point, denoted by $b_{k\sigma}, b_{k\sigma}^+$. In the linear approximation we may write our one-particle hamiltonian in the form

$$H_0 = \sum_{k,\sigma} v_F(k - k_F)a_{k\sigma}^+ a_{k\sigma} + \sum_{k,\sigma} v_F(-k - k_F)b_{k\sigma}^+ b_{k\sigma} \quad (3.6)$$

where we understand now that the cutoff $|\varepsilon(k) - \varepsilon_F| < E_0$ is implicit in the sum over modes.

3.2.1 Interactions

We apply now the scaling arguments introduced in the previous chapter to determine the form of the relevant or marginal interactions of the electronic system in $d = 1$. The expression (3.6) is nothing but the Dirac hamiltonian in $1 + 1$ dimensions with nonvanishing chemical potential. In terms of respective fermion fields of left and right chirality

$$\Psi_{1\sigma}(x) = \frac{1}{\sqrt{L}} \sum_k e^{ikx} b_{k\sigma} \quad (3.7)$$

$$\Psi_{2\sigma}(x) = \frac{1}{\sqrt{L}} \sum_k e^{ikx} a_{k\sigma} \quad (3.8)$$

we may write, apart from the chemical potential term,

$$H_0 = -iv_F \int dx \sum_{\sigma,j} (-1)^j \Psi_{j\sigma}^+ \partial_x \Psi_{j\sigma} \quad (3.9)$$

From this "free" hamiltonian we may read the scaling dimension of the fermion fields (in energy units) $[\Psi_j(x)] = 1/2$. We may now ask what kind of interaction

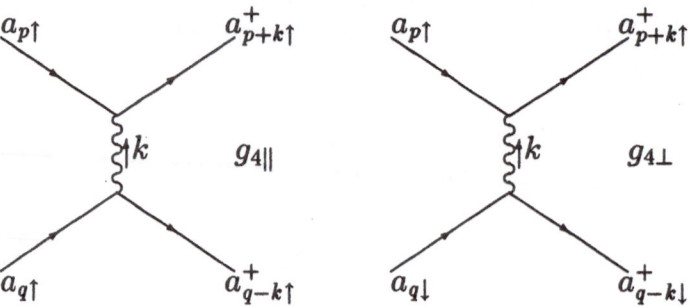

Fig. 3.2. Interactions between currents of like chirality.

hamiltonians can be built out of Ψ_1 and Ψ_2 if only dimensionless coupling constants are allowed. This catalog may still be too long, so that here we restrict it to interactions of density-times-density type. This amounts to the requirement of charge conservation in the processes under consideration. Then the only possible form of the interaction is (leaving aside subindices)

$$H_{int} \sim \int dx_1 \, dx_2 \, \Psi^+(x_1)\Psi(x_1)V(x_1, x_2)\Psi^+(x_2)\Psi(x_2) \qquad (3.10)$$

where the potential $V(x_1, x_2)$ must have dimensions of energy. Obviously, terms with higher content of fermion fields have to be irrelevant, in the sense of the previous chapter, since they have to enter through coupling constants with the dimensions of positive powers of some microscopic length scale.

The collection of all the interactions of the above type is most easily summarized by drawing the corresponding Feynman diagrams in momentum space. In these we represent right modes by a full line and left modes by a dashed line. A wavy line represents the exchange of momentum through the interaction potential. Its Fourier transform introduces some structure, that we suppose to be given by some smooth function of momentum, nonsingular at $k = 0$. There are, in general, four different kinds of four-fermion interactions, according to the type of incoming and outgoing modes[6]. We consider first the interaction between two densities of the same type of modes, corresponding to diagrams with two incoming right modes and two outgoing right modes, or the same construction with right replaced by left. Even then there are two different possibilities, as shown in Fig. 3.2, depending on whether the two densities have the same or opposite spin. The coupling constant in each case is conventionally denoted by $g_{4\parallel}$ and $g_{4\perp}$.

Another different situation is when the density of left modes interacts with the density of right modes. This is illustrated in Fig. 3.3 . Again the two densities may bear the same or different spin, making necessary to introduce respective coupling constants $g_{2\parallel}$ and $g_{2\perp}$.

Further on, a more ingenious interaction arises when two modes at different sides of the Fermi sea are excited to the respective opposite branches. As shown

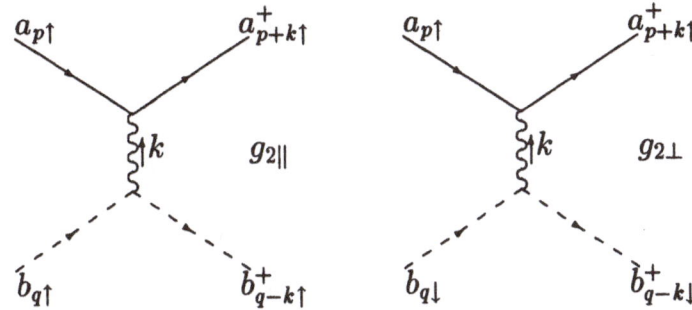

Fig. 3.3. Interactions between currents of different chirality.

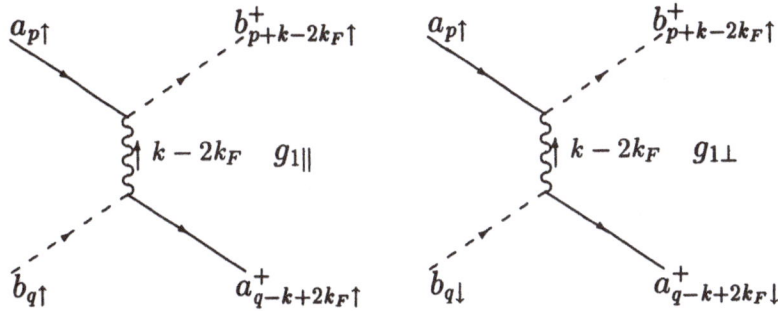

Fig. 3.4. Types of interaction with backscattering.

in Fig. 3.4, the typical momentum exchange has to be about $2k_F$. The coupling constant for this kind of interaction is now called $g_{1\parallel}$ or $g_{1\perp}$, for parallel or opposite spins at the two sides of the interaction, respectively. As we will see later, this process of backscattering introduces the main complication in the investigation of one-dimensional electron systems, as it spoils in general the integrability of the theory. The same can be said of the remaining interaction, which goes with the coupling constant g_3 and is built out of Umklapp processes. These are shown in Fig. 3.5, where the excess of momentum $4k_F$ is supposed to be absorbed by the lattice substrate. This interaction becomes relevant only at the point of half-filling, as we will see in the study of the harmonic chain in Chap. 5.

As we have already remarked, the momentum structure of the interaction is given by the Fourier transform of the potential. We have assumed that this has to be a nonsingular, smooth function for any value of k. In the rest of this chapter and for the sake of a clear exposition we will take it as a constant function of the momentum. Anyhow, our conclusions will not depend on this particular choice. In practice, it amounts to contract the interaction to a point in real space. Then, it is clear that $g_{1\parallel}$ and $g_{2\parallel}$ are describing, apart from a difference of sign, the same kind of process. We will fix this ambiguity by making the

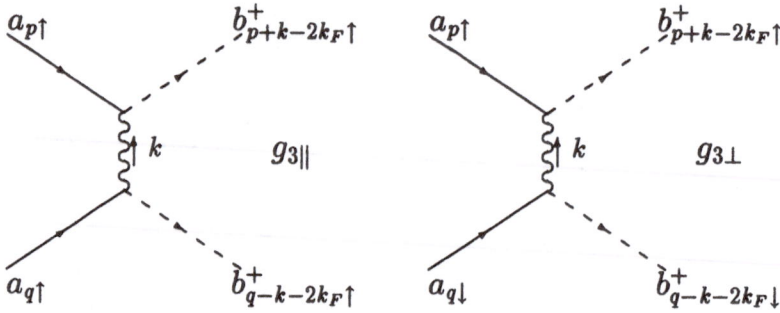

Fig. 3.5. Interactions originating from Umklapp processes.

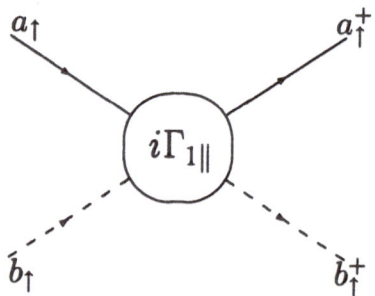

Fig. 3.6. Example of four-point vertex function.

choice $g_{2\parallel} = g_{2\perp} \equiv g_2$, so that only three independent couplings, $g_{1\parallel}$, $g_{1\perp}$ and g_2, remain[6]. We will use this convention to the end of this chapter.

3.2.2 Quantum Corrections

Now that we have written down the possible interaction terms in the hamiltonian, we want to investigate some of the properties of the quantum theory. We follow here the philosophy adopted in quantum field theory, in that the interaction corrects the "bare" coupling constants and promotes them to respective vertex functions. One of these, that we call $\Gamma_{1\parallel}$, is represented in Fig. 3.6, where the circle stands for all possible interactions connecting the external legs. The object can be computed in perturbation theory, the zeroth order being obviously the value of $g_{1\parallel}$ according to our above redefinition.

We undertake the computation of the first order corrections in $\Gamma_{1\parallel}$, from which one gains insight of the problems to define such object. Resorting to Feynman diagrams, there are essentially three different contributions, represented in Fig. 3.7 . In diagram (a) the two interactions have to be $g_{1\parallel}$, and in diagram (b) they both can be either $g_{1\parallel}$ or $g_{1\perp}$. It can be checked that, opposite to what happens with these two diagrams, the contribution of diagram (c) does

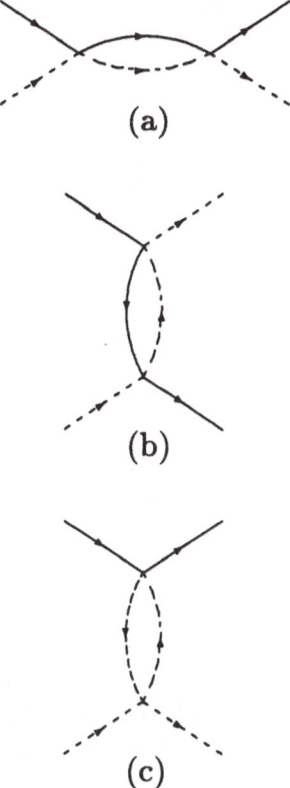

Fig. 3.7. Diagrams contributing to $\Gamma_{1\parallel}$ to the one-loop order.

not give rise to logarithmic divergences, so that it can be neglected in what follows.

The particular kinematics that we take for diagram (a) is indicated in Fig. 3.8 . The internal propagators for right modes and left modes are, respectively,

$$G_R^{(0)}(\omega, k) = \frac{1}{\omega - v_F(k - k_F) + i\epsilon \ \text{sgn}(k - k_F)} \tag{3.11}$$

$$G_L^{(0)}(\omega, k) = \frac{1}{\omega - v_F(-k - k_F) + i\epsilon \ \text{sgn}(-k - k_F)} \tag{3.12}$$

The contribution to $\Gamma_{1\parallel}$ from diagram (a) is therefore

$$\Gamma_{1\parallel}^{(a)} =$$

$$= -i2g_{1\parallel}^2 \int \frac{dk\,d\omega}{2\pi\,2\pi} G_R^{(0)}(\omega_1 + \omega, k_F + k) \ G_L^{(0)}(\omega_2 - \omega, -k_F - k)$$

$$= -i2g_{1\parallel}^2 \int \frac{dk\,d\omega}{2\pi\,2\pi} \frac{1}{\omega_1 + \omega - v_F k + i\epsilon \ \text{sgn}(k)} \frac{1}{\omega_2 - \omega - v_F k + i\epsilon \ \text{sgn}(k)}$$

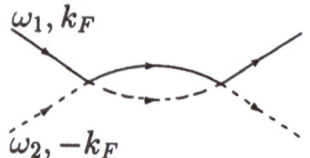

Fig. 3.8. Diagram (a) of the previous figure, with the kinematics used in its computation.

$$= i2g_{1\parallel}^2 \int \frac{dk\, d\omega}{2\pi\, 2\pi} \frac{1}{\overline{\omega} + \omega - v_F k + i\epsilon\ \text{sgn}(k)} \frac{1}{\omega + v_F k - i\epsilon\ \text{sgn}(k)} \qquad (3.13)$$

where we have called $\omega_1 + \omega_2 \equiv \overline{\omega}$. We can perform first the k integral by working in the complex k plane. If we take $\omega > 0$ (we suppose that $\overline{\omega} > 0$), we find a pole slightly over the real axis at $(\overline{\omega} + \omega)/v_F$ and another slightly under the real axis at $-\omega/v_F$. The result of the integral is, for $\omega > 0$,

$$-i2g_{1\parallel}^2 \int_{\omega > 0} \frac{d\omega}{2\pi} \frac{2\pi i}{2\pi} \frac{1}{v_F} \frac{1}{\overline{\omega} + 2\omega} = 2g_{1\parallel}^2 \frac{1}{2\pi v_F} \int_0^{E_0} d\omega \frac{1}{\overline{\omega} + 2\omega}$$

$$\approx -g_{1\parallel}^2 \frac{1}{2\pi v_F} \log \frac{\omega_1 + \omega_2}{E_0} \qquad (3.14)$$

where we have made use of the bandwith cutoff E_0. It can be checked that the contribution for $\omega < 0$ gives the same result. The total contribution from this diagram is therefore

$$\Gamma_{1\parallel}^{(a)} = -g_{1\parallel}^2 \frac{1}{\pi v_F} \log \frac{\omega_1 + \omega_2}{E_0} \qquad (3.15)$$

We turn now to the contribution of diagram (b). We evaluate it in the particular kinematics shown in Fig. 3.9 (in the case $\omega_1 - \omega_3 > 0$). Following the same steps as before we get

$$\Gamma_{1\parallel}^{(b)} =$$

$$= -i2(g_{1\parallel}^2 + g_{1\perp}^2) \int \frac{dk\, d\omega}{2\pi\, 2\pi} G_R^{(0)}(\omega, k_F + k)\ G_L^{(0)}(\omega_3 - \omega_1 + \omega, -k_F + k)$$

$$\approx (g_{1\parallel}^2 + g_{1\perp}^2) \frac{1}{\pi v_F} \log \frac{\omega_1 - \omega_3}{E_0} \qquad (3.16)$$

Finally, by adding up the contributions from diagrams (a) and (b) we get the result for the vertex function to first order in perturbation theory

$$\Gamma_{1\parallel} = g_{1\parallel} - g_{1\parallel}^2 \frac{1}{\pi v_F} \log \frac{\omega_1 + \omega_2}{E_0} + (g_{1\parallel}^2 + g_{1\perp}^2) \frac{1}{\pi v_F} \log \frac{\omega_1 - \omega_3}{E_0} + \dots \qquad (3.17)$$

There are two important observations to be made about this expression. The first of them is that, apparently, the value of $\Gamma_{1\parallel}$ depends on the particular value chosen for the cutoff E_0. This is actually inadmissible, since we may think of the

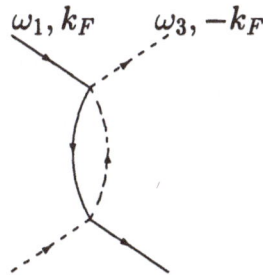

Fig. 3.9. Diagram (b) of Fig. 3.7, with the kinematics used in its computation.

vertex function as the "body" which allows us to obtain a four-point scattering amplitude just by inserting the appropriate modes in place of the external legs. This latter object is directly observable, and cannot depend on the cutoff as far as we are dealing with a predictive theory. The second remark concerns the divergence of (3.17) at small frequency values. This again cannot be a sensible effect, especially when this singularity appears even more pronounced at higher levels in perturbation theory. It can be checked, for instance, that the iteration of diagram (a) giving an n-loop contribution to the vertex function behaves like $\sim \log^n(\omega/E_0)$. Therefore what we are really observing is the breakdown of standard perturbation theory. We are going to see, however, how the first shortcoming is related, in a certain sense, to the second, and leads to the key idea giving physical meaning to the vertex function through the machinery of the renormalization group.

3.2.3 Renormalization Group

We have already learned in Chap. 2 how the ideas underlying the renormalization group can be applied to quantum statistical problems in which some energy scale can be varied by orders of magnitude. We face here one of these situations, since our aim is to make physical sense of a theory with a built-in cutoff E_0. This may start being of the same order than the bandwith W but much smaller than the typical energy of the electron interactions. Opposite to the approach to renormalization group based on a statistical physics description, we are going to place the emphasis on the philosophy usually adopted in quantum field theory. According to this point of view, our pretension will be to absorb the dependencies on the cutoff by applying some renormalization program. The outcome of this implementation is that the renormalization group is doing for us the partial sum of perturbation theory, in such a way that the above mentioned logarithmic divergences are cured. Actually, the problem of perturbation theory of electrons in one dimension is only one example of a recurrent logarithmic problem which appears in many different contexts, like the X-ray edge singularity problem, the Kondo effect, etc. In all these cases the recipe is to perform a partial sum of the perturbative expansion to get rid of the logarithms and obtain the right frequency (or energy) dependence.

Let us state the basic idea behind the renormalization group[7]. In a quantum (many-body) theory where there is an available energy scale, a change of the scale in posing a given problem leads to an effective hamiltonian description with different couplings at the new scale. Applied to our particular expression (3.17), this means that we may want perhaps to change the scale of energy E_0, but cannot then pretend to keep constant all the couplings. We have to allow for modifications of the "coupling constants" with E_0, so that a quantity like $\Gamma_{1\parallel}$ which may be used to compute an observable remains cutoff independent. The requirement of renormalizability is actually very strong for a quantum field theory. It means that all the dependence on the cutoff of physical quantities can be absorbed by redefinitions of the couplings and the scale of the fields[8]. To the extent in that it has been studied, our quantum theory of electrons with the given set of couplings has proven to be renormalizable, which gives us the right to adopt the quantum field theory point of view.

We illustrate the property of renormalizability in a very simple instance. In general we impose a relation of the couplings and the fields to renormalized partners which do not depend on the cutoff

$$g(E_0) = Z_g(E_0)g_R \tag{3.18}$$
$$\Psi(E_0) = Z_\Psi(E_0)\Psi_R \tag{3.19}$$

Usually, the knowledge that one has of Z_g and Z_Ψ is only perturbative. Working to the first order and collecting all the dependencies on E_0 we obtain the condition for the cutoff independence of the vertex function (3.17)

$$\frac{d}{dE_0}\left\{ g_{1\parallel}(E_0) - g_{1\perp}^2(E_0)\frac{1}{\pi v_F}\log E_0 + \ldots \right\} = 0 \tag{3.20}$$

In order to get a simple expression let us address the case of spin-independent interactions, that is $g_{1\parallel} = g_{1\perp}$. We obtain then

$$E_0\frac{d}{dE_0}g_1(E_0) = \frac{1}{\pi v_F}g_1^2(E_0) + \text{ higher orders} \tag{3.21}$$

This equation encodes at this level all the information from the renormalization group, since it tells us how do we have to change the coupling constant by a modification of the cutoff, in order to maintain invariant the vertex function. Now suppose that we do not want to describe a given problem at the scale E_0 but want to consider another scale E, which in general may be different by several orders of magnitude. The relation between the coupling constants at the two different scales is given by the integral of (3.21)

$$-\frac{1}{g_1(E)} + \frac{1}{g_1(E_0)} = \frac{1}{\pi v_F}\log\frac{E}{E_0} \tag{3.22}$$

or, finally,

$$g_1(E) = \frac{g_1(E_0)}{1 - (g_1(E_0)/\pi v_F) \log (E/E_0)} \tag{3.23}$$

Equation (3.23) displays the full insight of the renormalization group. We learn from it that, if we started at a scale E_0 with a coupling $g_1 > 0$ and turned to pose the problem at a smaller scale E, the effective coupling constant becomes reduced. Thus, if the original g_1 was already small (compared to πv_F) the predictions that we get from (3.23) have to be increasingly good. On the other hand, if we had started with a coupling $g_1 < 0$, by progressing to lower scales $E << E_0$, we would reach a point in which the effective coupling blows up. Well before this, however, our perturbative renormalization group approach must have lost its applicability. This accelerated growth of the coupling constant points at the onset of some instability which is beyond the framework of perturbation theory. We will comment ahead on the nature of the ground state in the repulsive ($g_1 > 0$) regime.

The main application of the renormalization group arises by trading the notion of the variable scale of energy E by the typical frequency at which a process is measured. If we interpret $g_1(E)$ as the effective value of the vertex function we get

$$\Gamma_1(\omega) \approx \frac{g_1(E_0)}{1 - (g_1(E_0)/\pi v_F) \log (\omega/E_0)} \tag{3.24}$$

This is the solution to the logarithmic problem posed by perturbation theory. Actually, it is known since long ago that the expression (3.24) arises also from the sum of certain class of diagrams in perturbation theory[9]. These are the so-called "parquet diagrams", which are obtained by repeated insertion of diagrams (a) and (b) of Fig. 3.7 into themselves. The above form of the vertex function has also the virtue of being independent of the value of the cutoff E_0 —the remaining problem we wanted to fix. This is not strange, since it is the starting point of the renormalization group. If a change of the cutoff is made to E_0' taking care of changing also the value of g_1, it can be checked that the value of Γ_1 in (3.24) remains invariant. From this expression we arrive again at the conclusion that, even in the neighborhood of vanishing coupling constant, the physics of the attractive interaction ($g_1 < 0$) is quite different to that of repulsive interaction ($g_1 > 0$), as we see that only in the latter case makes sense the notion of small perturbation.

The most interesting analysis, however, is that of the space of couplings with $g_{1\parallel} \neq g_{1\perp}$. In this case one has to solve a coupled set of differential renormalization group equations (one for each coupling constant). The form of the integrals as the value of the cutoff is decreased gives the flow of the renormalization group in coupling constant space. The most important issue is to determine the regions in which the flow is stable (bounded) and the regions in which it is not. The most illustrative flow diagram appears in the ($g_{1\perp}, g_{1\parallel}$) plane. It is represented in Fig. 3.10, as the result of solving the first order renormalization group equations.

This is a rather familiar phase diagram in condensed matter physics, since it appears also in quite different instances as the anisotropic Kondo problem[10]

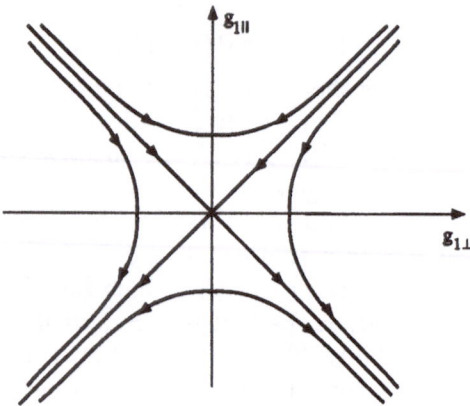

Fig. 3.10. Renormalization group flow in $(g_{1\perp}, g_{1\parallel})$ space.

or the Kosterlitz-Thouless phase transition[11]. We see that in the region $g_{1\parallel} \geq |g_{1\perp}|$ the flow is stable and leads always to a fixed point over the $g_{1\perp} = 0$ line. We remind that the meaning of the flow is that a problem with a certain set of coupling constants at a given scale is equivalent to another with lower energy scale and coupling constants found ahead in the corresponding orbit of the flow. This implies that a system in the region $g_{1\parallel} \geq |g_{1\perp}|$ must have physical properties related to another in which the backscattering mediated by $g_{1\perp}$ can be neglected. This is of great help, as we will see in the next chapter, since the model with $g_{1\perp} = 0$ is a gapless, integrable system. On the other hand, in the region $g_{1\parallel} < |g_{1\perp}|$ the flow is not bounded at this level, and the physical properties cannot be easily predicted. There is a horizontal line of points in the lower half-plane (Luther-Emery line) in which the theory can be solved, though this is not a line of fixed points[6]. There is evidence that in this regime the system ends up acquiring a gap in the spin excitation spectrum, which is the signal of an incipient superconductivity.

3.2.4 Ground State Properties

We now turn to the question of the nature of the ground state of the one-dimensional systems. This is studied by looking at the response functions, which give the hint of the kind of fluctuations which dominate at $T = 0$. In general one may expect instabilities of superconducting, charge density wave and spin density wave type. In these one-dimensional systems one cannot have true long-range order, yet it is possible to study the divergence of the response functions in ω at certain values of the momentum and determine in this competition which fluctuation prevails at low energy.

The response function related to a given charge, spin or superconductivity pairing operator $\mathcal{O}$ is defined by

$$R(\omega, k) = -i \int dt \, e^{i\omega t} \langle \mathcal{O}(t, k) \mathcal{O}^+(0, k) \rangle \tag{3.25}$$

In the case of the charge or the spin density operator, $R(\omega, k)$ follows at $k = 2k_F$ a power law behavior ω^α for small ω, which gives the measure of the instabilities present in the system[6]. One is led therefore to study the dynamic correlations for the charge density operator

$$\mathcal{O}_{CDW}(t, k + 2k_F) = \sum_p a^+_{p+k+2k_F\uparrow} b_{p\uparrow} + \sum_p a^+_{p+k+2k_F\downarrow} b_{p\downarrow} \tag{3.26}$$

the spin density operator

$$\mathcal{O}_{SDW}(t, k + 2k_F) = \sum_p a^+_{p+k+2k_F\uparrow} b_{p\uparrow} - \sum_p a^+_{p+k+2k_F\downarrow} b_{p\downarrow} \tag{3.27}$$

the singlet pairing operator

$$\mathcal{O}_{SP}(t, k) = \sum_p a_{p+k\uparrow} b_{-p\downarrow} + \sum_p b_{-p\uparrow} a_{p+k\downarrow} \tag{3.28}$$

and the triplet pairing operator

$$\mathcal{O}_{TP}(t, k) = \sum_p a_{p+k\uparrow} b_{-p\downarrow} - \sum_p b_{-p\uparrow} a_{p+k\downarrow} \tag{3.29}$$

The power law behavior ω^α of the response functions is consistent with the fact that under the renormalization group the correlators get anomalous dimensions at small ω. To see how this works, let us consider the correlation of the charge density operator, in the simple case of spin-independent interactions $g_{1\parallel} = g_{1\perp}$. In the noninteracting theory, the corresponding response function $R_{CDW}(\omega, k)$ is given by a single particle-hole loop of the same type that appears in Fig. 3.9 . We already know the result for this object from (3.16). To zeroth order in perturbation theory we have

$$R_{CDW}(\omega, 2k_F) = \frac{1}{\pi v_F} \log \frac{\omega}{E_0} + O(g^2) \tag{3.30}$$

where ω is now the external frequency injected into the loop. The first perturbative corrections appear by iterating the loop by means of g_1 interactions or by inserting a g_2 interaction in the middle of it. Adding up both contributions we have

$$R_{CDW}(\omega, 2k_F) = \frac{1}{\pi v_F} \log \frac{\omega}{E_0} \left(1 + \frac{2g_1 - g_2}{2\pi v_F} \log \frac{\omega}{E_0} + \dots \right) \tag{3.31}$$

We observe that the iteration of the above operations in the loop produces more severe logarithmic divergences to higher orders in the interaction. This is an indication that the breakdown of perturbation theory arises from the wrong expansion of an exponential dependence on the coupling constants. Renormalization group methods allow to reproduce again the correct dependence by exploiting the scaling properties of the correlator with respect to changes in the cutoff E_0. If we take the derivative of $R_{CDW}(\omega, 2k_F)$ with respect to E_0, we have to bear in mind the explicit dependences at the right-hand-side of (3.31) as well as the implicit dependences of the coupling constants on the cutoff. However, if we apply the derivative on g_1 or g_2, this produces higher order terms according to (3.21). Therefore, to first order we have

$$\frac{\partial}{\partial E_0} R_{CDW} = -\frac{1}{\pi v_F} \frac{1}{E_0} \left(1 + \frac{2g_1 - g_2}{\pi v_F} \log \frac{\omega}{E_0} + \dots \right) \tag{3.32}$$

Within the same approach of the perturbative renormalization group, we may replace $(\pi v_F)^{-1} \log(\omega/E_0)$ in the second term of the derivative by the full response function R_{CDW}, the difference being of higher order in the coupling constants. Finally we get the differential equation

$$\frac{\partial}{\partial E_0} R_{CDW} = -\frac{1}{\pi v_F} \frac{1}{E_0} \left(1 + (2g_1 - g_2) R_{CDW} + \dots \right) \tag{3.33}$$

In order to obtain the dominant behavior of R_{CDW}, we observe that $g_1(E_0)$ and $g_2(E_0)$ are both regular at small values of the cutoff (for $g_1 > 0$). The only singularities may arise due to the $1/E_0$ factor at the right-hand-side of (3.33) and g_1 and g_2 can be set to their respective fixed-point values, 0 and g_2^*. The solution of (3.33) has therefore the leading behavior

$$R_{CDW} \sim \left(\frac{\omega}{E_0} \right)^{\alpha_{CDW}} \tag{3.34}$$

with $\alpha_{CDW} = -g_2^*/(\pi v_F)$. We remark that the combination $g_2 - g_1/2$ is a renormalization group invariant, to the order we are working here. Therefore, in terms of the original couplings we have $\alpha_{CDW} = -(g_2 - g_1/2)/(\pi v_F)$. By means of similar calculations, one may check that the rest of exponents for the spin density, singlet pairing and triplet pairing response functions are given respectively by

$$\alpha_{SDW} = -(g_2 - g_1/2)/(\pi v_F) \tag{3.35}$$
$$\alpha_{SP} = (g_2 - g_1/2)/(\pi v_F) \tag{3.36}$$
$$\alpha_{TP} = (g_2 - g_1/2)/(\pi v_F) \tag{3.37}$$

Characterizing the nature of the ground state by the most singular behavior among the mentioned response functions, one may draw the phase diagram in the coupling constant space. This is shown in Fig. 3.11 for the region $g_1 > 0$, which is susceptible of perturbative treatment. For $g_2 > g_1/2$ we may expect a tendency to the formation of a charge density wave or a spin density wave in the ground state of the theory. On the contrary, for $g_2 < g_1/2$ the superconductivity pairing correlations are supposed to dominate the system. Though we have only

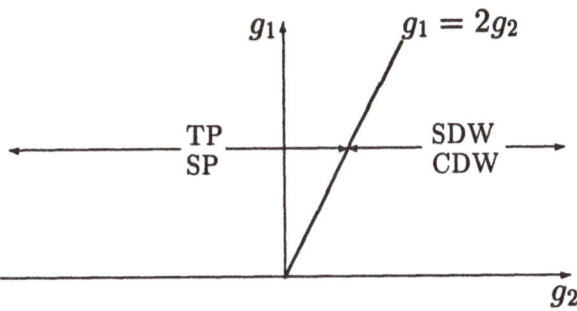

Fig. 3.11.   Phase diagram of the one-dimensional electron system with spin-independent interactions.

worried about the power-law dependence of the response functions, these are affected by logarithmic corrections, which enhance the spin density or the triplet pairing instability over the charge density or the singlet pairing instability in the corresponding situation.

We may think of the power-law dependence of the response functions as a reflection of the critical behavior of the system, though the exponents that we find are non-universal, i.e. they depend on the coupling constants of the model. On the other hand, the correlations are enhanced with respect to those of the noninteracting theory. The logarithmic behavior in (3.30) translates into a power-law decay in real space (t, x) of the form $\sim 1/x^2$, while a scaling of the type (3.34) gives rise to a long-distance behavior $\sim 1/x^{2+\alpha}$. For the corresponding response function with $\alpha < 0$, this slower decay reflects a tendency to order, which cannot be attained anyhow as far as in two space-time dimensions a continuous symmetry cannot be spontaneously broken. We will find again the phenomenon of enhancement at $2k_F$ when discussing the correlation functions of the Hubbard model in Chap. 5.

We close this chapter by remarking that the above physical picture has to be taken with the reserve inherent to a perturbative treatment. In particular, it is not guaranteed that modifications of the flow and new physics may not arise for sufficiently large values of the couplings. In the unstable phase $g_1 < 0$ we even lack a perturbative description of the low-energy physics. What would be needed is actually some specific modelling of the large attractive interaction. These and other questions, as the effects of band curvature, are still open problems in the topic of one-dimensional electron systems.

exercise 3.1 *Compute the contribution to $\Gamma_{1\parallel}$ from the diagram in Fig. 3.9 .*

exercise 3.2 *Check that the value of $\Gamma_1(\omega)$ in (3.24) is invariant under the change to a different cutoff E_0'*

References

[1]A. J. Heeger, S. Kivelson, J. R. Schrieffer and W. P. Su, *Rev. Mod. Phys.* (1988) 781.

[2]D. Jérome and H. J. Schulz, *Adv. Phys.* (1982) 299. See also D. Jérome and L. G. Caron, *Low-Dimensional Conductors and Superconductors* (Plenum, New York, 1987).

[3]C. L. Kane and M. P. A. Fisher, *Phys. Rev. B* (1992) 15233.

[4]F. P. Milliken, C. P. Umbach and R. A. Webb, to be published in *Phys. Rev. Lett.*

[5]J. Hubbard, *Proc. R. Soc. A* (1963) 238.

[6]J. Sólyom, *Adv. Phys.* (1979) 201.

[7]L. P. Kadanoff, *Rev. Mod. Phys.* (1977) 267; K. G. Wilson, *Rev. Mod. Phys.* (1983) 583.

[8]P. Ramond, *Field Theory. A Modern Primer.* (Addison-Wesley, Reading, 1989).

[9]Yu. A. Bychkov, L. P. Gorkov and I. E. Dzyaloshinsky, *Soviet Phys. JETP* (1966) 489. In the context of the Kondo effect, see also A. A. Abrikosov, *Physics* (1965) 5. Regarding the X-ray edge singularity problem, see B. Roulet, J. Gavoret and P. Nozières, *Phys. Rev.* (1969) 1072.

[10]P. W. Anderson, G. Yuval and D. R. Hamann, *Phys. Rev. B* (1970) 4464.

[11]J. M. Kosterlitz, *J. Phys. C* (1974) 1046.

4. Bosonization. Luttinger Liquid

4.1 Luttinger Model. Bosonization

In this section we are going to focus our attention on the line of critical points $g_{1\perp} = 0$ in the upper half-plane of Fig. 3.10 . These are worth of study since, as stated in the previous chapter, the properties of the theory on that line give the low-energy physics on the whole region $g_{1\parallel} \geq |g_{1\perp}|$, where the backscattering is an irrelevant perturbation. The model with $g_{1\perp} = 0$ is called the Tomonaga model[1]. A further simplification is achieved if one introduces an infinite linear dispersion relation for both left and right channels, as shown in Fig. 4.1 . It is argued that the influence of the deeper, spurious electronic states can be neglected if one only cares about low-energy processes. The consequences of this variant are important since the model with the infinite linear dispersion relation and the interactions given in Figs. 3.2 and 3.3 is exactly solvable. This is called the Luttinger model[2]. Quite remarkably, this is a quantum field theory in which the complete summation of diagrams in perturbation theory can be achieved[3]. This can be done using the great degree of symmetry of the model: the dynamics conserves the number of particles of given spin in a given channel.

We will not adopt, though, the perturbative approach to solve the Luttinger model. Instead we will introduce bosonization techniques which are well-suited for more general problems in $1+1$ dimensions. Haldane has made plausible that certain properties that one describes by means of bosonization are robust, in the sense that they should be shared also by the more realistic one-dimensional electron systems (whenever backscattering can be neglected). These properties configure a kind of universality class, referred to by the name of Luttinger liquid[4]. Its most prominent feature is the absence of quasiparticles with the same quantum numbers of the electron, together with the possibility of classifying all the excitations into boson-like objects. These account, in general, for charge and spin degrees of freedom, whose dynamics becomes completely independent.

4.1.1 Bosonic Excitations

We begin by classifying the excitations of the non-interacting Luttinger model. Since in the present development the two possible orientations of spin play

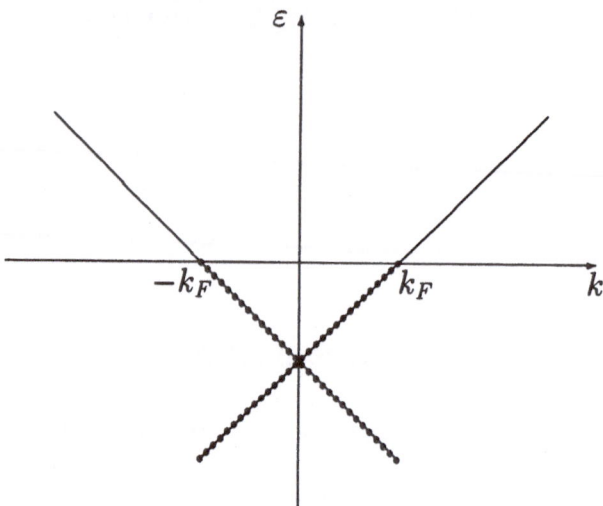

Fig. 4.1. Electronic dispersion relation in the Luttinger model.

parallel roles we will simply omit them, instead of duplicating all the expressions (spin requires an explicit consideration, however, when the interactions are turned on). Our "free" hamiltonian reads then

$$H_0 = \sum_k v_F(k - k_F)a_k^+ a_k + \sum_k v_F(-k - k_F)b_k^+ b_k \tag{4.1}$$

where, according to the infinite dispersion relation, no cutoff is implied in the sum over modes.

If we think of all the possible ways in which one can confer energy to the Fermi sea represented in Fig. 4.1, we find three essentially different operations in which this can be done.

i) Density excitations

The first may consist of taking an electron of the Fermi sea, in a given channel, and promoting it to an empty state above the Fermi level, in the same branch of the dispersion relation. In practice one considers the linear combination of operations involving the same excitation energy, for instance

$$\rho_{kR} = \sum_q a_{q+k}^+ a_q \tag{4.2}$$

for the right branch, and

$$\rho_{kL} = \sum_q b_{q+k}^+ b_q \tag{4.3}$$

for the left branch.

Now, it is clear that for $k > 0$ ρ_{kR} creates an excitation while ρ_{kL} desexcites the Fermi sea. For $k < 0$ the roles are reversed. These operators cannot be interpreted yet as true creation and annihilation operators, since their commutation relations do not have the proper normalization. It can be shown that

$$[\rho_{-k'R}, \rho_{kR}] = \delta_{kk'} k \frac{L}{2\pi} \tag{4.4}$$

$$[\rho_{-k'L}, \rho_{kL}] = -\delta_{kk'} k \frac{L}{2\pi} \tag{4.5}$$

We have supposed for convenience that the electrons are confined in a compact dimension of length L. The proof of the commutation relations (4.4), (4.5) has actually some subtle points that can be conveniently dealt with taking into account the infinite linear dispersion relation[5]. If one only cares about low-energy excitations, though, the same results can be obtained by using a linear dispersion relation with bandwith cutoff. We give a sketch of the proof of (4.4) and (4.5) in this latter case. Suppose that we consider the region of momentum space between $-\Lambda + k_F \equiv k_1$ and $\Lambda + k_F \equiv k_2$. The commutator of ρ_{-kR} and ρ_{kR}, for instance, reads

$$
\begin{aligned}
[\rho_{-kR}, \rho_{kR}] &= \left[\sum_{k_1 < q-k, q < k_2} a^+_{q-k} a_q, \sum_{k_1 < r+k, r < k_2} a^+_{r+k} a_r \right] \\
&= \sum_{k_1 < r, r+k < k_2} \delta_{q, r+k}\, a^+_{q-k} a_r \\
&\quad - \sum_{k_1 < q, q-k < k_2} \delta_{r, q-k}\, a^+_{r+k} a_q
\end{aligned}
\tag{4.6}
$$

It is important to have in mind that in these sums all the subindices run between k_1 and k_2. For this reason, provided that $0 < k < \Lambda$, one can see that the first sum in (4.6) has k over $2\pi/L$ more contributions than the other from the lower part of the band, while the second gets the same number in excess from the upper part of the band. We may now replace the right-hand-side of (4.6) by its vacuum expectation value on the ground state, which is an increasingly good approximation as $\Lambda \to \infty$ or, equivalently, at low-energies. Since the number operator $a^+_r a_r$ gives zero acting on states above k_F, we get $kL/(2\pi)$ for the final result of the commutator.

At this point, we form boson creation and annihilation operators by defining

$$B^+_k = \sqrt{\frac{2\pi}{L|k|}} \rho_{kR} \quad k > 0 \quad , \quad B^+_k = -\sqrt{\frac{2\pi}{L|k|}} \rho_{kL} \quad k < 0$$

$$B_k = \sqrt{\frac{2\pi}{L|k|}} \rho_{-kR} \quad k > 0 \quad , \quad B_k = -\sqrt{\frac{2\pi}{L|k|}} \rho_{-kL} \quad k < 0 \tag{4.7}$$

These obey the canonical commutation relations

$$[B_k, B^+_{k'}] = \delta_{kk'} \tag{4.8}$$

In this boson representation, the unexcited Fermi sea is equivalent to the boson ground state, defined as the state which is annihilated by all the B_k. The energy associated to the creation of each boson is obviously $v_F |k|$.

ii) Current excitations

We may imagine, furthermore, a second type of excitation process of the Fermi sea. It arises when a number of electrons of one of the branches are taken from below the Fermi level and shifted to the other branch. For a given number of shifted electrons, we are going to be interested in the *minimal* energy needed in the process. Thus, in the case of one electron we have to take it at the Fermi level, say from the left channel, and place it on top of the highest occupied level in the right channel. One has to remember that the system has spatial periodicity equal to L, so that the distance between neighboring states in momentum space is $2\pi/L$. The energy involved in shifting one electron is therefore this quantity times v_F.

In the process of transferring two electrons from one channel to the other, we have to realize that, after having shifted the first, the second electron is pulled from a state which is at a distance $2\pi/L$ of the Fermi level. As it has to be placed on top of the previously transferred electron, the total energy involved in the process is $2\pi/L$ times $v_F(1+3)$. It is not hard to see that the energy needed in the case of three electrons is $2\pi/L$ times $v_F(1+3+5)$, in the case of four electrons $2\pi/L$ times $v_F(1+3+5+7)$, etc. In general, the energy needed to transfer a number J of electrons can be expressed in compact form as $(2\pi/L)v_F J^2$ [4]. What is important to realize is that J is also half the difference of the number of electrons of one channel with respect to the other. In the Luttinger model the dynamics conserves the number of electrons N_R in the right channel and the number of electrons N_L in the left channel. Therefore it is really appropriate to take J as a number, since it is a *conserved charge* of the problem. The energy for the kind of process we have just described is therefore

$$\frac{\pi}{2L}v_F(N_R - N_L)^2 \tag{4.9}$$

iii) Charge excitations

Finally, we are left with only one more possibility to excite the Fermi sea of Fig. 4.1 . This is the addition of net charge to the system. In other words, we have to evaluate the dependence of the energy of the system on charge variations. Since the preceding operation involved the asymmetry in the distribution of electrons in the two branches, we will suppose for convenience that the same charge is confered to each channel. We take the Fermi level as a reference when measuring the energy balance. Thus, after addition of one electron to each channel the energy gained by the system is $(2\pi/L)$ times $2v_F$. The energy needed to modify by two units the charge per channel is $(2\pi/L)$ times $2v_F(1+2)$, to modify it by three $(2\pi/L)$ times $2v_F(1+2+3)$, etc. The sequence of integer numbers appears as the result of placing each new electron on top of the previous one. In general, when the total charge added to the system is N, the balance of energy becomes $(2\pi/L)v_F(N/2+1)N/2$. In this operation, however, the Fermi level is risen by $v_F \delta k_F = (2\pi/L)v_F N/2$. Since we are measuring the energy with respect to this reference level we have to discount this shift. The energy gained by addition of charge becomes then[4]

$$\frac{\pi}{2L}v_F(N_R + N_L)^2 \tag{4.10}$$

where now N_R and N_L are normal ordered quantities, i.e. measured after subtraction of the charge of the Fermi sea.

4.1.2 Bosonization

The remarkable fact is that, in one dimension, the set of bosonic excitations we have just described spans the whole Hilbert space of the original fermion problem. Thus, the basis of one-particle electron states and that of boson occupation numbers are nothing but different representations for the same quantum system. The boson excitations are all independent by construction. That they also provide a complete set of states can be shown in several ways. The most sophisticated consists of computing the partition function of the grandcanonical ensemble

$$\mathcal{Z}(\beta) = \text{tr} \exp\{-\beta H_0\} \tag{4.11}$$

in both bases and checking that the result is independent of taking the trace over fermion or boson excitations[4]. Actually, the identity between the expressions one obtains in the two cases is highly nontrivial. A quicker way of showing the equivalence of the two representations manages to compare the thermal energy for the two descriptions, in the infinite volume limit $L \to \infty$ [6]. Then, the sums over modes can be replaced by integrals which are easier to evaluate. The thermal energy in the fermion representation is

$$E_F = \sum_{k > k_F} 2\frac{\varepsilon(k) - \varepsilon_F}{e^{\beta(\varepsilon(k)-\varepsilon_F)} + 1} \approx \frac{1}{\beta^2}\frac{L}{2\pi v_F} \int_0^\infty dx \frac{2x}{e^x + 1} \tag{4.12}$$

where the factor of 2 takes into account the energy of the holes. The same quantity in the boson representation reads

$$E_B = \sum_{k > 0} \frac{\varepsilon(k)}{e^{\beta\varepsilon(k)} - 1} \approx \frac{1}{\beta^2}\frac{L}{2\pi v_F} \int_0^\infty dx \frac{x}{e^x - 1} \tag{4.13}$$

It can be checked that the two integrals in (4.12) and (4.13) give the same result, which completes the proof in the limit $v_F\beta/L \ll 1$.

As a consequence of the mentioned equivalence, we may also express the hamiltonian (4.1) in terms of the bosons B_k, B_k^+ and the conserved charges N_L, N_R. Recalling the energy of each mode evaluated before, we have

$$
\begin{aligned}
H_0 &= \sum_{k \neq 0} v_F |k| B_k^+ B_k + \frac{\pi}{2L}v_F(N_R - N_L)^2 + \frac{\pi}{2L}v_F(N_R + N_L)^2 \\
&= \sum_{k \neq 0} v_F |k| B_k^+ B_k + \frac{\pi}{L}v_F N_L^2 + \frac{\pi}{L}v_F N_R^2
\end{aligned} \tag{4.14}
$$

In fact, one can organize more efficiently the boson excitations into a pair of boson fields in $1+1$ dimensions. As stated before, the conserved charges N_L and

N_R count additional fermions to the Fermi sea and are therefore represented by the normal ordered form of the respective zero momentum boson modes. We have, for instance,

$$N_R = \sum_q a_q^+ a_q - \langle \sum_q a_q^+ a_q \rangle \qquad (4.15)$$

along with a similar expression for the left charge. It is possible to arrange ρ_{kR} and N_R into a genuine chiral boson field

$$\Phi_R(x) = \frac{2\pi}{L} \left(x N_R + i \sum_{k \neq 0} \frac{e^{-ikx}}{k} \rho_{kR} \right) \qquad (4.16)$$

This field has a definite chirality since, in the sum over modes, creation operators, for instance, enter only for $k > 0$. The field with the opposite chirality is given by

$$\Phi_L(x) = \frac{2\pi}{L} \left(x N_L + i \sum_{k \neq 0} \frac{e^{-ikx}}{k} \rho_{kL} \right) \qquad (4.17)$$

We have obtained this chiral decomposition quite naturally, since we have started with a problem in which the two branches of the dispersion relation are completely disconnected.

In field theory one usually deals with the complete boson field

$$
\begin{aligned}
\Phi(x) &= \Phi_L(x) + \Phi_R(x) \\
&= \frac{2\pi}{L} x \left(N_L + N_R \right) + i \sum_{q \neq 0} \sqrt{\frac{2\pi}{L|q|}} \left(e^{-iqx} B_q^+ - e^{iqx} B_q \right) \quad (4.18)
\end{aligned}
$$

The decomposition into left and right parts requires the use of the momentum field operator

$$
\begin{aligned}
\Pi(x) &= \frac{1}{4\pi} \left(\partial_x \Phi_L(x) - \partial_x \Phi_R(x) \right) \\
&= \frac{1}{4\pi} \left(\frac{2\pi}{L} \left(N_L - N_R \right) - \sum_{q \neq 0} \sqrt{\frac{2\pi}{L|q|}} |q| \left(e^{-iqx} B_q^+ + e^{iqx} B_q \right) \right) (4.19)
\end{aligned}
$$

which enters in the canonical commutation relation

$$[\Phi(x), \Pi(y)] = i\delta(x - y) \qquad (4.20)$$

We have actually the relations

$$\Phi_L(x) = \frac{1}{2} \left(\Phi(x) + 4\pi \int^x dy\, \Pi(y) \right) \qquad (4.21)$$

$$\Phi_R(x) = \frac{1}{2} \left(\Phi(x) - 4\pi \int^x dy\, \Pi(y) \right) \qquad (4.22)$$

that will be needed in the next section for the boson representation of the fermion field operator. The representation of the chiral boson fields in terms of the fermions is straightforward and reads, from (4.16) and (4.17),

$$\partial_x \Phi_L(x) = \frac{2\pi}{L} \rho_L(x) \tag{4.23}$$

$$\partial_x \Phi_R(x) = \frac{2\pi}{L} \rho_R(x) \tag{4.24}$$

In terms of the boson fields the hamiltonian (4.14) can also be expressed in integral compact form. From the above expressions it can be proven that

$$
\begin{aligned}
H_0 &= \int_0^L dx \, v_F \frac{\pi}{L^2} \left(: \rho_L(x)\rho_L(x) : + : \rho_R(x)\rho_R(x) : \right) \\
&= \frac{v_F}{2} \int_0^L dx \, : \left(4\pi \Pi^2(x) + \frac{1}{4\pi} \left(\partial_x \Phi(x) \right)^2 \right) :
\end{aligned}
\tag{4.25}
$$

Normal ordering in the above expression is needed to deal with singularities arising from the product of fields at the same point. This representation of the hamiltonian as a quadratic form on the density fields stems from a more general property for Kac-Moody algebras (Sugawara construction), of which ours is a trivial example[7]. As we are going to see, it plays also an essential role in the integrability of the Luttinger model.

4.1.3 Interacting Theory

As mentioned before, the Luttinger model describes points of the critical line in which only the g_2 and g_4 types of interaction are present, in the notation of Figs. 3.2 and 3.3 . This makes possible to express the interacting hamiltonian, as well as the "free" hamiltonian (4.25), as a form quadratic in boson creation and annihilation operators. The $g_{1\perp}$ type of interaction cannot be written in the form "density" times "density", which explains why the backscattering term spoils in general the integrability of the model.

According to the description of Figs. 3.2 and 3.3, the interaction hamiltonian now reads

$$
\begin{aligned}
H_{int} &= \frac{1}{L} \sum_{q>0} g_2(q) \left(\rho_{qR}\rho_{-qL} + \rho_{-qR}\rho_{qL} \right) \\
&\quad + \frac{1}{L} \sum_{q>0} \frac{g_4(q)}{2} \left(\rho_{qR}\rho_{-qR} + \rho_{-qR}\rho_{qR} + \rho_{qL}\rho_{-qL} + \rho_{-qL}\rho_{qL} \right)
\end{aligned}
\tag{4.26}
$$

In what follows we let g_2 and g_4 depend on the momentum transfer q. We will see that this level of generality is needed since some constraints on the "coupling constants" arise for the strict integrability of the Luttinger model.

Aside from the zero momentum modes, we write the total hamiltonian of the interacting model in terms of the boson operators (4.7)

$$\begin{aligned}
H &= H_0 + H_{int} \\
&= v_F \sum_{q \neq 0} |q| \, B_q^+ B_q - \sum_{q>0} \frac{g_2}{2\pi} |q| \left(B_q^+ B_{-q}^+ + B_q B_{-q} \right) \\
&\quad + \sum_{q>0} \frac{g_4}{4\pi} |q| \left(B_q^+ B_q + B_q B_q^+ + B_{-q}^+ B_{-q} + B_{-q} B_{-q}^+ \right)
\end{aligned} \tag{4.27}$$

It is convenient to rewrite the first term in the form

$$v_F \sum_{q \neq 0} |q| \, B_q^+ B_q = \frac{v_F}{2} \sum_{q \neq 0} |q| \left(B_q^+ B_q + B_q B_q^+ \right) - \frac{v_F}{2} \sum_{q \neq 0} |q| \tag{4.28}$$

where we suppose that the last sum is conveniently cut off in the ultraviolet. It turns out that the g_4 interaction can be absorbed into a redefinition of the "free" hamiltonian

$$\begin{aligned}
H &= \frac{1}{2} \sum_{q \neq 0} \left(v_F + \frac{g_4}{2\pi} \right) |q| \left(B_q^+ B_q + B_q B_q^+ \right) - \frac{v_F}{2} \sum_{q \neq 0} |q| \\
&\quad - \sum_{q \neq 0} \frac{g_2}{4\pi} |q| \left(B_q^+ B_{-q}^+ + B_q B_{-q} \right)
\end{aligned} \tag{4.29}$$

We see therefore that the effect of g_2 and that of g_4 are quite different on the electronic system. The effect of g_4 is to "renormalize" the value of the Fermi velocity

$$v_F \rightarrow v_F + \frac{g_4}{2\pi} \equiv v_F' \tag{4.30}$$

while that of g_2 is more drastic since it changes the ground state.

In fact, the total hamiltonian can be diagonalized by means of a Bogoliubov transformation[8]

$$\tilde{B}_q^+ = \cosh \phi(q) \, B_q^+ - \sinh \phi(q) \, B_{-q} \tag{4.31}$$
$$\tilde{B}_{-q} = \cosh \phi(q) \, B_{-q} - \sinh \phi(q) \, B_q^+ \tag{4.32}$$

With this parametrization, the new operators $\tilde{B}_q$ and $\tilde{B}_q^+$ continue satisfying canonical commutation relations. Writing the hamiltonian in terms of them and requiring the cancellation of the $\tilde{B}_q^+ \tilde{B}_{-q}^+$ and $\tilde{B}_q \tilde{B}_{-q}$ contributions, we get the condition on $\phi(q)$

$$\tanh 2\phi(q) = \frac{1}{v_F'} \frac{g_2}{2\pi} \tag{4.33}$$

The diagonal hamiltonian reads then

$$\begin{aligned}
H &= \frac{1}{2} \sum_{q \neq 0} \tilde{v}_F |q| \left(\tilde{B}_q^+ \tilde{B}_q + \tilde{B}_q \tilde{B}_q^+ \right) - \frac{v_F}{2} \sum |q| \\
&= \sum_{q \neq 0} \tilde{v}_F |q| \, \tilde{B}_q^+ \tilde{B}_q + \frac{1}{2} \sum (\tilde{v}_F - v_F) |q|
\end{aligned} \tag{4.34}$$

where

$$\tilde{v}_F \;=\; v_F' \cosh 2\phi(q) - \frac{g_2}{2\pi} \sinh 2\phi(q)$$

$$=\; \sqrt{\left(v_F + \frac{g_4}{2\pi}\right)^2 - \left(\frac{g_2}{2\pi}\right)^2} \tag{4.35}$$

We find that the consistency of the above procedure demands a first constraint to be satisfied

$$(I) \qquad\qquad |g_2| \le |2\pi v_F + g_4| \tag{4.36}$$

On the other hand, removing the ultraviolet cutoff implicit in the last term of (4.34) may lead, in general, to a divergent boson ground state energy, unless $\tilde{v}_F - v_F$ goes appropriately to zero at large $|q|$. This finiteness or independence of the ground state energy on the cutoff is not usually seen, though, as very relevant, since even a divergent result may be interpreted as a global shift of the energy scale. It cannot be said the same regarding the normalization of the boson ground state. This is now annihilated by all the "free" operators $\tilde{B}_q$ and is related to the state $|O\rangle_B$ annihilated by the original B_q operators through the expression

$$|O\rangle_{\tilde{B}} \sim \exp\left(\sum_{q>0} \tanh \phi(q)\, B_q^+ B_{-q}^+\right) |O\rangle_B \tag{4.37}$$

It can be shown that this state is normalizable only if the following constraint is satisfied[4]

$$(II) \qquad\qquad \lim_{|q|\to\infty} \sqrt{|q|}\,\frac{g_2}{2\pi v_F + g_4} = 0 \tag{4.38}$$

We should impose therefore conditions (I) and (II) to ensure the integrability of the Luttinger model. This implies the existence of some length scale which separates the short and long distance ranges in the system. Finally, the condition that both g_2 and g_4 are finite at $q = 0$ is also usually imposed within the context of the Luttinger model. Though it does not arise from the above development, it is needed to maintain the physical properties which characterize the universality class of the model. Its violation, as it happens in the case of the Coulomb interaction, leads in general to new physics, announced by the abnormal behavior of the response functions[9].

To summarize, we have been discussing the Luttinger model, which is actually a class of electronic systems with "density" times "density" type of interactions. The main property of them is that all their excitations can be classified in terms of boson degrees of freedom. Though we have been carrying out the discussion without taking into account the spin, its inclusion leads to the same essential result. This is treated in the following section, where we also complete the bosonization program giving the translation of the electron field in terms of boson operators.

exercise 4.1 *Check the equivalence of the representations (4.14) and (4.25) of H_0.*

exercise 4.2 *Find the norm of the state (4.37) and the condition for its finiteness.*

4.2 Charge-Spin Separation. Luttinger Liquid

In this section we consider the Luttinger model of one-dimensional electrons with spin. The discussion carried out in the previous section for the noninteracting theory can be now entirely reproduced, but the introduction of the spin gives rise to the genuine physical properties that characterize the interacting theory. The most important of them, namely the complete separation of the charge and the spin excitations, builds up the concept of Luttinger liquid. In this universality class should fall one-dimensional systems which may deviate slightly from the Luttinger model description, but share with it the absence of excitations with the same quantum numbers of the electron. This property alone establishes a neat difference with the Fermi liquid behavior characteristic of the higher dimensions.

In order to discuss the above points, the main technical achievement of this section will be the representation of the fermions fields in terms of the boson operators. This involves the rather nontrivial concept of how fermions may arise in a theory built out of bosons[10, 11]. The most clear explanation of this fact is perhaps given by Mandelstam[12]. Once we are able to express every field in terms of the boson fields, we may compute all the correlators of the interacting theory and, in particular, the one-particle electron Green function. Its inspection will clearly show that it lacks the characteristic pole structure of Fermi liquid behavior, proving that in the interacting theory, and no matter how small the strength of the couplings may be, there are no physical states with spin 1/2 and the electron charge. We will finally provide some intuitive picture trying to understand how this deconfinement of the charge and the spin is a plausible phenomenon in one dimension.

4.2.1 Charge-Spin Separation in a Simple Case

The introduction of the spin just amounts to make a parallel counting of the boson excitations for the two spin orientations, in the noninteracting theory. Taking into account (4.25) the expression of the "free" hamiltonian becomes

$$H_0 = \int_0^L dx \; v_F \frac{\pi}{L^2} \left(: \rho_{R\uparrow}(x)\rho_{R\uparrow}(x) : + : \rho_{R\downarrow}(x)\rho_{R\downarrow}(x) : + \quad L \leftrightarrow R \right) \quad (4.39)$$

It is convenient to introduce the respective charge density and spin density fields (we omit the subindices L, R in this formula)

$$\rho(x) = \frac{\rho_\uparrow(x) + \rho_\downarrow(x)}{\sqrt{2}} \quad , \quad \sigma(x) = \frac{\rho_\uparrow(x) - \rho_\downarrow(x)}{\sqrt{2}} \quad (4.40)$$

The separation of charge and spin is then manifest in the "free" hamiltonian since

$$H_0 = \int_0^L dx \; v_F \frac{\pi}{L^2} \left(: \rho_R(x)\rho_R(x) : + : \sigma_R(x)\sigma_R(x) : + \quad L \leftrightarrow R \right) \quad (4.41)$$

It can be also shown that in the interacting hamiltonian of the Luttinger model (g_2 and g_4 interactions alone) charge and spin always decouple. Suppose a simple instance in which the interaction is local (apart from some momentum transfer cutoff which we do not write explicitly), involving only the charge density

$$
\begin{aligned}
H_{int} &= g\frac{1}{L^2}\int_0^L dx \; : (\rho_L(x) + \rho_R(x))\,(\rho_L(x) + \rho_R(x)) : \\
&= g\frac{1}{L^2}\int_0^L dx \; (: \rho_L(x)\rho_L(x) : + : \rho_R(x)\rho_R(x) :) \\
&\quad +2g\frac{1}{L^2}\int_0^L dx \; \rho_L(x)\rho_R(x)
\end{aligned}
\tag{4.42}
$$

The combination $H_0 + H_{int}$ can be diagonalized, as in the previous section, by means of the "pseudorotation"

$$
\begin{pmatrix} \rho_L \\ \rho_R \end{pmatrix} = \begin{pmatrix} \cosh\phi & \sinh\phi \\ \sinh\phi & \cosh\phi \end{pmatrix} \begin{pmatrix} \tilde\rho_L \\ \tilde\rho_R \end{pmatrix}
\tag{4.43}
$$

For the value

$$
\tanh 2\phi = -\frac{g/\pi}{v_F + g/\pi}
\tag{4.44}
$$

we obtain the diagonal form

$$
\begin{aligned}
H &= \tilde v_F \frac{\pi}{L^2}\int_0^L dx \; (: \tilde\rho_R(x)\tilde\rho_R(x) : + L \leftrightarrow R) \\
&\quad + v_F \frac{\pi}{L^2}\int_0^L dx \; (: \sigma_R(x)\sigma_R(x) : + L \leftrightarrow R)
\end{aligned}
\tag{4.45}
$$

In this example it is quite trivial that the only eigenstates of the hamiltonian are given by spin waves and plasmons. The response functions for charge and spin show the poles corresponding to these physical states, for instance,

$$
\begin{aligned}
\chi(\omega, k) &\equiv -i\int dt \, e^{i\omega t} \langle \sigma(t, k)\sigma(0, -k) \rangle \\
&= \frac{v_F^2 k^2}{\omega^2 - v_F^2 k^2 + i\epsilon}
\end{aligned}
\tag{4.46}
$$

The above is just an illustration of the charge-spin separation which *always* takes place in the Luttinger model. It is a good exercise to show that for arbitrary values of $g_{2\parallel}, g_{2\perp}, g_{4\parallel}$ and $g_{4\perp}$ the change of variables (4.40) always decouples the charge and the spin excitations[13]. Then one can perform two independent "pseudorotations" in the respective sectors to bring the hamiltonian to diagonal form, showing that the velocity of spin excitations is different than that of charge excitations, and both different, in general, to v_F.

4.2.2 Boson Representation of Fermion Operators

At first sight it may appear intriguing that a fermion-like object may be built in a theory which, as we have already seen, is made out of bosons. This point is perhaps best understood following Mandelstam[12] and his argumentation about the sine-Gordon model[11]. This is the model for a boson field $\Phi(x)$ with an interaction hamiltonian

$$H_{int} \sim \int dx \, \cos \, \Phi(x) \qquad (4.47)$$

Together with trivial static solutions like $\Phi(x) = \pm\pi$, the sine-Gordon model has *soliton* solutions which interpolate between different wells of the interaction potential. A single soliton, in particular, complies to the boundary conditions

$$\Phi(x) \to \pi \quad , \quad x \to \infty$$
$$\Phi(x) \to -\pi \quad , \quad x \to -\infty \qquad (4.48)$$

In the quantum theory, therefore, we have two different types of fluctuations. We have particle excitations, which appear by quantizing about the solution $\Phi(x) = \pi$, for instance, and we have also the quanta which arise from the soliton solution. It can be shown that the soliton and the corresponding antisoliton bear nonvanishing fermion number, being therefore the manifestation of the fermion-like object in the theory. This is consistent with our representation (4.23) and (4.24) of the density fluctuations. The total charge of the system (which for the particular sine-Gordon model (4.47) is a conserved charge, see Ref. [12]) is

$$Q = \frac{1}{2\pi} \int_{-\infty}^{+\infty} dx \, \partial_x \Phi \qquad (4.49)$$

which gets nonvanishing (integer) values for the soliton solutions like (4.48). From the point of view of the particles (bosons), the fermion is an object which interpolates between different topological sectors of the model. We will reach a more intuitive understanding of this phenomenon when describing the harmonic chain in Chap. 5.

Following with this argumentation, Mandelstam represents the fermion field $\Psi(x)$ as a soliton *annihilation* operator. That is, it must be an operator which shifts the value of $\Phi(y)$ by 2π to the left of x, and leaves it unmodified to the right. This is expressed in the form

$$[\Phi(y), \Psi(x)] = 2\pi\Psi(x) \quad , \quad y < x \qquad (4.50)$$
$$[\Phi(y), \Psi(x)] = 0 \quad , \quad y > x \qquad (4.51)$$

A possible representation of $\Psi(x)$ which satisfies these commutation relations is (see (4.20))

$$\Psi(x) =: \mathcal{O}(x) \exp \left(-i2\pi \int^x dy \, \Pi(y) \right) : \qquad (4.52)$$

where the operator $\mathcal{O}(x)$ is still undetermined. If we now require that $\Psi(x)$ and $\Psi(y)$ anticommute at different points x and y, we are led to take the two possibilities[12]

$$\Psi_1(x) \;=\; : \exp\left(-i\frac{1}{2}\Phi(x) - i2\pi \int^x dy\, \Pi(y)\right) : \qquad (4.53)$$

$$\Psi_2(x) \;=\; : \exp\left(i\frac{1}{2}\Phi(x) - i2\pi \int^x dy\, \Pi(y)\right) : \qquad (4.54)$$

At this point we recognize in the argument of the exponentials the two chiral boson fields (4.21), (4.22). It is therefore appropriate to identify (4.53) and (4.54), respectively, with the two fermion components of given chirality

$$\Psi_R(x) \;=\; : \exp\left(i\Phi_R(x)\right) : \qquad (4.55)$$
$$\Psi_L(x) \;=\; : \exp\left(-i\Phi_L(x)\right) : \qquad (4.56)$$

Quite satisfactorily, it can be also shown that, in a given channel, $\Psi_i(x)$ and $\Psi_i^+(y)$ also satisfy the canonical anticommutation relations

$$\{\Psi_i(x), \Psi_i^+(y)\} = \delta(x-y) \qquad i = L, R \qquad (4.57)$$

In all these considerations it is implicit the assumption that the fields $\Phi_L(x)$ and $\Phi_R(x)$ are the very same that those defined from the density fields $\rho_L(x)$ and $\rho_R(x)$. There is, however, a nontrivial self-consistency check to be carried out, that is, to verify that the computation of the fermion density for a given channel, $: \Psi_i^+(x)\Psi_i(x) :$, with the above boson representation of the fermion fields actually matches the definitions (4.23) and (4.24). This is left as an exercise at the end of the section.

The above boson representation of the chiral fermion fields makes possible the computation of any correlator of the electron system. The full fermion field operator can be decomposed in the form

$$\begin{aligned}
\Psi(x) \;=\;& \frac{1}{\sqrt{L}}\sum_k e^{ikx}a_k + \frac{1}{\sqrt{L}}\sum_k e^{ikx}b_k \\
=\;& e^{ik_Fx}\frac{1}{\sqrt{L}}\sum_k e^{i(k-k_F)x}a_k + e^{-ik_Fx}\frac{1}{\sqrt{L}}\sum_k e^{i(k+k_F)x}b_k \\
=\;& e^{ik_Fx}\Psi_R(x) + e^{-ik_Fx}\Psi_L(x) \qquad (4.58)
\end{aligned}$$

Correlation functions involving any number of Ψ fields can be translated into expectation values on the boson vacuum of products of exponentials of boson fields. The computation requires the knowledge, in general, of the object

$$\langle : e^{i\alpha\Phi(x)} : : e^{-i\alpha\Phi(y)} : \rangle \qquad (4.59)$$

In the case $\alpha = 1$, we should recover the free propagator for the chiral fermion with the well-known behavior $\sim (x-y)^{-1}$. As we will see in the next subsection, though, in the interacting theory we get to know the expression (4.59) for arbitrary values of α.

The calculation, for instance, of

$$\langle : e^{i\alpha\Phi_R(x)} : : e^{-i\alpha\Phi_R(y)} : \rangle \tag{4.60}$$

can be accomplished in the following way. In all the pertinent instances, the expectation value refers to the vacuum free of bosons. The way to proceed, then, is to place to the right all the boson annihilation operators. We have already the normal ordered form

$$: \exp\left(i\alpha\Phi_R(x)\right) := \exp\left(-\alpha \sum_{k>0} \sqrt{\frac{2\pi}{kL}} e^{-ikx} B_k^+\right) \exp\left(\alpha \sum_{k>0} \sqrt{\frac{2\pi}{kL}} e^{ikx} B_k\right) \tag{4.61}$$

After use of the formula

$$e^A e^B = e^{B+A} e^{\frac{1}{2}[A,B]} = e^B e^A e^{[A,B]} \tag{4.62}$$

we get

$$\langle : e^{i\alpha\Phi_R(x)} : : e^{-i\alpha\Phi_R(y)} : \rangle =$$

$$= \langle \prod_{k>0} \exp\left(-\alpha\sqrt{\frac{2\pi}{kL}} e^{-ikx} B_k^+\right) \prod_{p>0} \exp\left(\alpha\sqrt{\frac{2\pi}{pL}} e^{ipx} B_p\right) \times$$

$$\prod_{q>0} \exp\left(\alpha\sqrt{\frac{2\pi}{qL}} e^{-iqy} B_q^+\right) \prod_{r>0} \exp\left(-\alpha\sqrt{\frac{2\pi}{rL}} e^{iry} B_r\right) \rangle$$

$$= \langle \prod_{k>0} \exp\left(-\alpha\sqrt{\frac{2\pi}{kL}} e^{-ikx} B_k^+\right) \prod_{q>0} \exp\left(\alpha\sqrt{\frac{2\pi}{qL}} e^{-iqy} B_q^+\right) \times$$

$$\prod_{p>0} \exp\left(\alpha\sqrt{\frac{2\pi}{pL}} e^{ipx} B_p\right) \prod_{r>0} \exp\left(-\alpha\sqrt{\frac{2\pi}{rL}} e^{iry} B_r\right) \rangle \times$$

$$\exp\left(\alpha^2 \frac{2\pi}{pL} e^{ip(x-y)}\right)$$

$$= \exp\left(\alpha^2 \frac{2\pi}{L} \sum_{p>0} \frac{1}{p} e^{ip(x-y)}\right) \tag{4.63}$$

The last expression is more easily evaluated in the infinite volume limit $L \to \infty$. The sum can be traded then by an integral, taking into account that it is actually regulated in the infrared by the minimum value of the momentum $\Delta \equiv 2\pi/L$,

$$\langle : e^{i\alpha\Phi_R(x)} : : e^{-i\alpha\Phi_R(y)} : \rangle \approx$$

$$\approx \exp\left\{\alpha^2 \int_\Delta^\infty dk \, \frac{1}{k} e^{ik(x-y)}\right\}$$

$$\approx \exp\left\{-\alpha^2 \left[\log\left(\Delta(x-y)\right) - i\frac{\pi}{2}\right]\right\} = \left(\frac{L}{2\pi} \frac{i}{(x-y)}\right)^{\alpha^2} \tag{4.64}$$

This is the final result. In the case of the free electron correlator $\alpha = 1$, it shows the correct spatial behavior, apart from normalization factors. The boson representation of the fermion field needs anyhow to be regulated in the ultraviolet, and the way we have done it here by the normal order prescription is the usual in quantum field theory[14]. There are, however, more possibilities. In condensed matter physics a more sophisticated prescription is usually taken[10], which allows to reproduce also the imaginary part of the electron Green function (the $i\epsilon$ prescription). This will be used in the description of the discrete models in Chap. 5.

In general, time dependent correlators can be infered from their static limit, by enforcing the manifest Galilean invariance of the theory. Thus, in the case of a right-handed field the time dependence is obtained replacing $x - x'$ by $x - x' - v_F(t - t')$, while in the case of a left-handed field it amounts to replace $x - x'$ by $x - x' + v_F(t - t')$. This of course can be checked by direct computation with operators in the interaction representation.

4.2.3 Electron Green Function

We illustrate the computation of the electron Green function in the Luttinger model, in the case of the charge density interactions already quoted in (4.42). The general case can be dealt with by means of the same method. We begin by dissociating charge and spin fields (we omit subindices L, R at this point)

$$\Phi_c = \frac{\Phi_\uparrow + \Phi_\downarrow}{\sqrt{2}} \quad , \quad \Phi_s = \frac{\Phi_\uparrow - \Phi_\downarrow}{\sqrt{2}} \tag{4.65}$$

We have, for instance, depending on the orientation of the spin $\uparrow$ or $\downarrow$

$$\Psi_{R\uparrow} = e^{i(\Phi_{cR} + \Phi_{sR})/\sqrt{2}} \tag{4.66}$$

$$\Psi_{R\downarrow} = e^{i(\Phi_{cR} - \Phi_{sR})/\sqrt{2}} \tag{4.67}$$

We may interpret therefore the electron field as being composed of two completely independent fields, one that carries charge but no spin (the holon) and other that carries spin but no charge (the spinon). This picture is correct to the extent that the hamiltonian is built out of disjoint charge and spin excitations. We may say then that the holon and the spinon are deconfined in one dimension.

What we want to obtain, on the other hand, is a correlator like

$$G_{R\uparrow}(t - t', x - x') = -i\langle T \, \Psi_{R\uparrow}(t, x)\Psi_{R\uparrow}^+(t', x')\rangle \tag{4.68}$$

in the interacting theory. We could apply for that purpose the expression (4.64), were not for the fact that it is valid only for free fields, as implied in its derivation. It is possible to resort, however, to the linear change of variables (4.43) that decouples left and right modes in the hamiltonian and renders it a diagonal quadratic form in the boson fields. In our case we have to make the canonical transformation

$$\begin{pmatrix} \Phi_{cL} \\ \Phi_{cR} \end{pmatrix} = \begin{pmatrix} \cosh\phi & \sinh\phi \\ \sinh\phi & \cosh\phi \end{pmatrix} \begin{pmatrix} \tilde{\Phi}_{cL} \\ \tilde{\Phi}_{cR} \end{pmatrix} \qquad (4.69)$$

to obtain free fields $\tilde{\Phi}_{cL}$ and $\tilde{\Phi}_{cR}$. Φ_{sL} and Φ_{sR} are already decoupled, but in the general case with spin-dependent interactions another "pseudorotation" to free fields may be needed in the spin sector. By playing with the holon-spinon decomposition and the decoupling of left and right modes, the interacting electron Green function reduces to a product of correlators of the type (4.64).

Without paying attention to normalization factors, we have

$$\begin{aligned}
G_{R\uparrow}(t - t', x - x') &= \\
&= \langle e^{i(\Phi_{cR}(t,x)+\Phi_{sR}(t,x))/\sqrt{2}}\, e^{-i(\Phi_{cR}(t',x')+\Phi_{sR}(t',x'))/\sqrt{2}} \rangle \\
&= \langle e^{i\Phi_{cR}(t,x)/\sqrt{2}}\, e^{-i\Phi_{cR}(t',x')/\sqrt{2}} \rangle \langle e^{i\Phi_{sR}(t,x)/\sqrt{2}}\, e^{-i\Phi_{sR}(t',x')/\sqrt{2}} \rangle \\
&= \langle e^{i\Phi_{cR}(t,x)/\sqrt{2}}\, e^{-i\Phi_{cR}(t',x')/\sqrt{2}} \rangle \frac{1}{|x - x' - v_F(t - t')|^{1/2}} \qquad (4.70)
\end{aligned}$$

where we have made use of (4.64) for the spinon operators. The charge correlator requires first the transformation to decoupled holon fields. Abbreviating $\sinh\phi \equiv s$ and $\cosh\phi \equiv c$, we have

$$\begin{aligned}
G_{R\uparrow}(t - t', x - x') &= \\
&= \langle e^{i(s\tilde{\Phi}_{cL}(t,x)+c\tilde{\Phi}_{cR}(t,x))/\sqrt{2}}\, e^{-i(s\tilde{\Phi}_{cL}(t',x')+c\tilde{\Phi}_{cR}(t',x'))/\sqrt{2}} \rangle \times \\
&\qquad \frac{1}{|x - x' - v_F(t - t')|^{1/2}} \\
&= \langle e^{is\tilde{\Phi}_{cL}(t,x)/\sqrt{2}}\, e^{-is\tilde{\Phi}_{cL}(t',x')/\sqrt{2}} \rangle \langle e^{ic\tilde{\Phi}_{cR}(t,x)/\sqrt{2}}\, e^{-ic\tilde{\Phi}_{cR}(t',x')/\sqrt{2}} \rangle \times \\
&\qquad \frac{1}{|x - x' - v_F(t - t')|^{1/2}} \\
&= \frac{1}{|x - x' + \tilde{v}_F(t - t')|^{s^2/2}} \frac{1}{|x - x' - \tilde{v}_F(t - t')|^{c^2/2}} \times \\
&\qquad \frac{1}{|x - x' - v_F(t - t')|^{1/2}} \qquad (4.71)
\end{aligned}$$

where $\tilde{v}_F$ is, from (4.35),

$$\tilde{v}_F = \sqrt{\left(v_F + \frac{g}{\pi}\right)^2 - \left(\frac{g}{\pi}\right)^2} \qquad (4.72)$$

The final result can be written in the form

$$\begin{aligned}
G_{R\uparrow}(t - t', x - x') &= \\
&= \frac{1}{|x - x' - \tilde{v}_F(t - t')|^{1/2}} \frac{1}{|x - x' - v_F(t - t')|^{1/2}} \times \\
&\qquad \frac{1}{|(x - x')^2 - \tilde{v}_F^2(t - t')^2|^{s^2/2}} \qquad (4.73)
\end{aligned}$$

The effect of the separation of charge and spin is manifest in the electron Green function, since the two different velocities v_F and $\tilde{v}_F$ account for the different propagation speed of charge and spin excitations. It is also remarkable that the propagation of the fermion field $\Psi_{R\uparrow}$ is not purely in the right direction, as far as $s^2 \neq 0$. The same admixture is also found in the propagator of the left-handed fermion, which is obtained from (4.73) by replacing $v_F \to -v_F$ and $\tilde{v}_F \to -\tilde{v}_F$. This lack of perfect propagation in a given direction can be interpreted on physical grounds by the fact that fermions of a given chirality excite, in its propagation, the fermions of the opposite chirality. Thus, there is a phenomenon of "reflection", by which we always find the superposition of two waves travelling in opposite directions.

Though the electron Green function has the simplest expression in (t, x) space, the physical properties related to the spectrum have to be addressed in momentum representation. The above propagator (4.73) does not admit a simple expression in (ω, k) space, but one may still obtain the Fourier transform of the factors and write that of the product as a convolution operation. The Fourier transform of (4.73) becomes then[15]

$$G_{R\uparrow}(\omega, k) \sim \frac{1}{|(\tilde{v}_F k - \omega)(v_F k - \omega)|^{1/2}} * (\tilde{v}_F^2 k^2 - \omega^2)^{s^2/2 - 1} \tag{4.74}$$

where $*$ symbolically denotes the convolution product. The corresponding expression for the left-handed fermion is obtained again by the replacements $v_F \to -v_F$ and $\tilde{v}_F \to -\tilde{v}_F$. The conclusion that we get by inspection of (4.74) is that we do not find fermion excitations after looking for poles of the fermion propagator. Actually, for a given value of the momentum this object has branch cuts instead of poles. This absence of fermion states in the spectrum is not surprising since, after all, we had already classified the physical states into bosonic charge and spin excitations. These certainly do appear as poles of the corresponding response functions for charge and spin operators.

The above result is very important since it shows that the Luttinger model is the prototype for a quite different liquid than that dealt with in Landau's Fermi liquid theory. In fact, we may characterize this universality class, called by the name of Luttinger liquid[4], by the breakdown of the quasiparticle picture. No matter how small the strength of the interaction may be, all trace of quasiparticle poles disappears in the electron Green function, and no physical states remain with the same quantum numbers of the electron.

Another distinctive signal not shared by the Fermi liquid arises from the behavior of the momentum distribution function $n(k)$ at the Fermi level. This is again a consequence of the particular form of the correlator (4.74). $n(k)$ is defined about k_F by

$$n(k) = -i \lim_{t \to 0^-} \int_{-\infty}^{+\infty} dx \, e^{-i(k - k_F)x} G_R(t, x) \tag{4.75}$$

In Fermi liquid theory the small imaginary part of the Green function at the quasiparticle pole leads to the well-known discontinuity of $n(k)$ at the Fermi level. Now the presence of the branch cut changes the nature of the singularity. In order to apply correctly (4.75) we need the precise knowledge of the complex structure of $G_R(t, x)$. This can be found in Ref. [3]. The limit $t \to 0^-$ fixes the imaginary part of the Green function

$$G_R(0^-, x) = \frac{1}{2\pi} \frac{1}{x - i\delta} \frac{1}{\left(\Lambda^2 \left(x - i\Lambda^{-1}\right)\left(x + i\Lambda^{-1}\right)\right)^{s^2/2}} \tag{4.76}$$

where δ^{-1} is a bandwith cutoff and Λ is a momentum transfer cutoff in the interaction. By introducing this expression in (4.75) we find

$$
\begin{aligned}
n(k) &= -i\frac{1}{2\pi} \int_{-\infty}^{+\infty} dx \, e^{-i(k-k_F)x} \frac{1}{x - i\delta} \frac{1}{\left(\Lambda^2 x^2 + 1\right)^{s^2/2}} \\
&= \frac{1}{2} \int_{-\infty}^{+\infty} dx \, e^{-i(k-k_F)x} \delta(x) \frac{1}{\left(\Lambda^2 x^2 + 1\right)^{s^2/2}} \\
&\quad - \frac{1}{2\pi} \, \mathrm{PP} \int_{-\infty}^{+\infty} dx \, \frac{\sin(k - k_F)x}{x} \frac{1}{\left(\Lambda^2 x^2 + 1\right)^{s^2/2}} \\
&\approx \frac{1}{2} - \mathrm{const.} \, |k - k_F|^{s^2} \, \mathrm{sgn}(k - k_F)
\end{aligned}
\tag{4.77}
$$

A symmetric expression holds about the other Fermi point at $-k_F$. We see that the jump in the momentum distribution function, which in the Fermi liquid equals the residue of the quasiparticle pole, has disappeared. Instead we have a continuous function, though with infinite slope at the Fermi points. This allows us, in a certain sense, to maintain the notion of Fermi level in the Luttinger liquid, as the level at which the derivative of $n(k)$ becomes infinite.

4.2.4 Intuitive Picture of Charge-Spin Separation

To close this section and the chapter, we try to give here an intuitive explanation of the phenomenon of separation of charge and spin in one dimension. Suppose, for instance, that we have a periodic disposition of electrons, with antiferromagnetic order, as shown in Fig. 4.2(a). This cannot represent a physical state of our one-dimensional theory but we may consider it as a local picture, pertinent to the ground state of systems like the one-dimensional Hubbard model. It represents a state free of spin and charge excitations. Suppose now that we introduce a hole in the system, which amounts to remove the charge and the spin at a given site of the periodic disposition (Fig. 4.2(b)). If we exchange the empty site with its neighbor to the left, after repeated application of this operation we observe that two different kinds of perturbation have been introduced in the system. We end up with the empty site, which does not disturb the antiferromagnetic order, and with a "domain wall" (the two consecutive spins pointing up) representing a frustration of the order (Fig. 4.2(c)). Obviously this frustration can be also propagated by exchanging neighboring spins at one side of the wall. Thus, we have the picture of two different perturbations of the

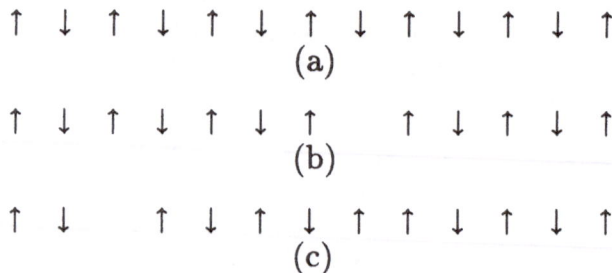

Fig. 4.2. Sequence of events showing the creation of a hole and a domain wall of the antiferromagnetic order.

original configuration, free to evolve one independently of the other. The first corresponds to what we called "holon" in the mathematical framework, while the second is the image of the "spinon" excitation. This simple picture makes plausible that the "holon" and the "spinon" may be disjoint excitations, able to propagate each of them with its own velocity. It also stresses to what extent the dimensionality of the system is crucial to understand the phenomenon of charge and spin separation. One can readily see that any attempt to implement this picture in two dimensions is not going to work out with the simplicity shown here. The question of whether there can be two-dimensional systems with the property of charge-spin separation similar to that of the Luttinger liquid is nowadays the subject of a big debate.

exercise 4.3 *Check that in the most general Luttinger model with $g_{2\parallel}$, $g_{2\perp}$, $g_{4\parallel}$ and $g_{4\perp}$ interactions the hamiltonian splits into two respective parts for charge and spin operators.*

exercise 4.4 *Compute the fermion density operator* $: \Psi_R^+(x)\Psi_R(x) :$ *using the boson representation of the fermion field and compare with the former definition (4.24).*

References

[1]S. Tomonaga, *Prog. Theor. Phys.* **5** (1950) 349.

[2]J. M. Luttinger, *J. Math. Phys.* **4** (1963) 1154.

[3]I. E. Dzyaloshinsky and A. I. Larkin, *Soviet Phys. JETP* **38** (1974) 202.

[4]F. D. M. Haldane, *J. Phys. C* **14** (1981) 2585.

[5]G. D. Mahan, *Many Particle Physics* (Plenum, New York, 1981).

[6]A. Luther and K. D. Schotte, *Nucl. Phys. B* **242** (1984) 407.

[7]D. Minic, *Mod. Phys. Lett. B* **7** (1993) 641.

[8]D. C. Mattis and E. H. Lieb, *J. Math. Phys.* **6** (1965) 304.

[9]H. J. Schulz, *Phys. Rev. Lett.* **71** (1993) 1864.

[10]A. Luther and I. Peschel, *Phys. Rev. B* **9** (1974) 2911.

[11]S. Coleman, *Phys. Rev. D* **11** (1975) 2088.

[12]S. Mandelstam, *Phys. Rev. D* **11** (1975) 3026.

[13]J. Sólyom, *Adv. Phys.* **28** (1979) 201.

[14]A. M. Polyakov, *Gauge Fields and Strings*, Chap. 9 (Harwood, London, 1989).

[15]X. G. Wen, *Phys. Rev. B* **42** (1990) 6623.

5. Correspondence from Discrete to Continuum Models

5.1 Introduction

We have seen in the preceding chapter that the bosonization technique is a very powerful tool for studying one-dimensional electron systems. We have uncovered with it the properties of a continuous set of models showing critical behavior and governing the low-energy physics of a wide region of the parameter space of all the theories. The very fact of describing a number of critical properties is actually what justifies the study accomplished in the continuum limit. One may think that the assumption of an infinite linear dispersion relation for the electrons is a crucial step in the rigorous proof of the boson-fermion correspondence and wonder to what extent the results obtained by means of the bosonization program may apply to more realistic models. In particular, any model of fermions hopping on a lattice must have a compact dispersion relation of the kind shown in Fig. 3.1 . There is empirical evidence, though, that the presence of the lattice does not change essentially the low-energy properties and that the Luttinger liquid paradigm continues applying to a wide class of discrete models. Some of these can be studied analytically by means of the Bethe ansatz technique, which provides exact results for the energy spectrum and some thermodynamic quantities. The case of the Hubbard model, for instance, is treated with detail in Chap. 10. From the exact resolution of the models one also gets information about the mapping to the Luttinger liquid universality class. This amounts to the knowledge of the parameters which characterize the model within the line of critical points obtained in Chap. 3. Thus, different systems like the antiferromagnetic Heisenberg chain, the spinless fermion model into which the former can be mapped or the Hubbard model at weak coupling have a low-energy behavior dictated by points over the critical line. The effect of the lattice only shows up in the form of certain renormalization of the parameters determining the critical point. Most significantly, it turns out that the correspondence to the mentioned universality class holds even in the presence of strong interactions, as is the case of the harmonic chain or the Hubbard model in the limit of large on-site repulsion. We will discuss these two models in what follows. The correspondence with the continuum approach has important consequences from a practical point of view, since it allows to ascertain low-energy properties like the charge-spin separation and the long-distance behavior of the correlation functions in the discrete models. These are not obtained by the exact Bethe ansatz resolution and, in most cases, are not susceptible either of an accurate numerical determination with the present computational techniques.

5.2 The Harmonic Chain

The first example of discrete one-dimensional model that we consider is the harmonic chain, which is essentially a system of particles tied by a nearest-neighbor harmonic oscillator potential[1]. This model does not seem to bear, at first sight, a direct relation with the continuum electron systems discussed previously, but it actually provides a good description of the electron liquid when there are strong correlation effects. This is the case of the large-U limit of the Hubbard model that we will discuss further on. In that extreme situation only the charge dynamics becomes nontrivial in the low-energy limit, and it can be modeled by some elastic coupling provided by the holes between adjacent electrons in the lattice. A most important point is that the continuum limit of the harmonic chain adopts the same form that the boson representation of the Luttinger model. This gives a clear indication that systems in the strong coupling regime may fall into the Luttinger universality class. The harmonic chain will also serve us to reach a more intuitive understanding of the boson description of one-dimensional electron systems.

5.2.1 Continuum Limit

We start with a system of N particles whose first-quantized hamiltonian is

$$H = \sum_{n=1}^{N} \left\{ \frac{P_n^2}{2m} + \frac{\lambda}{2} \left(X_{n+1} - X_n - d \right)^2 \right\} \tag{5.1}$$

Thus, the particles X_n and X_{n+1} suffer a repulsive interaction at short distances, while they reach a most stable configuration at a relative distance d. The quantum properties of the model are best investigated in the continuum limit $d \to 0$ [1]. Then the label n is promoted to a continuous variable $x = nd$ while the position of each particle is expanded about the classical configuration, so that

$$nd \;\; \to \;\; x \tag{5.2}$$

$$X_n \;\; \to \;\; nd + \frac{1}{2\pi} d\, \Phi(x) \tag{5.3}$$

$$P_n \;\; \to \;\; 2\pi\, \Pi(x) \tag{5.4}$$

The fields $\Phi(x)$ and $\Pi(x)$ satisfy the canonical commutation relation

$$[\Phi(x), \Pi(y)] = i\delta(x - y) \tag{5.5}$$

On the other hand, we have

$$X_{n+1} - X_n - d \rightarrow \frac{1}{2\pi} d^2\, \partial_x \Phi \qquad (5.6)$$

The hamiltonian in the continuum limit becomes

$$
\begin{aligned}
H &= \frac{1}{2m}\frac{1}{d}4\pi^2 \int dx\, \Pi^2(x) + \frac{1}{2}\lambda d^3 \frac{1}{4\pi^2} \int dx\, (\partial_x \Phi(x))^2 \\
&= \frac{v}{2} \int dx\, \left\{ 4\pi\mu \Pi^2(x) + \frac{1}{4\pi\mu}(\partial_x\Phi(x))^2 \right\}
\end{aligned}
\qquad (5.7)
$$

with the parameters

$$v = d\sqrt{\frac{\lambda}{m}} \quad , \quad \mu = \frac{1}{d^2}\frac{\pi}{\sqrt{\lambda m}} \qquad (5.8)$$

The model is therefore completely specified by the velocity v and the dimensionless parameter μ. We will see afterwards that the latter gives a measure of the strength of the quantum fluctuations, the classical limit being approached as $\mu \rightarrow 0$.

In order to bring the hamiltonian to the canonical free boson form (4.25), one may perform the transformation

$$
\begin{aligned}
\Phi(x) &= \sqrt{\mu}\tilde{\Phi}(x) \\
\Pi(x) &= \frac{1}{\sqrt{\mu}}\tilde{\Pi}(x)
\end{aligned}
\qquad (5.9)
$$

which, obviously, preserves the canonical commutation relations. In fact, this is nothing but the real-space version of the Bogoliubov transformation in (4.31) and (4.32), that we introduced to diagonalize the boson expression of the Luttinger model hamiltonian. Recalling (4.18) and (4.19), we may cast the above transformation in the already familiar form

$$
\begin{pmatrix} \Phi_L \\ \Phi_R \end{pmatrix} = \begin{pmatrix} \frac{1}{2}(\sqrt{\mu}+\frac{1}{\sqrt{\mu}}) & \frac{1}{2}(\sqrt{\mu}-\frac{1}{\sqrt{\mu}}) \\ \frac{1}{2}(\sqrt{\mu}-\frac{1}{\sqrt{\mu}}) & \frac{1}{2}(\sqrt{\mu}+\frac{1}{\sqrt{\mu}}) \end{pmatrix} \begin{pmatrix} \tilde{\Phi}_L \\ \tilde{\Phi}_R \end{pmatrix} \qquad (5.10)
$$

which draws the relation between μ and the rotation angle ϕ of the Bogoliubov transformation

$$\cosh\phi \leftrightarrow \frac{1}{2}\left(\sqrt{\mu}+\frac{1}{\sqrt{\mu}}\right) \qquad (5.11)$$

In the present model one is mainly interested in the correlators of the density operator $\rho(x)$. In terms of the discrete variables, this is written in the form

$$
\begin{aligned}
\rho(x) &= \sum_{n=1}^{N} \delta(x - X_n) \\
&= \int_{-\infty}^{+\infty} \frac{dq}{2\pi} \sum_{n=1}^{N} e^{iq(x-X_n)}
\end{aligned}
\qquad (5.12)
$$

If we make the passage to the $\Phi(x)$ field variable we get

$$\rho(x) = \int_{-\infty}^{+\infty} \frac{dq}{2\pi} e^{iqx} \sum_{n=1}^{N} e^{-iqdn} \, e^{-iqd\Phi/(2\pi)} \tag{5.13}$$

A crucial assumption is that the Φ field accounts for a smooth quantum fluctuation effect. In that case the sum over n has a significant value only for $q \approx 2\pi m/d, m = 0, \pm 1, \pm 2, \ldots$ In the continuum limit we get a contribution about each of these dominant momenta

$$\rho(x) = \sum_{m=-\infty}^{+\infty} e^{i2\pi mx/d} \, \rho_m(x) \tag{5.14}$$

For $m \neq 0$ we have

$$\rho_m(x) \approx \frac{1}{d} e^{im\Phi(x)} \tag{5.15}$$

The $m = \pm 1$ contributions correspond to the well-known chirality-mixing density fluctuations near $2k_F$. In fact, for fermions without spin this quantity equals 2π over the mean separation between particles. What we learn here is that under strong correlation effects, as they occur in the harmonic chain, we may expect density fluctuations about any integer multiple of $2k_F$. We may think of this phenomenon in the same way we understand the Bragg difraction in a lattice, though it is clear that in the present model we cannot have long-range order regarding the position of the particles. The square of the relative displacement between these diverges logarithmically with their distance, according to the two-point correlator of the boson field.

   On the other hand, for $m = 0$, $\rho_0(x)$ is the uniform density of the chain. In the original discrete variables this quantity is $1/(X_{n+1} - X_n)$, which in the continuum limit becomes

$$\rho_0(x) \approx \frac{1}{d + d^2 \frac{1}{2\pi} \partial_x \Phi} \approx \frac{1}{d} - \frac{1}{2\pi} \partial_x \Phi \tag{5.16}$$

We recover in this way the expression of the uniform density $\rho_L + \rho_R$ in terms of the boson field $\Phi = \Phi_L + \Phi_R$. We see, moreover, the explicit appearance of the constant mode $1/d$, which is nothing but the mean particle density of the chain. It corresponds to the contribution from the Fermi sea in the Luttinger model, which we conveniently subtracted out in the definition of the charges N_L and N_R. To summarize, the above description makes clear that there is a correspondence of the dynamics and the observables of the harmonic chain to those of the boson representation of the Luttinger model[1]. There are higher harmonic effects from the contributions (5.15), which arise from the discrete character of the model and survive in the continuum limit, but we may conclude that the physical properties of the harmonic chain are essentially the same as those configuring the Luttinger liquid universality class.

5.2.2 Correlation Functions

The correlators of the density field $\rho(x)$ are significant observables in the harmonic chain. According to the decomposition (5.14), in the two-point density correlator we may expect oscillations of any multiple of $2\pi/d$. Each of these contributions is given by the vacuum expectation value of the product of two exponentials of the type (5.15). After making the transformation (5.9) to free fields, one could resort again to the computational technique developed for the electron Green function in the Luttinger model. Here, however, we want to follow the same steps but implementing a different regularization, better suited to the discrete character of the model. Instead of dealing with the normal order prescription for the exponential of the field, we make use of the fact that the model describes fluctuations about a mean separation d between neighboring particles. Thus, in the mode expansion of the boson field $\Phi(x)$ there is a built-in short-distance cutoff, which we can make explicit by introducing a factor $\exp(-\epsilon|k|/2)$, with $\epsilon \equiv d/(2\pi)$, in the sum over momenta. Following the same strategy as in (4.63), we get for a typical, say right-handed, free-field correlator

$$
\begin{aligned}
\langle e^{i\alpha\Phi_R(x)} \, e^{-i\alpha\Phi_R(y)} \rangle & = \\
& = \langle \exp\left(-\alpha \int_\Delta^\infty \frac{dk}{\sqrt{k}} e^{-\epsilon k/2} e^{-ikx} B_k^+\right) \exp\left(\alpha \int_\Delta^\infty \frac{dp}{\sqrt{p}} e^{-\epsilon p/2} e^{ipx} B_p\right) \times \\
& \quad \exp\left(\alpha \int_\Delta^\infty \frac{dq}{\sqrt{q}} e^{-\epsilon q/2} e^{-iqy} B_q^+\right) \exp\left(-\alpha \int_\Delta^\infty \frac{dr}{\sqrt{r}} e^{-\epsilon r/2} e^{iry} B_r\right)\rangle \times \\
& \quad \exp\left(-\frac{\alpha^2}{2} \int_\Delta^\infty \frac{dp}{p} e^{-\epsilon p}\right) \exp\left(-\frac{\alpha^2}{2} \int_\Delta^\infty \frac{dr}{r} e^{-\epsilon r}\right) \\
& = \langle \exp\left(-\alpha \int_\Delta^\infty \frac{dk}{\sqrt{k}} e^{-\epsilon k/2} e^{-ikx} B_k^+\right) \exp\left(\alpha \int_\Delta^\infty \frac{dq}{\sqrt{q}} e^{-\epsilon q/2} e^{-iqy} B_q^+\right) \times \\
& \quad \exp\left(\alpha \int_\Delta^\infty \frac{dp}{\sqrt{p}} e^{-\epsilon p/2} e^{ipx} B_p\right) \exp\left(-\alpha \int_\Delta^\infty \frac{dr}{\sqrt{r}} e^{-\epsilon r/2} e^{iry} B_r\right)\rangle \times \\
& \quad \exp\left(\alpha^2 \int_\Delta^\infty \frac{dp}{p} e^{-\epsilon p} e^{ip(x-y)}\right) \exp\left(-\alpha^2 \int_\Delta^\infty \frac{dp}{p} e^{-\epsilon p}\right) \\
& = \exp\left(-\alpha^2 \int_\Delta^\infty \frac{dp}{p} e^{-\epsilon p} \left(1 - e^{ip(x-y)}\right)\right)
\end{aligned}
\tag{5.17}
$$

We have performed the computation in the limit of an infinite chain ($L \to \infty$). Consequently, we can set $\Delta = 0$ as the integral inside the exponential is well-behaved in the infrared. The ϵ parameter regulates it in the ultraviolet, giving the result

$$
\begin{aligned}
\langle e^{i\alpha\Phi_R(x)} \, e^{-i\alpha\Phi_R(y)} \rangle & = \exp\left(-\alpha^2 \log\left(1 - i(x-y)/\epsilon\right)\right) \\
& = (\epsilon)^{\alpha^2} \left(\frac{i}{x - y + i\epsilon}\right)^{\alpha^2}
\end{aligned}
\tag{5.18}
$$

It is worthwhile to compare the result (5.18) with that in (4.64) obtained with the field theory prescription. Using the latter, the exponentials of the boson fields get effective dimensions with respect to the length L of the system. By using the discrete approach, the dimensions are given in terms of the chain spacing $d \equiv 2\pi\epsilon$. One may think that one method is dual of the other, but the agreement between the spatial dependences in both cases shows that they are simply different ways of extracting the long-distance behavior of the correlator.

Putting together the left-handed and right-handed factors, we finally have for a typical contribution to the density-density correlator

$$
\begin{aligned}
\langle \rho_m(x)\rho_{-m}(y)\rangle &= \frac{1}{d^2}\langle e^{im\Phi(x)}\, e^{-im\Phi(y)}\rangle \\
&= \frac{1}{d^2}\langle e^{im\sqrt{\bar\mu}\tilde\phi(x)}\, e^{-im\sqrt{\bar\mu}\tilde\phi(y)}\rangle \\
&= \left(d^2\right)^{m^2\mu-1}\left(\frac{1}{4\pi^2}\frac{1}{(x-y)^2}\right)^{m^2\mu}
\end{aligned}
\tag{5.19}
$$

The power-law decay of the correlators is a reflection of the critical behavior of the model. The critical properties are however non-universal since they depend on the interaction strength. We are thus describing a continuous line of critical points labeled by the μ parameter.

The space-time dependence of the correlators can be inferred from that of the chiral factors to obtain

$$
\langle \rho_m(t,x)\rho_{-m}(t',y)\rangle = \left(d^2\right)^{m^2\mu-1}\left(\frac{1}{4\pi^2}\frac{1}{(x-y)^2-v^2(t-t')^2}\right)^{m^2\mu}
\tag{5.20}
$$

The Fourier transform of the density-density correlator is the dynamical structure factor of the model $S(\omega,k)$ [1]. About $k = 2\pi m/d$, this is given by the Fourier transform of the expression (5.20), that is,

$$
S(\omega, 2\pi m/d + q) \sim \left(d^2\right)^{m^2\mu-1}\left|\omega^2 - v^2 q^2\right|^{m^2\mu-1}
\tag{5.21}
$$

On the other hand, the Fourier transform of the equal-time density-density correlator gives the X-ray structure factor $S(k)$ [1], which is made of a sum of contributions about $2\pi m/d$

$$
S(2\pi m/d + q) \sim \left(d^2\right)^{m^2\mu-1}|q|^{2m^2\mu-1}
\tag{5.22}
$$

In this object we find the first divergence only for $\mu < 1/2$. In the classical model, $S(k)$ is given at zero temperature by a sum of delta function peaks, reflecting the perfect order of the chain. In the quantum theory, this long-range order is replaced by the tendency to order ("quasi" long-range order) that the power-law singularity in equation (5.22) shows. It becomes clear that the parameter μ provides a measure of the strength of the quantum fluctuations. The correlations become weaker for greater values of μ, when the singularities in $S(k)$ disappear and the system resembles more a gas. On the contrary, the classical limit is attained when $\mu \to 0$, as more and more peaks appear in the X-ray structure factor.

5.2.3 Massive Interactions

The harmonic chain in the continuum limit is governed by the line of critical points forming the Luttinger liquid universality class. In fact, it has helped us to give a more concrete meaning to some of the objects that we introduced in the description of the Luttinger model. Thus, we have visualized the $\Phi(x)$ field as a displacement of the particles about the equilibrium position, and we have obtained the $2k_F$ oscillation in the density correlator as an effect due to the quasiperiodic character of the system. As in the Luttinger model, it is clear that the harmonic chain becomes a system without a gap in the excitation spectrum, in the thermodynamic limit $N \to \infty$. One may ask, however, whether there is any relevant interaction which may drive the model out of criticality.

In order to answer this question one has to incorporate the lattice substrate which is supposed to support the chain of particles. That is, the description of the chain has not implied up to now any consideration about the crystal lattice underlying the system. The periodicity of this lattice may not bear any relation with the mean distance d between the particles. However, when there is commensuration, i.e. when $2\pi m/d$ equals the length G of a vector of the reciprocal lattice, Umklapp processes appear in which the momentum of the particles is absorbed by the crystal. In the first instance $m = 1$, we have an interaction of the type

$$
\begin{aligned}
H_U &= \frac{1}{2}\frac{g_U}{d} \int dx \ (\rho_1(x) + \rho_{-1}(x)) \\
&= \frac{g_U}{d^2} \int dx \ \cos \Phi(x)
\end{aligned}
\tag{5.23}
$$

The field theory takes the form of the well-known sine-Gordon model, that we already introduced in Chap. 4. In the quantum theory all the excited states of this model are massive. In particular, the low-energy excitations are either small fluctuations about any of the minima of the potential or soliton-like excitations in which the $\Phi(x)$ field goes from one to other of the minima.

We have already seen in Chap. 4 that the soliton-like objects have fermionic character. In the present context we may understand this in the following way. A soliton which complies with the boundary conditions

$$
\begin{aligned}
\Phi(x) &\to \pi \ , \quad x \to \infty \\
\Phi(x) &\to -\pi \ , \quad x \to -\infty
\end{aligned}
\tag{5.24}
$$

is an object that, according to (5.3), represents a progressive relative displacement of the particles of the chain going from left to right. The particle placed to the extreme right is shifted a relative distance d to the right with respect to that at the extreme left. This is like having effectively one particle less in the chain. Similarly, the antisoliton complying with the boundary conditions

$$
\begin{aligned}
\Phi(x) &\to -\pi \ , \quad x \to \infty \\
\Phi(x) &\to \pi \ , \quad x \to -\infty
\end{aligned}
\tag{5.25}
$$

appears like having in the average one more particle in the chain.

As stated before, the soliton-like excitations have a nonvanishing mass in the quantum theory. After performing the canonical transformation (5.9), the complete hamiltonian of the field theory becomes

$$H = \frac{v}{2} \int dx \; \left\{ 4\pi \tilde{\Pi}^2(x) + \frac{1}{4\pi}(\partial_x \tilde{\Phi}(x))^2 \right\} + \frac{g_U}{d^2} \int dx \; \cos\left(\sqrt{\mu}\tilde{\Phi}(x) \right) \quad (5.26)$$

The quantum theory is unambiguously defined only for $\mu < 2$ (check this constraint with the different normalization used in Ref. [2]). At $\mu = 1$, for instance, the model is equivalent to a theory of free Dirac massive fermions. We may arrive at this conclusion by using the boson representation (4.55),(4.56) of the two fermion chiralities (see also (5.34) and (5.35)). For the above value we have

$$\frac{1}{d^2} \int dx \; \cos \tilde{\Phi}(x) = \frac{1}{2}\frac{1}{d^2} \int dx \; \left(e^{i(\tilde{\Phi}_L + \tilde{\Phi}_R)} + \text{h. c.} \right)$$

$$\sim \frac{1}{2}\frac{1}{d} \int dx \; \left(\Psi_L^+ \Psi_R + \Psi_R^+ \Psi_L \right) \quad (5.27)$$

which is nothing but the mass term of a one-dimensional relativistic fermion theory. For $\mu < 2$, the last term in (5.26) is in general a relevant perturbation of the critical theory. This can be seen by noticing that, if one were to treat it as a weak interaction, one would face in a perturbative computation the appearance of correlators of the form

$$\frac{1}{d^4}\langle \cos\left(\sqrt{\mu}\tilde{\Phi}(x) \right) \cos\left(\sqrt{\mu}\tilde{\Phi}(y) \right) \rangle \sim \frac{1}{d^4} \left(\frac{d^2}{(x-y)^2} \right)^\mu \quad (5.28)$$

which diverge in the limit $d \to 0$ as far as $\mu < 2$.

As a consequence of the gap in the excitation spectrum, systems in which the charge is commensurate with the lattice spacing are genuine insulators. Only upon doping do fermion kinks (the soliton-like objects) appear in the ground state configuration, that is the way in which the system may transport charge. The brief account carried out here stresses the relevance of the lattice effects in the metal-insulator transition in one-dimensional systems, which is a topic extensively treated in the literature[3].

5.3 The Hubbard Model

The Hubbard model is the prototype of a system in which the emphasis is placed on correlation effects between the electrons. In one dimension the model can be solved exactly[4], the meaning of this being that the energy of the ground state and of the excited states can be obtained through a system of coupled nonlinear equations. The exact resolution of the model by the Bethe ansatz technique is reviewed in Chap. 10. Here we rather want to work out the continuum limit of the model, in the instances in which this may be feasible. The reason of this approach to the problem is that, being the interaction purely on-site on the lattice, we should expect a simple field theory governing the continuum limit and, therefore, the low-energy properties of the discrete model. This description in terms of continuous field variables is of the utmost interest, since the Bethe ansatz technique does not shed light on the long-distance behavior of the correlation functions. It turns out, however, that the continuum limit of the model is not so straightforward as in the case of the harmonic chain, for instance, and that it can be easily established only in the extreme situations of weak coupling[5] and large on-site repulsion[6, 7]. In the latter case, the procedure is somewhat heuristic and combines information from the ground state wavefunction in the large-U limit and from numerical results. A different line of research has been pursued by applying finite size scaling to the low-energy data obtained by the Bethe ansatz approach[8, 9]. In this way, a correspondence between the low-energy excitations in the spectrum and the fields of two independent $c = 1$ conformal field theories has been drawn. This development has been also used in the determination of several critical exponents, with apparent success.

5.3.1 Weak Coupling

The Hubbard model in one dimension is given by the hamiltonian

$$H = -t \sum_{i,\sigma} \left(a_{i,\sigma}^+ a_{i+1,\sigma} + \text{h. c.} \right) + U \sum_i n_{i\uparrow} n_{i\downarrow} \qquad (5.29)$$

where $a_{i,\sigma}^+, a_{i,\sigma}$ are fermion creation and annihilation operators and $n_{i,\sigma}$ is the number operator at site i for spin $\sigma = \uparrow, \downarrow$. We will be concerned all the time with the case of repulsive interaction ($U > 0$). When the on-site repulsion U is not very large (compared to t), we may expect not very strong correlation effects and that a sensible continuum limit can be worked out from the microscopic hamiltonian.

The kinetic part of the hamiltonian poses no problem in that respect. Following the same steps as in the beginning of Chap. 3, we may introduce a fermion field at each of the two Fermi points in the limit in which the lattice spacing a goes to zero, that is,

$$a_n \rightarrow e^{-ik_F na} \Psi_L(na) + e^{ik_F na} \Psi_R(na) \qquad (5.30)$$

Some momentum cutoff of order $1/a$ is implied in the definition of the fields Ψ_L, Ψ_R, which represents no drawback since we are only interested in the low-energy properties of the model. Inserting (5.30) in the kinetic part of (5.29) we get, in the limit $a \rightarrow 0$,

$$-t \sum_{i,\sigma} \left(a_{i,\sigma}^+ a_{i+1,\sigma} + \text{h. c.} \right) \rightarrow$$
$$-iv_F \tfrac{1}{a} \int dx \left(-\Psi_{L\sigma}^+(x) \partial_x \Psi_{L\sigma}(x) + \Psi_{R\sigma}^+(x) \partial_x \Psi_{R\sigma}(x) \right) \qquad (5.31)$$
$$v_F \equiv 2ta \, \sin(k_F a)$$

The dimensions of the fermion fields are obtained simply by scaling them according to the lattice spacing a

$$a^{-1/2}\Psi_{L,R} \rightarrow \Psi_{L,R} \tag{5.32}$$

With regard to the interaction term in (5.29), the proposal is to promote the on-site interaction in the lattice to a delta-function type of interaction between density fields in continuous space[5]. The density operators built up from (5.30) have a divergent limit $a \rightarrow 0$, though, as a consequence of the infinite Fermi sea which appears in the continuum limit. We may regularize the product of Fermi fields by the normal order prescription, so that

$$a_{n\sigma}^{+} a_{n\sigma} \quad \rightarrow \quad :\Psi_{L\sigma}^{+}(x)\Psi_{L\sigma}(x): + :\Psi_{R\sigma}^{+}(x)\Psi_{R\sigma}(x): \\ + \left(e^{i2k_F x}\Psi_{L\sigma}^{+}(x)\Psi_{R\sigma}(x) + \text{ h. c. }\right) \tag{5.33}$$

The computation of $:\Psi_{L\sigma}^{+}\Psi_{L\sigma}:$ and $:\Psi_{R\sigma}^{+}\Psi_{R\sigma}:$ is now a little bit different in the discrete regularization of the theory, compared to the field theory approach of Chap. 4. The sensible bosonization formulas for the chiral fermion fields read (we omit spin indices for the time being)

$$\Psi_L(x) \quad = \quad \frac{1}{\sqrt{a}}e^{-i\Phi_L(x)} \tag{5.34}$$

$$\Psi_R(x) \quad = \quad \frac{1}{\sqrt{a}}e^{i\Phi_R(x)} \tag{5.35}$$

where we do not apply the boson normal order prescription but assume that the chiral boson fields have a momentum cutoff of order $1/\epsilon \equiv 2\pi/a$, i.e.

$$\Phi_R(x) \quad = \quad i \int_0^\infty \frac{dk}{\sqrt{k}} e^{-\epsilon k/2} \left(e^{-ikx}B_k^+ - e^{ikx}B_k\right) \tag{5.36}$$

$$\Phi_L(x) \quad = \quad i \int_{-\infty}^0 \frac{dk}{\sqrt{|k|}} e^{-\epsilon|k|/2} \left(e^{-ikx}B_k^+ - e^{ikx}B_k\right) \tag{5.37}$$

With this regularization the calculation of an electron correlator can be done following similar steps as in (5.17). For free boson fields we have, for instance,

$$\frac{1}{2\pi\epsilon}\langle e^{i\Phi_R(x)}e^{-i\Phi_R(y)}\rangle \quad = \quad \frac{1}{2\pi\epsilon} \exp\left(-\int_0^\infty \frac{dp}{p} e^{-\epsilon p}\left(1 - e^{ip(x-y)}\right)\right)$$

$$= \quad \frac{1}{2\pi}\frac{i}{x - y + i\epsilon} \tag{5.38}$$

which gives the free electron Green function with the correct causal prescription. By using the same method and a point-splitting technique[5], one may also check that

$$\Psi_L^{+}(x)\Psi_L(x) \quad = \quad \frac{1}{2\pi\epsilon} + \frac{1}{2\pi}\partial_x\Phi_L(x) \tag{5.39}$$

$$\Psi_R^{+}(x)\Psi_R(x) \quad = \quad \frac{1}{2\pi\epsilon} + \frac{1}{2\pi}\partial_x\Phi_R(x) \tag{5.40}$$

Normal ordering with reference to the Fermi level just gets rid of the divergent contributions in the limit $\epsilon \to 0$.

In terms of the chiral boson fields, the Hubbard hamiltonian becomes in the limit $a \to 0$

$$H = \frac{v_F}{4\pi} \int dx \; (: \partial_x \Phi_{L\uparrow} \partial_x \Phi_{L\uparrow} : + : \partial_x \Phi_{R\uparrow} \partial_x \Phi_{R\uparrow} : + \; \uparrow \leftrightarrow \downarrow)$$
$$+ Ua \int dx \; \left(\frac{1}{4\pi^2} (\partial_x \Phi_{L\uparrow} + \partial_x \Phi_{R\uparrow})(\partial_x \Phi_{L\downarrow} + \partial_x \Phi_{R\downarrow}) \right.$$
$$\left. + (\Psi_{L\uparrow}^+ \Psi_{R\uparrow} \Psi_{R\downarrow}^+ \Psi_{L\downarrow} + \text{ h. c. }) \right) \tag{5.41}$$

In spite of its appearance, the interaction entering in (5.41) has a sensible limit $a \to 0$. We recall that its effective strength is given by $\sim Ua/v_F$, which is a finite quantity in the continuum limit as long as k_F is appropriately scaled with a. What is not so evident is the scaling of the backscattering term in the last line of (5.41) and, therefore, if the model may fall into the Luttinger liquid universality class. We may still give a complete transcription of the hamiltonian in terms of boson fields, using formulas (5.34) and (5.35),

$$H = \frac{v_F}{2} \int dx \; \left(4\pi \Pi_\uparrow^2(x) + \frac{1}{4\pi}(\partial_x \Phi_\uparrow(x))^2 + \; \uparrow \leftrightarrow \downarrow \right)$$
$$+ Ua \int dx \; \left(\frac{1}{4\pi^2}(\partial_x \Phi_\uparrow(x))(\partial_x \Phi_\downarrow(x)) \right.$$
$$\left. + \frac{1}{(2\pi\epsilon)^2}(e^{i\Phi_\uparrow(x)} e^{-i\Phi_\downarrow(x)} + e^{i\Phi_\downarrow(x)} e^{-i\Phi_\uparrow(x)}) \right) \tag{5.42}$$

with

$$\Phi_\sigma(x) = \Phi_{L\sigma}(x) + \Phi_{R\sigma}(x) \tag{5.43}$$

Introducing charge and spin fields

$$\Phi_c = \frac{1}{\sqrt{2}}(\Phi_\uparrow + \Phi_\downarrow) \quad , \quad \Phi_s = \frac{1}{\sqrt{2}}(\Phi_\uparrow - \Phi_\downarrow) \tag{5.44}$$

we see that the dynamics of these two kind of variables completely decouples in (5.42). The hamiltonian takes the form

$$H = \frac{v_F}{2} \int dx \; \left(4\pi \Pi_c^2(x) + \frac{1}{4\pi}(\partial_x \Phi_c(x))^2 \right)$$
$$+ Ua \frac{1}{8\pi^2} \int dx \; (\partial_x \Phi_c(x))^2$$
$$+ \frac{v_F}{2} \int dx \; \left(4\pi \Pi_s^2(x) + \frac{1}{4\pi}(\partial_x \Phi_s(x))^2 \right)$$
$$- Ua \frac{1}{8\pi^2} \int dx \; (\partial_x \Phi_s(x))^2$$
$$+ Ua \frac{2}{(2\pi\epsilon)^2} \int dx \; \cos(\sqrt{2}\Phi_s(x)) \tag{5.45}$$

The charge sector has the same boson expression as that of the Luttinger model, while the backscattering term gives rise to an interaction of the sine-Gordon type in the spin sector. Following the same procedure as in (5.7), we bring the hamiltonian to a form easy to diagonalize, in the case that the sine-Gordon interaction were switched off,

$$
\begin{aligned}
H = \; & \frac{v_c}{2} \int dx \left(4\pi\mu \Pi_c^2(x) + \frac{1}{4\pi\mu}(\partial_x \Phi_c(x))^2 \right) \\
& + \frac{v_s}{2} \int dx \left(4\pi\eta \Pi_s^2(x) + \frac{1}{4\pi\eta}(\partial_x \Phi_s(x))^2 \right) \\
& + U a \frac{2}{(2\pi\epsilon)^2} \int dx \; \cos(\sqrt{2}\Phi_s(x))
\end{aligned}
\tag{5.46}
$$

where

$$
v_c = v_F\sqrt{1 + Ua/(\pi v_F)} \qquad v_s = v_F\sqrt{1 - Ua/(\pi v_F)} \tag{5.47}
$$

$$
\mu = \frac{1}{\sqrt{1 + Ua/(\pi v_F)}} \qquad \eta = \frac{1}{\sqrt{1 - Ua/(\pi v_F)}} \tag{5.48}
$$

The charge dynamics can be mapped to a free field theory by a canonical transformation of the same type as (5.9). A similar transformation in the spin sector

$$
\begin{aligned}
\Phi_s(x) &= \sqrt{\eta}\,\tilde{\Phi}_s(x) \\
\Pi_s(x) &= \frac{1}{\sqrt{\eta}}\tilde{\Pi}_s(x)
\end{aligned}
\tag{5.49}
$$

leads to a hamiltonian H_s for the spin variables

$$
\begin{aligned}
H_s = \; & \frac{v_s}{2} \int dx \left(4\pi \tilde{\Pi}_s^2(x) + \frac{1}{4\pi}(\partial_x \tilde{\Phi}_s(x))^2 \right) \\
& + U a \frac{2}{(2\pi\epsilon)^2} \int dx \; \cos(\sqrt{2\eta}\tilde{\Phi}_s(x))
\end{aligned}
\tag{5.50}
$$

In order to ascertain if the spin dynamics corresponds to that of a Luttinger liquid (and, in particular, if it is gapless) one has to check the relevance or irrelevance of the last term in (5.50). This can be done by looking at the scaling dimension of this operator as $\epsilon \to 0$. We recall once again that the effective coupling constant is given by $\sim Ua/v_F$, so that when taking the limit we do not mind about the factor Ua in front of the interaction term. The scaling dimension can be extracted from the correlator of the cosine operator[5], as we did at the end of the previous section. We get

$$
\frac{1}{\epsilon^4}\langle\cos(\sqrt{2\eta}\tilde{\Phi}_s(x))\cos(\sqrt{2\eta}\tilde{\Phi}_s(y))\rangle \sim \frac{1}{\epsilon^4}\langle e^{i\sqrt{2\eta}\tilde{\Phi}_s(x)}e^{-i\sqrt{2\eta}\tilde{\Phi}_s(y)}\rangle
$$

$$
\sim \frac{1}{\epsilon^4}\left(\frac{\epsilon^2}{(x-y)^2}\right)^{2\eta}
\tag{5.51}
$$

For the case of repulsive interaction that we are treating here, it is clear that $\eta > 1$, so that the sine-Gordon interaction scales to zero in the continuum limit $\epsilon \to 0$.

Thus, we may conclude that the Hubbard model at weak coupling is a system with a gapless spectrum of charge and spin excitations. In the above considerations it is also implicit that the states of the system can be divided in two disjoint sectors of charge and spin, respectively, at least in the low-energy region of the spectrum. Unfortunately, it is also clear from the above treatment that the applicability of our continuum approach has a limit at a point in which U/t is $\sim O(1)$, where formulas like (5.47) and (5.48) become meaningless. It seems therefore that a sufficiently large on-site repulsion in the lattice becomes too rough to admit a smooth description in terms of our field variables and that the inability to capture strong correlation effects in the model is signaled by the shortcoming of our continuum theory. The singularity in equations (5.47) and (5.48) cannot have any physical meaning in any event, as long as the exact solution of the model does not show any discontinuity for $U \neq 0$.

5.3.2 Large-U Limit. Correlation Functions

We investigate at this point the other extreme situation in the Hubbard model, namely that in which the on-site repulsion U is sent to infinity. The interest in considering this limit is that in such case the Bethe ansatz equations simplify considerably and more information is then available from them. We are not going into the analysis of the equations since we can easily take the main conclusions of their study. To begin with, the first important fact is that, in the limit $U \to \infty$, the ground state wavefunction of the system splits into a factor for the charge variables times another factor for the spin variables[10]. That is, the separation of charge and spin is an exact property of the Hubbard model in the limit $U \to \infty$. Moreover, the wavefunction of the spin variables corresponds to that of a spin-1/2 antiferromagnetic Heisenberg chain, getting rid of the holes in the system. The isotropic antiferromagnet belongs to the Luttinger liquid universality class and its low-energy properties are given by a Luttinger model hamiltonian

$$H_s = \frac{v_s}{2} \int dx \left(4\pi \Pi_s^2(x) + \frac{1}{4\pi}(\partial_x \Phi_s(x))^2 \right) \tag{5.52}$$

The dynamics of the charge sector is more subtle. In the limit of very large on-site repulsion, it becomes very unfavorable for the system to have two fermions on the same site, so that it mainly remains in a configuration in which each particle is confined between its nearest neighbors. These strong correlation effects are of the kind we have already seen in the harmonic chain. Therefore, we may tentatively introduce the hypothesis that the charge dynamics is governed by a hamiltonian like (5.7)

$$H_c = \frac{v_c}{2} \int dx \left(4\pi \mu \Pi_c^2(x) + \frac{1}{4\pi\mu}(\partial_x \Phi_c(x))^2 \right) \tag{5.53}$$

In the above expression v_c and μ are parameters to be determined. They can be fixed from the information that one can extract in the Bethe ansatz resolution of the model. In particular, v_c can be obtained from the spectrum of the low-energy charge excitations. The value of μ can be determined from the compressibility of the system. We recall that the constant mode in $\sqrt{2}\partial_x \Phi_c/(2\pi)$ is to be identified with the particle density n in the model. Therefore, taking the derivative of (5.53) with respect to n and applying the Hellman-Feynman theorem we get[6]

$$\frac{1}{L}\frac{\partial^2 E_0}{\partial n^2} = \frac{\pi}{2}\frac{v_c}{\mu} \tag{5.54}$$

where E_0 is the ground state energy of the system. The left-hand-side of (5.54) is nothing but the inverse of the compressibility and can be obtained numerically from Lieb and Wu exact solution. Nice curves of μ as a function of the on-site repulsion U and the particle density n obtained by this method are shown in Ref. [6]. From them one gets clear evidence that in the limit of very large repulsion $\mu \to 1/2$, irrespective of the value of the particle density.

The knowledge of the μ parameter is of particular interest, since it dictates the long-distance behavior of a number of correlators. The general structure of the correlation functions of the discrete model is, at large x,

$$\langle \Psi(x)\Psi^+(0)\rangle = a_0\frac{1}{x^{\alpha_0}}\cos(k_F x) + a_1\frac{1}{x^{\alpha_1}}\cos(3k_F x) + \dots \tag{5.55}$$

$$\langle \rho(x)\rho(0)\rangle = \frac{\mu}{\pi^2}\frac{1}{x^2} + b_1\frac{1}{x^{\beta_1}}\cos(2k_F x) + b_2\frac{1}{x^{\beta_2}}\cos(4k_F x) + \dots \tag{5.56}$$

$$\langle \sigma(x)\sigma(0)\rangle = \frac{1}{\pi^2}\frac{1}{x^2} + c_1\frac{1}{x^{\gamma_1}}\cos(2k_F x) + \dots \tag{5.57}$$

$\rho(x)$ being the charge density operator and $\sigma(x)$ the spin density operator. We are omitting at this point logarithmic corrections to the above power-law decays, which affect in particular to the $2k_F$ harmonic contributions to the charge density and spin density correlators[11]. The $4k_F$ and higher harmonics in these functions, as well as the $3k_F$ and higher harmonics in the electron Green function, arise due to the discrete character of the model[12], as we already learned from the study of the harmonic chain. These contributions cannot be obtained sticking strictly to the computational framework of the Luttinger model and their consideration requires a modification of the standard bosonization formulas incorporating the lattice effects[13].

The critical exponents giving the dominant behavior in equations (5.55)-(5.57) have been estimated numerically in the large-U limit by several authors. We are now in a position to test whether the Luttinger model description synthesized by (5.52) and (5.53) is able to predict the correct values of the exponents. Focusing first on the electron Green function, the determination of α_0 follows the same steps as in (4.71). We have to remind the correspondence (5.11) between μ and $\cosh\phi$ in that equation, which gives in the present case $c^2 = 9/8$. Thus, we are predicting an exponent $\alpha_0 = 9/8$ for the dominant behavior of the electron Green function. In a numerical computation one usually measures some observable defined from that object as, for instance, the momentum distribution function. In our Luttinger model description, this must have a singular behavior at k_F of the type (4.77), with $s^2 = 1/8$. The numerical estimates range between 0.13 and 0.15 [10], and a refined finite size analysis of the results gives an exponent ≈ 0.126 [14]. This is in very good agreement with the theoretical prediction, taking into account that the extrapolations are made from measures over relatively small systems.

Regarding the charge density and spin density correlators, the $1/x^2$ contributions stem from the correlations of the first two terms in (5.33), while the $2k_F$ oscillations arise from the hybridization of left and right modes in the density fields. In the case of the charge density, for instance, we have

$$\rho(x) = \frac{\sqrt{2}}{2\pi}\partial_x \Phi_c + \frac{1}{2\pi\epsilon}\left(e^{i2k_F x}e^{i\Phi_\uparrow} + \text{h. c. }\right) + \frac{1}{2\pi\epsilon}\left(e^{i2k_F x}e^{i\Phi_\downarrow} + \text{h. c. }\right) \quad (5.58)$$

We need to compute the correlators between the $2k_F$ harmonics in (5.58). This can be done again making first the transformation to free fields and using the formula (5.18). We have, for instance,

$$
\begin{aligned}
\langle e^{i\Phi_\uparrow(x)}e^{-i\Phi_\uparrow(0)}\rangle &= \\
&= \langle e^{i\sqrt{\mu}\tilde{\Phi}_c(x)/\sqrt{2}}\, e^{i\Phi_s(x)/\sqrt{2}}\, e^{-i\sqrt{\mu}\tilde{\Phi}_c(0)/\sqrt{2}}\, e^{-i\Phi_s(0)/\sqrt{2}}\rangle \\
&\sim \frac{1}{(x^2)^{\mu/2}}\frac{1}{(x^2)^{1/2}} = \frac{1}{|x|^{\mu+1}}
\end{aligned}
\quad (5.59)
$$

It can be seen that our continuum approach gives the same prediction for the exponents β_1 in (5.56) and γ_1 in (5.57), which according to (5.59) is $\beta_1 = \gamma_1 = 3/2$. On the other hand, the spin correlation function has been the object of a thorough numerical investigation[10, 14]. The results of Ref. [10] place the exponent γ_1 between 1.3 and 1.45, while the finite size analysis of the same data seems to give $\gamma_1 \approx 1.48$ [14]. Again, this is a clear confirmation of the applicability of the Luttinger model description to the Hubbard model at very large on-site repulsion.

It has to be stressed that, contrary to the procedure followed in the weak coupling regime, we have not been able to formulate an analytical correspondence from the microscopic Hubbard hamiltonian in the large-U limit to the continuum field theory description of the model. In this sense, our approach has been rather heuristic. There have been also other attempts to develop a picture of the low-energy excitations of the model, at arbitrary U, either classifying the different states which appear in finite systems[15] or establishing the correspondence with conformal field theory through finite size scaling[8, 9]. The conclusion that one reaches is that the low-energy limit of the model may be given by a simple field theory (two $c = 1$ conformal field theories, one of them with varying radius of compactification), yet it is not known how to bring explicitly the strongly correlated system into the framework of continuum field theory.

References

[1] See, for instance, V. J. Emery, *The Many-Body Problem in One Dimension*, in *Correlated Electron Systems*, Vol. 9, ed. V. J. Emery (World Scientific, Singapore, 1993).

[2] S. Coleman, *Phys. Rev. D* **11** (1975) 2088.

[3] For a review, see H. J. Schulz, *The Metal-Insulator Transition in One Dimension*, Lectures given at Los Alamos Meeting on Strongly Correlated Electron Systems, December 1993, database report cond-mat/9412036.

[4] E. H. Lieb and F. Y. Wu, *Phys. Rev. Lett.* **20** (1968) 1445.

[5] R. Shankar, *Bosonization: How to Make It Work for You in Condensed Matter*, Lectures given at the BCSPIN School, Katmandu, May 1991.

[6] H. J. Schulz, *Phys. Rev. Lett.* **64** (1990) 2831.

[7] P. W. Anderson and Y. Ren, *The One-Dimensional Hubbard Model with Repulsive Interaction* (unpublished).

[8] N. Kawakami and S.-K. Yang, *Phys. Lett. A* **148** (1990) 359.

[9] H. Frahm and V. E. Korepin, *Phys. Rev. B* **42** (1990) 10553.

[10] M. Ogata and H. Shiba, *Phys. Rev. B* **41** (1990) 2326.

[11] T. Giamarchi and H. J. Schulz, *Phys. Rev. B* **39** (1989) 4620.

[12] F. D. M. Haldane, *Phys. Rev. Lett.* **47** (1981) 1840.

[13] J. Ferrer, J. González and M.-A. Martín-Delgado, *Phys. Rev. B* **51** (1995) 4807.

[14] S. Sorella, A. Parola, M. Parrinello and E. Tosatti, *Europhys. Lett.* **12** (1990) 721.

[15] H. J. Schulz, *Interacting Fermions in One Dimension: From Weak to Strong Correlation*, in *Correlated Electron Systems*, Vol. 9, ed. V. J. Emery (World Scientific, Singapore, 1993).

Part III

6. From the Cuprate Compounds to the Hubbard Model

6.1 Phenomenology of the Cuprate Compounds

The Hubbard model in all its variants is the microscopic model which supposedly describes the unusual properties of the cuprate compounds in the normal state and in the superconducting state as well.

From the title of this chapter we do not want the readers to draw the conclusion that it is meant to be a "proof" that the Hubbard model is the theoretical scenario where all the parts of the high-T_c superconductivity puzzle fit together; we are far from the solution of the problem, not only from the theoretical viewpoint but also because the experimental data are not conclusive enough as to discriminate among several models. The main purpose of this chapter is to explain how we may arrive to the Hubbard hamiltonian starting from the lattice structure of the cuprates in a fashion which may be illuminating to the theorists who study the properties of the model and quite oftenly take its connection to the cuprates for granted and obviate how intrincate that possible connection is.

When Anderson [1] proposed this model he pointed out that the strong Coulomb repulsion among the holes in the Copper $3d$-orbitals are responsible for the known antiferromagnetic properties of these materials in the absence of doping as well as for the superconductivity upon doping.

Definition 6.1 Cuprate Compounds *We will call a cuprate compound to any member of the family of materials having planes of CuO_2 as its basic building blocks.*

Most of the high-T_c superconductors are cuprates, but there are other materials like Nb_3Ge and the fullerenes like K_3C_{60} which also exhibit high-T_c superconductivity. For an excellent review on this topic see [2].

6.1.1 Lattice Structure of the Cuprates

The lattice structure of the cuprates is found to be strongly anisotropic. The basic unit cell is made up of one or a few planes of CuO_2 atoms on top of which

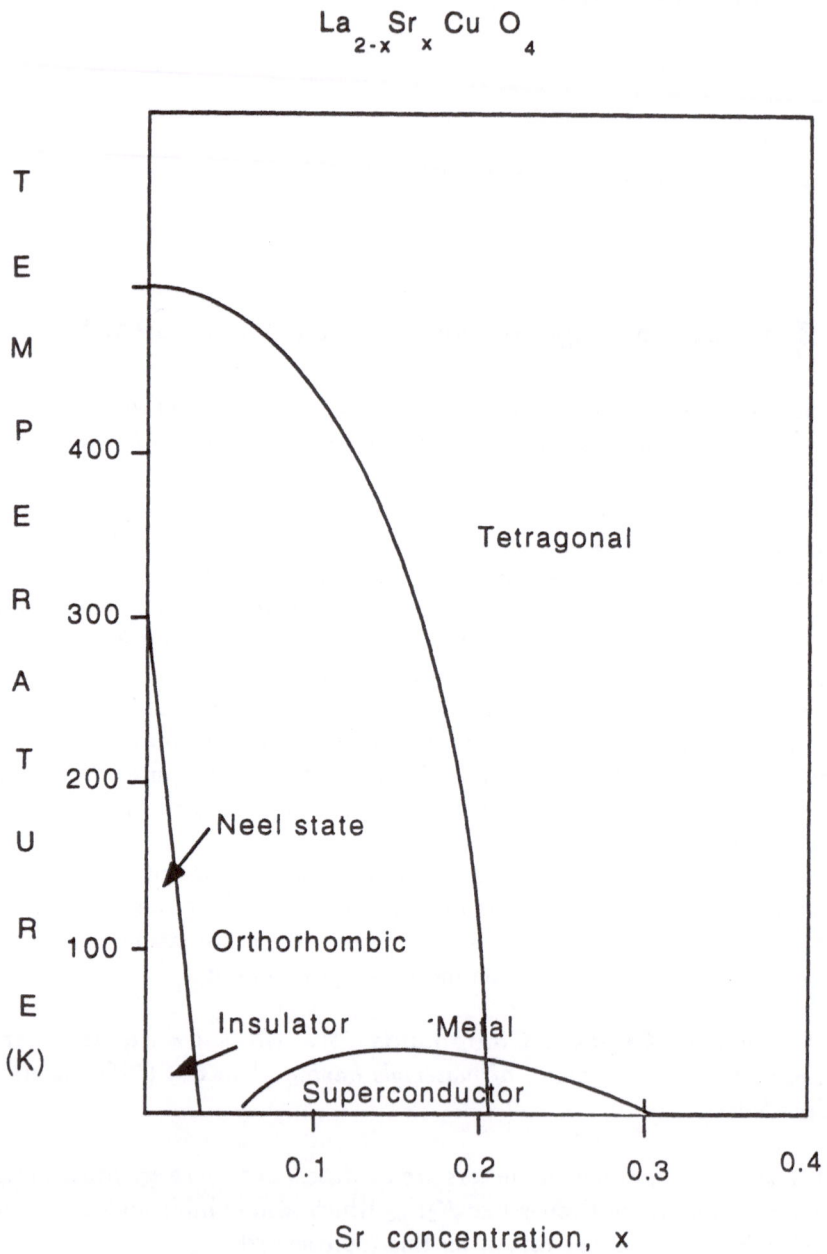

Fig. 6.1. Phase diagram of a cuprate compound ($La_{2-x}Sr_xCuO_4$) as a function of temperature.

there are layers of other atoms (Ba, La, O ...). This unit cell is repeated all over the z-direction.

The current belief is that superconductivity is confined to the CuO_2 planes, while the surrounding layers of other atoms play the only role of providing the carriers (whether electrons or holes) which produce the superconductivity. This is why the neighbouring layers to the CuO_2 planes are called "charge reservoirs". The assumption that high-T_c superconductivity is a two-dimensional phenomenon is based on the fact that the distance between two CuO_2 planes is longer than the Copper-Oxigen interspacing so that the electrons (or holes) shared by these atoms are more likely to hoppe in these planes than off the planes.

The mechanism which permits to transfer charge from the charge reservoirs to the conducting planes is the *doping*. It consists in the substitution of atoms from the charge reservoirs by others in a different ionization state in such a way that electrons are taken out of the CuO_2 planes or are donated to them. In the former case the carriers of the superconducting current are holes while they are electrons in the latter. The hole or electron nature of the carriers is detected measuring the sign of the Hall effect coefficient.

Definition 6.2 Definition of Doping *We will denote the amount of doping by x which is the ratio of substituted atoms to the originally present atoms in the charge reservoirs.*

We may be tempted to think that doping is a trivial issue, that we can easily and straightfowardly give and substract electrons at will thereby controlling the superconducting current. This is not the case but for the moment we will not dwell upon this point until we are not more acquainted with the basics features of high-T_c superconductors.

Varying the chemical composition of the charge reservoirs, thereby changing x, it is possible to change the charge density of the carriers in the CuO_2 planes and for a given temperature, this determines the metallic or insulating state of the material. Then changing the temperature we may obtain the phase diagram of the cuprate compounds (See figure 6.1).

This phase diagram is very characteristic of the high-T_c superconductors and any acceptable theory of high-T_c superconductivity should be capable of explaning that diagram. The phase diagram provides us with very helpful information about the possible mechanisms that governs the physics of the cuprates. Hereby we summerize the main features exhibited in the phase diagram which happend to be quite universal qualitatively and to a large extent quantitatively.

6.1.2 Basic Features of the Cuprates Phase Diagrams

- To very low doping $x \sim 0$ the material is in an *antiferromagnetic* (AFM) ordered state with small quantum fluctuations. Whenever this is the case, the material is an *insulator* (if the temperature is high enough, thermal

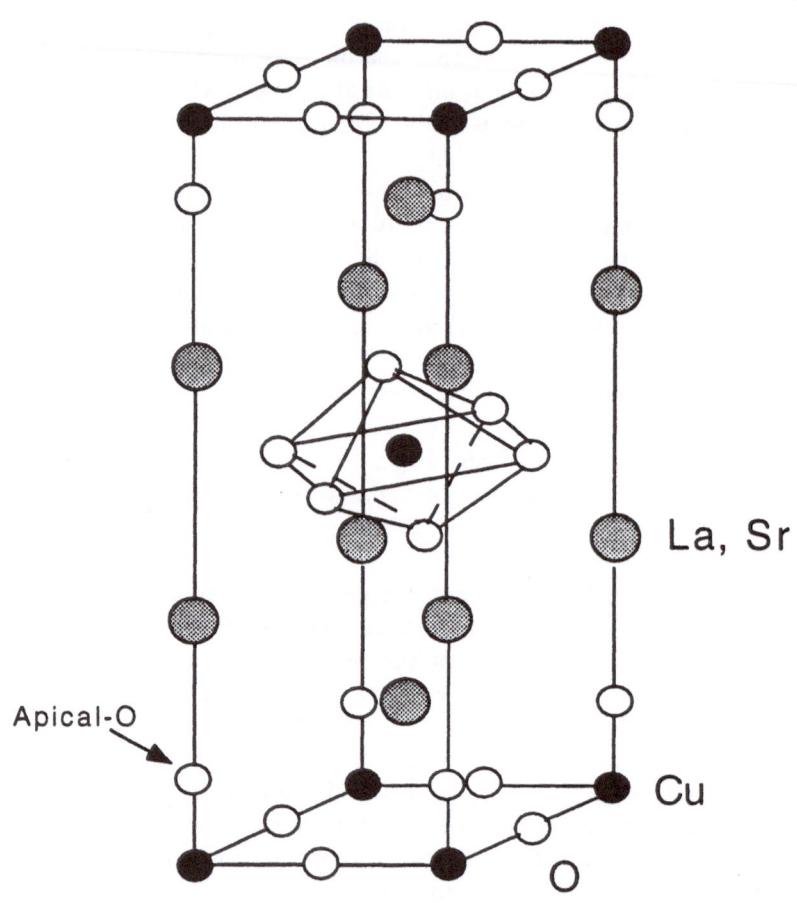

Fig. 6.2. The body centered tetragonal structue of the $La_{2-x}Sr_xCuO_4$ crystal.

fluctuations may destroy the magnetic order and the cuprate becomes conductor).

- When the doping is approximately $x \sim 0.15 \equiv$ *optimal doping* and the temperature is low (but "high" compared to ordirary superconductors) the material is a *superconductor*. The optimal doping means that the critical temperature T_c of the superconductor-conductor phase transition is the highest possible. When we depart below or above the optimal doping the critical temperature decreases thereby bounding the superconducting region in the phase diagram. This region usually extends from $x = 0.10$ to $x = 0.20$.

- When the temperature is above the critical value $T > T_c$ the material is in the *normal metal state* (NMS). The word "normal" may be misleading for it only means that the material shows ordinary conductivity but not superconductivity. Apart from this fact, we will see below that even in this non-superconducting state these materials may exhibit quite unusual metallic properties (anomalous resistivity, susceptivility etc).

Before carry on with more details on the phenomenology of the cuprates let us get down to some examples of cuprates and see what is its microscopic lattice structure. We will consider three cases which are prototypes of cuprate compounds.

Example 6.3 $La_{2-x}Sr_xCuO_2$

This is a typical cuprate which exhibits superconductivity where the carriers are *holes*. It was one of the first high-T_c superconductors discovered. Its crystalline structure is body centered tetragonal as show in figure 6.2 . Each Copper atom is surrounded by 6 Oxygen atoms, the ones below and above each Copper atom in the z-direction are called *apical* Oxygens.

The doping mechanism consits in trading Lanthanum atoms for Stronthium atoms:

$$La \xrightarrow{\text{Doping}} Sr$$

$x = 0$ At zero doping the La_2CuO_2 is *insulator* and exhibits *antiferromagnetic* long range order.

$x_{op} = 0.15$ At the optimal doping the $La_{1.85}Sr_{0.15}CuO_2$ is a *superconductor* with a critical temperature of $T_c = 39K$.

In the following table are shown the electronic configurations of the elements in the Lanthanum cuprates which are needed to explain its undoped structure and the doping mechanism.

Table 6.1. Electronic configurations of the elements in the Lanthanum cuprates.

Electronic Configurations	Crystal ($x = 0$)	Doped ($x \neq 0$)
Cu: [Ar] $3d^{10}4s$	Cu^{2+}	Cu^{2+}
O: [He] $2s^2 2p^4$	O^{2-}	O^{1-}, O^{2-}
La: [Xe] $5d\ 6s^2$	La^{3+}	La^{3+}
Sr: [Kr] $5s^2$	$\varnothing$	Sr^{2+}

$x = 0$ In the crystal La_2CuO_2 the Oxygen atoms are in a O^{2-} state so that they achieve a complete p-shell. In the same fashion, Lanthanum loses 3 electrons and becomes La^{3+} which is a stable configuration. Once this is favored by low energy cost, electrical neutrality forces the Copper to adopt the Cu^{2+} configuration:

$$2 \times (Cu^{2+}) + 3 \times 2\,(La^{3+}) = 4 \times 2\,(O^{2+})$$

Losing one electron makes the Copper to achieve shell completeness but the lost of the second electron causes the creation of a *hole* in the d-shell. Thus Cu^{2+} has a net spin of $1/2$ in the crystal. These holes carrying spin but not charge are responsible for the *antiferromagnetic* long range order observed in the undoped insulating state.

$x_{op} \neq 0$ $La_{2-x}Sr_xCuO_2$ The doping mechanism consists in randomly replacing La^{3+} by Sr^{2+} in the charge reservoirs. Consequently, there are less electrons available in the conducting CuO_2 planes and we have *doping by holes*.

There are two possible alternatives in principle but one of them is energetically more favorable:

1. $Cu^{2+} \longrightarrow Cu^{3+}$ This yields a 3 d^8 configuration which means two holes in the same d-shell. The outcome is a strong coulomb repulsion which makes this configuration non-favorable energetically.

2. $O^{2-} \longrightarrow O^{1-}$ This yields a 2 p^5 configuration which means one hole in th p-shell. Altogether we have two holes in different shells which is a situation more favorable energetically than before.

Example 6.4 $Nd_{2-x}Ce_xCuO_4$

This is a typical cuprate which exhibits superconductivity where the carriers are *electrons*. Its crystalline structure is body centered tetragonal as show in figure 6.3. The difference with the Lanthanum cuprates is that there are not apical Oxygen atoms in the charge reservoirs.

The doping mechanism consits in trading Neodymium atoms for Cesium atoms:

$$Nd \xrightarrow{\text{Doping}} Ce$$

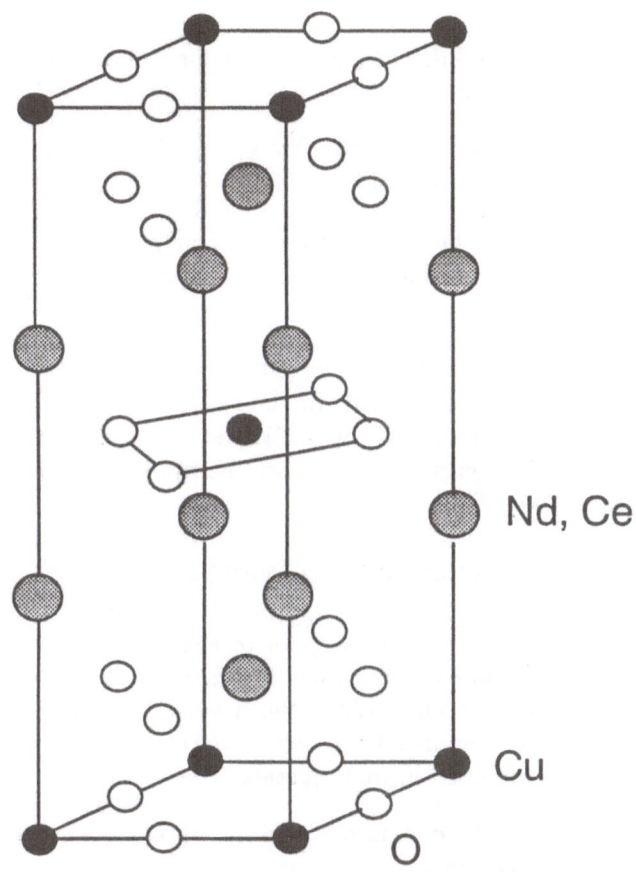

Fig. 6.3. The body centered tetragonal structue of the $Nd_{2-x}Ce_xCuO_4$ crystal.

Table 6.2. Electronic configurations of the elements in the Neodymium cuprates.

Electronic Configurations	Crystal ($x = 0$)	Doped ($x \neq 0$)
Cu: [Ar] $3d^{10}4s$	Cu^{2+}	Cu^{1+}
O: [He] $2s^2 2p^4$	O^{2-}	O^{2-}
Nd: [Xe] $4f^4\ 6s^2$	Nd^{3+}	Nd^{3+}
Ce: [Xe] $4f\ 5d\ 6s^2$	$\varnothing$	Ce^{4+}

$x = 0$ At zero doping the Nd_2CuO_4 is an *insulator* and exhibits *antiferromagnetic* long range order.

$x_{op} = 0.15$ At the optimal doping the $Nd_{1.85}Ce_{0.15}CuO_4$ is a *superconductor* with a critical temperature of $T_c = 24K$.

In the following table are shown the electronic configurations of the elements in the Neodymium cuprates which are needed to explain its undoped structure and the doping mechanism.

$x = 0$ In the crystal Nd_2CuO_4 the Oxygen atoms are in a O^{2-} state so that they achieve a complete p-shell. In the same fashion, Neodymium loses 3 electrons and becomes Nd^{3+} which is a stable configuration. Once this is favored by low energy cost, electrical neutrality forces the Copper to adopt the Cu^{2+} configuration:

$$1 \times 1\ (Cu^{2+}) + 3 \times 2\ (Nd^{3+}) = 4 \times 2\ (O^{2+})$$

Losing one electron makes the Copper to achieve shell completeness but the lost of the second electron causes the creation of a *hole* in the d-shell. Thus Cu^{2+} has a net spin of $1/2$ in the crystal. These holes carrying spin but not charge are responsible for the *antiferromagnetic* long range order observed in the undoped insulating state.

$x \neq 0$ $Nd_{2-x}Ce_xCuO_4$ The doping mechanism consists in randomly replacing Nd^{3+} by Ce^{4+} in the charge reservoirs. Consequently, there are more electrons available in the conducting CuO_2 planes and we have *doping by electrons*.

Now it is believed that there is only one alternative to locate the excess of electrones, namely, to occupy the hole in the d-shell of Copper completing the shell configuration. Thus the net spin for the Copper ion is $S = 0$ in this case.

In figure 6.4 we may see the phase diagram of the Neodymium cuprates (electron doping) as compared to the Lanthanum cuprates (hole doping). It is worthwhile to pointing out the qualitatively mirror symmetry exhibited by the phase diagrams. The only difference being the Spin Glass phase present in

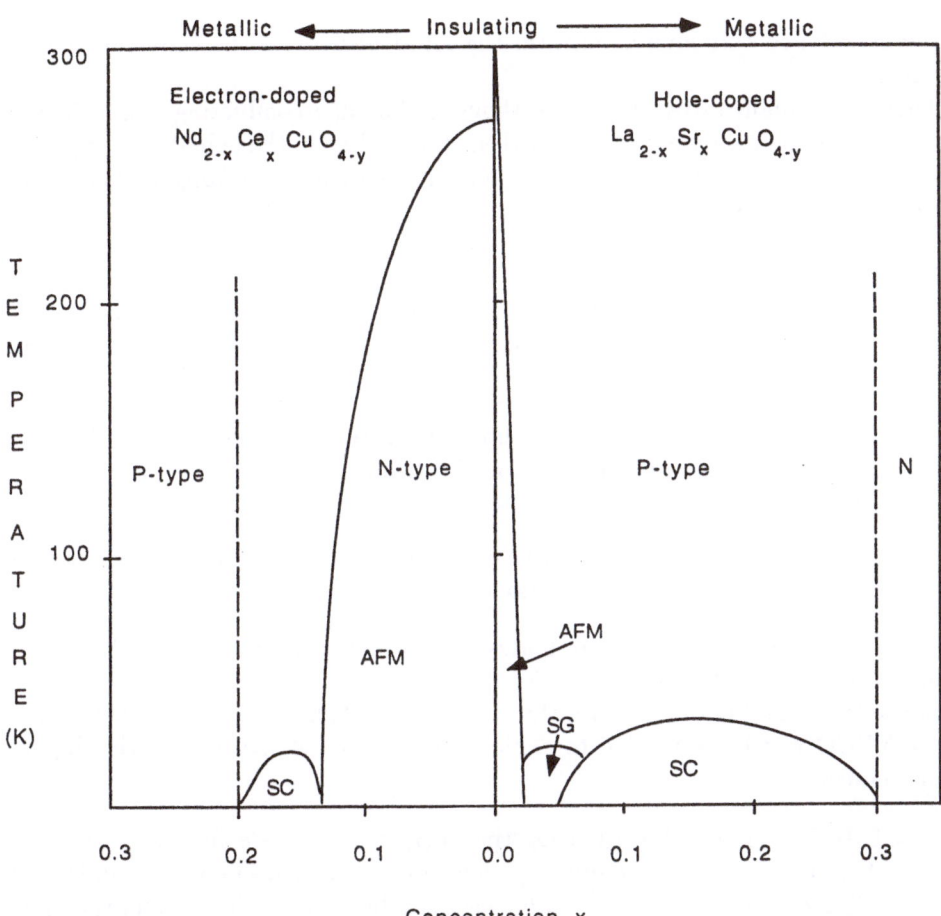

Fig. 6.4. Phase diagram of the Neodymium cuprates (electron doping) as compared to the Lanthanum cuprates.

Table 6.3. Electronic configurations of the elements in the YBCO cuprates.

Electronic Configurations	Crystal ($x = 0$)	Doped ($x \neq 0$)
Cu: [Ar] $3d^{10}4s$	Cu^{2+} (pl), Cu^{1+} (ch)	$Cu^{2.2+}$ (pl), $Cu^{2.5+}$ (ch)
O: [He] $2s^2 2p^4$	O^{2-}	O^{2-}
Y: [Kr] $5s^2\ 4d$	Y^{3+}	Y^{3+}
Ba: [Xe] $6s^2$	Ba^{2+}	Ba^{2+}

the hole-doped compounds between the AFM phase and the superconducting phase.

There is a quantitative difference though, the superconducting phase in the electron-doped compouds is smaller (less stable) than in the hole-doped cuprates meaning that in the former the antiferromagnetism is a stronger effect than superconductivity.

Example 6.5 $YBa_2Cu_3O_{6+x}$

It is oftenly denoted by $YBCO$. The lattice structure of this cuprate compound is clearly more complicated than the previous examples as shown in figure 6.5.

- There are 2 CuO_2 planes per unit cell instead of only one and they are separated by an additional plane of Ytrium atoms from the charge reservoir.

- The charge reservoirs consists of Barium atoms and chains of Copper-Oxygen CuO_x in the y-direction.

The presence of more than one CuO_2 planes makes increase the critical temperatures to $T_c = 92K$.

In the following table are shown the electronic configurations of the elements in the YBCO which are needed to explain its undoped structure and the doping mechanism.

$x = 0$ In the crystal $YBa_2Cu_3O_6$ the Oxygen atoms are in a O^{2-} state so that they achieve a complete p-shell. In the same fashion, Ytrium loses 3 electrons and becomes Y^{3+} which is a stable configuration and for the same reason Barium becomes Ba^{2+}. Once this is favored by low energy cost, electrical neutrality forces the Copper to adopt the Cu^{2+} configuration in the planes and Cu^{1+} in the chains:

$$2 \times 2\ (Cu^{2+}) + 1 \times 1\ (Cu^{1+}) + 3 \times 1\ (Y^{3+}) + 2 \times 2\ (Ba^{2+}) = 6 \times 2\ (O^{2+})$$

As a matter of fact, no $Cu - O$ chains exist at zero doping.

Losing one electron makes the Copper to achieve shell completeness but the lost of the second electron causes the creation of a *hole* in the d-shell.

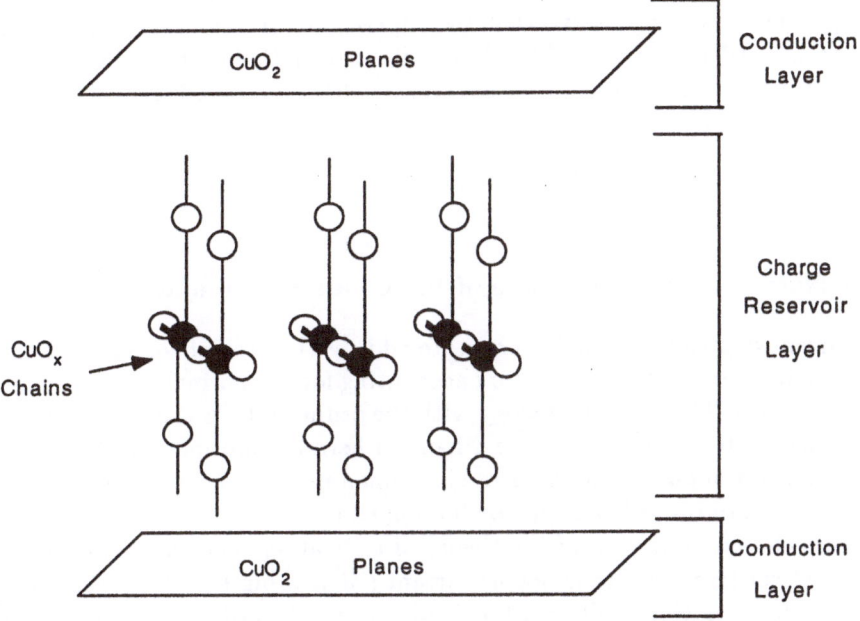

Fig. 6.5. Schematic lattice structure of the $YBCO$ compound.

Thus Cu^{2+} in the planes has a net spin of $1/2$ in the crystal. These holes carrying spin but not charge are responsible for the *antiferromagnetic* long range order observed in the undoped insulating state. The Cu^{1+} in the chains do not exhibit magnetism for they have net spin zero.

$x \neq 0$ $YBa_2Cu_3O_{6+x}$ The doping mechanism consists in changing the Oxygen content in the charge reservoirs. It is believed that this amounts to adding *holes* to the conducting planes since the new oxygens are O^{2-} substracting electrons from the CuO_2 planes. However, this picture is not yet confirmed experimentally by measuring the sign of the Hall coefficient which in this compound is fairly temperature dependent.

Notice that only at maximum doping $x = 1$ the $Cu - O$ chains are completely formed.

In figure 6.6 it is shown the phase diagram of the $YBCO$ cuprates (hole doping) where we can observe that the superconducting phase is appreciately large (same order than the AFM phase). We also notice that the SC phase do not show a marked maximum but apparently there are two. Optimal doping is achieved at $x_{op} = 1$

6.1.3 Normal State Properties of the Oxide Superconductors

The current strategy (due to Anderson) to address the problem of high-T_c superconductivity is to try to find a theory accounting for the normal state properties of the cuprates. This goes in analogy with the ordinary BCS theory where firsly the normal state is identified to be a Landau Fermi liquid metal and then it is found an electron pairing mechanism via phonon interactions which destabilizes the normal state towards a superconducting state.

From the point of view of the theory of critical phenomena the magnitude of the critical temperature is not important (it is a big deal for technological reasons though!), what really matters is to identify the critical exponents which define the possible universality classes or phases. In the problem of high-T_c superconductivity there exists the impossibility of reaching these high temperatures within the standard BCS theory. But a major issue is to explain the unusual critical exponents in the normal state of the cuprate compounds.

We are not going to make an exhaustive enumeration of all the properties of the normal state but we quote hereby some of them, making special emphasis in the resistivity.

- Temperature dependence in the magnetic susceptibility contrary to what is expected from canonical Landau Fermi liquid metals.

- Temperature dependence in the Hall coefficient contrary to what is expected from canonical Landau Fermi liquid metals.

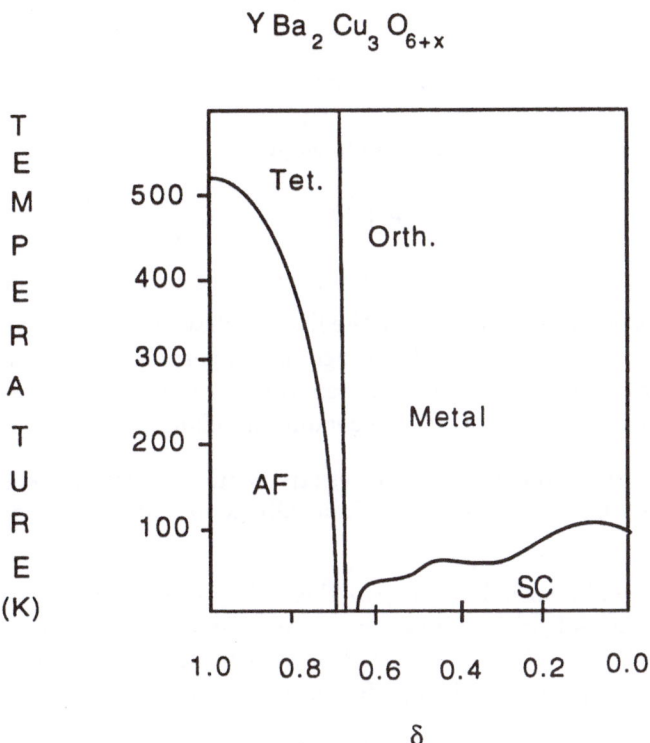

$Y Ba_2 Cu_3 O_{6+x}$

Fig. 6.6. Phase diagram of the $YBCO$ compounds (electron doping).

- The Nuclear Magnetic Resonance (NMR) relaxation is proportional to temperature $1/T_1 \sim T$ contrary to the usual temperature independence in Fermi liquids.

- Strong anisotropies are observed in the resistivity signaling the two-dimensional nature of the superconducting mechanism. Namely, the resistivity in the perpendicular direction to the CuO_2 planes is much bigger than the in-plane resistivity.

- As for the resistivity in the conducting planes, it exhibits an unusual linear behaviour with temperature

$$\rho_{xy} \sim a + bT$$

instead of the canonical Fermi liquid behaviour

$$\rho_{xy} \sim a + bT^5$$

where the fifth exponent comes from scattering of electrons with phonons.

In figure 6.7 it is possible to see clearly the linear behaviour of the conductivity for several materials at a wide range of temperatures. It is worth noting the similar slope of the curves denoting an universal scattering mechanism for carrier transport in the conducting planes.

Regarding the unusual properties of the resistivity in the normal state we want to stress the following two remarks which usually go unnoticed among the theorists:

remark 1 The observed linear behaviour $\rho \sim T$ is only present at the optimal doping. Off this special point, the curves of the resistivity depart from the simple linear regimen and become more involved the larger the departure form the optimal doping. This is manifest in figure 6.8.

remark2 There is an intriguing lack of universality between electron-doped materials and hole-doped materials. It is the latter which exhibit the linear behaviour in ρ_{xy} while the former behave quadratically $\rho_{xy} \sim T^2$. This behaviour in the electron-doped compounds might be explained with Fermi liquid theory, although there are also logarithmic corrections which might imply a modification of that theory called Marginal Fermi Liquid (MFL).

These anomalies in the normal state properties of the high-T_c superconductors have led Anderson [1] to draw the conclusion that any attempt to explain these properties within conventional Landau Fermi liquid theory is doomed to failure. Moreover, he asserts that this normal state is a novel quantum liquid that he identifies with the Luttinger liquid introduced by Haldane [3] in the description of one dimensional interacting electron systems. However, the validity of this approach as well as other more conventional, are still open questions.

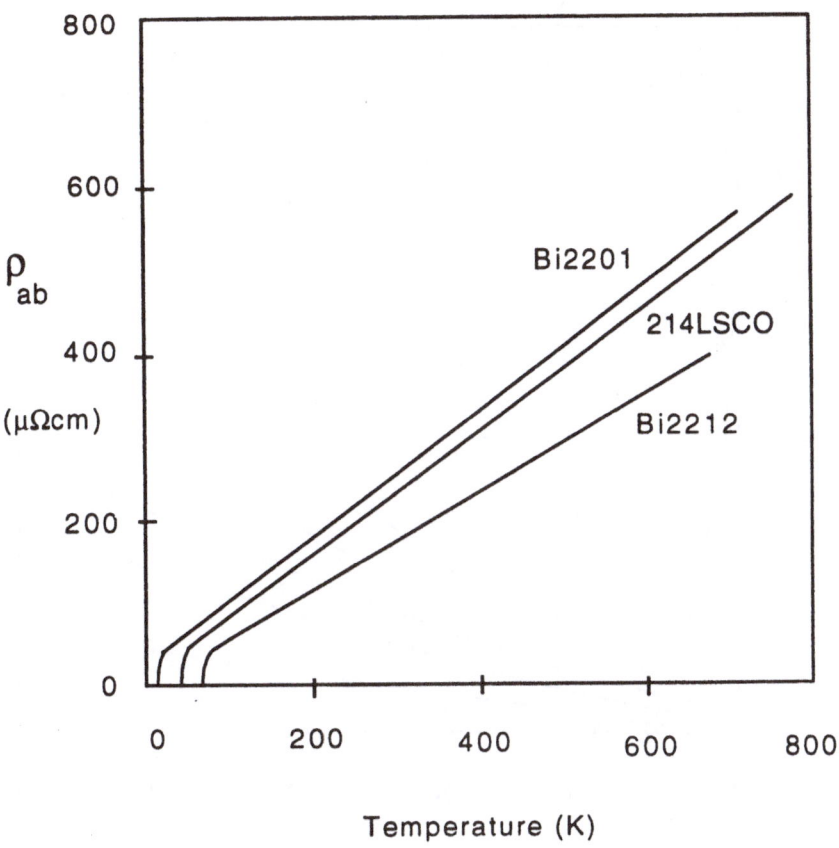

Fig. 6.7. Linear behaviour of the conductivity for several materials at a wide range of temperatures.

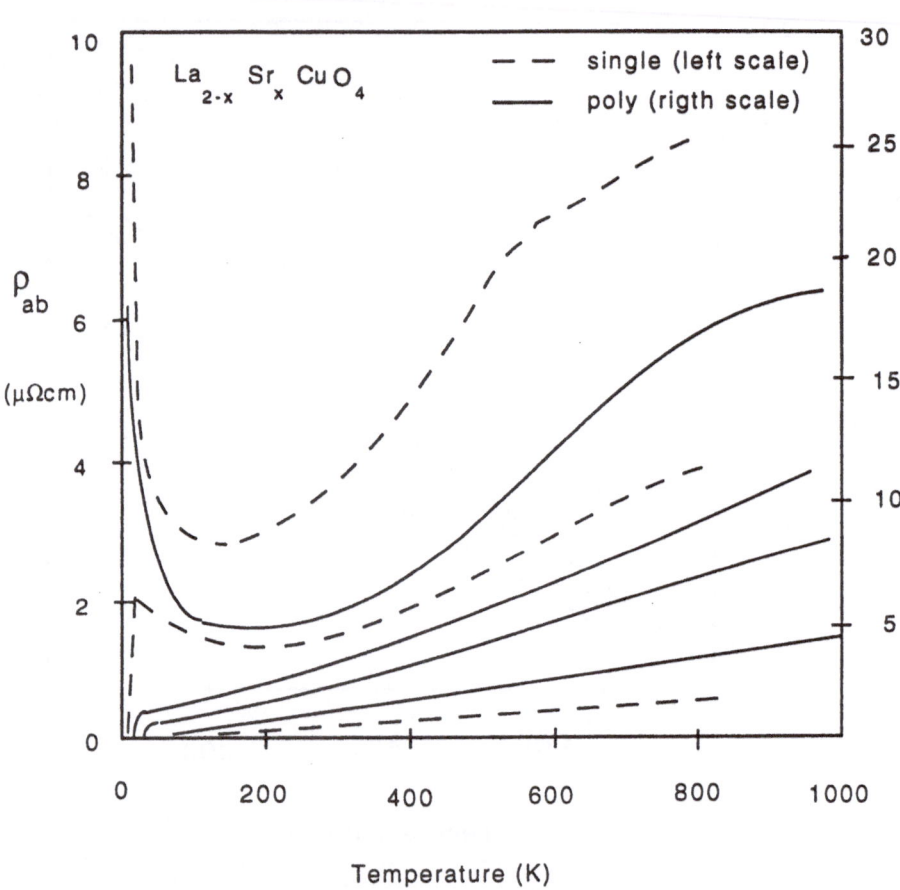

Fig. 6.8. Resistivity curves in $La_{2-x}Sr_xCuO_4$ for several dopings.

6.2 Hubbard Model Description of the Cuprate Compounds

From the description of the cuprate properties in the previous section and according to Anderson, we can state the following *paradigms* which help to make the connection between the Hubbard model and the cuprate compounds.

paradigm I All the relevant spin and charge carriers reside in the CuO_2 planes.

paradigm II Magnetism and high-T_c superconductivity are intimately related, in the sense that the electrons responsible for the magnetism are also responsible for the observed superconductiivity.

paradigm III The dominant interactions are repulsive and the carriers are strongly correlated.

According to Anderson, the model fulfilling these properties and others is the *2 dimensional Hubbard model* which we hereby present.
To formulate a microscopic model we have to take into account the orbital structure in the conducting CuO_2 planes. Mainly, we have to concentrate on the orbitals subject to hybridization with other orbitals. Hence, as far as the hybridization in the conducting planes is concerned, there are 3 relevants orbitals (see figure 6.9)

Cu $3d_{x^2-y^2}$.

O $2p_x$, $2p_y$.

The hybridization mechanism is squematically ilustrated in figure 6.10.
Should we want to have into account other planes, the most interesting situation is for instance that of Lanthanum cuprates where we already saw that there exist apical oxygens in the z-direction. In the case the relevant orbitals are 5,

Cu $3d_{x^2-y^2}$, $3d_{3z^2-r^2}$.

O $2p_x$, $2p_y$.

The effect of hybridization is to break up the degeneracy in the p- and d-orbitals in such a fashion that the state of maximum energy is $3d_{x^2-y^2}$ (see figure 6.9).

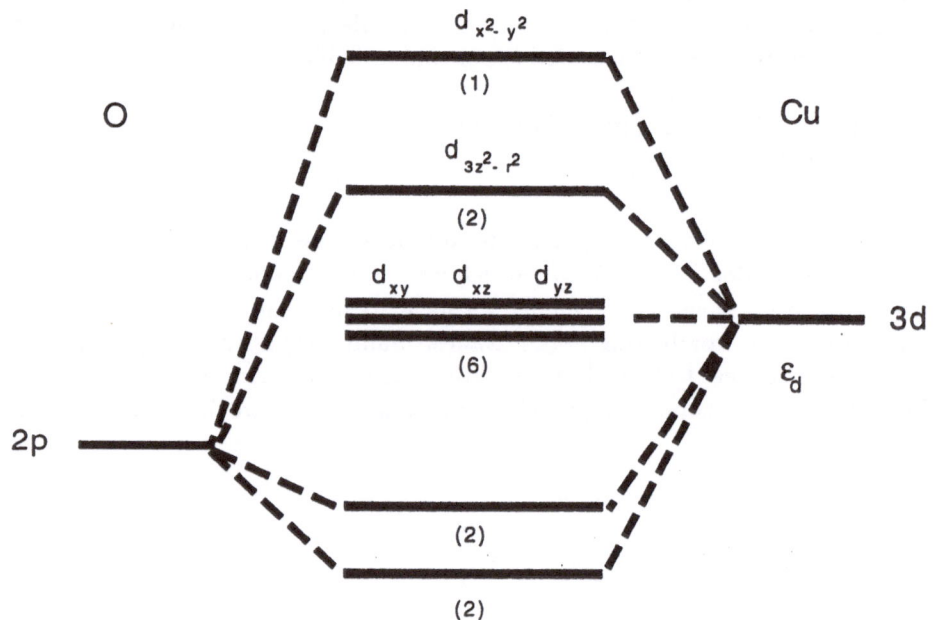

Fig. 6.9. Orbital structure of the conducting CuO_2 planes.

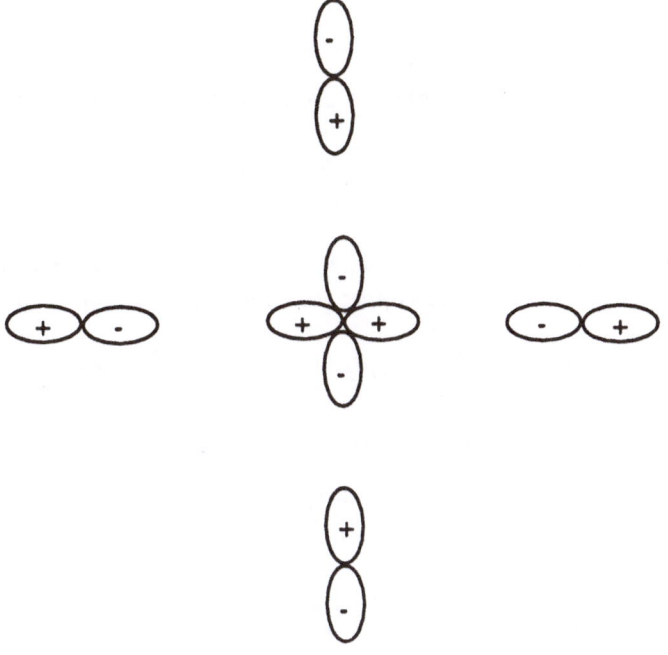

Fig. 6.10. Hybridization of orbitals in the conducting CuO_2 planes.

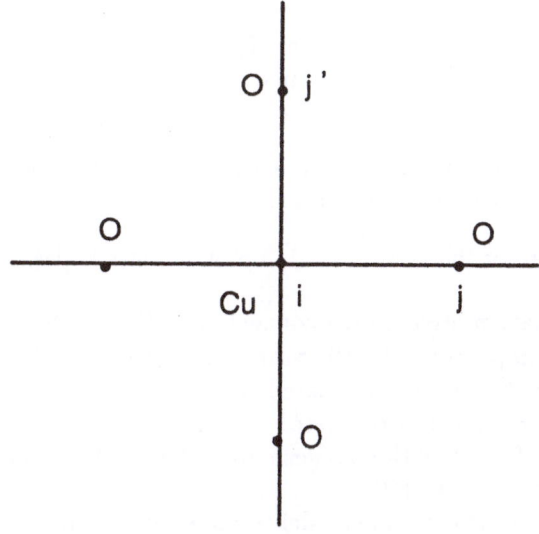

Fig. 6.11. Unit cell of couplings in Hubbard model description of the conducting CuO_2 planes.

6.2.1 One Band Hubbard Model

In a first approximation, possibly too oversimplified, we could make a description of the conducting CuO_2 planes based upon only one hybridized orbital, namely, $3d_{x^2-y^2} - 2p_\sigma$ which would be localized at the copper atoms, discarding the effect of the oxygen atoms except for the fact of being considered when hybridizing the only orbital.

Thus, in the one band Hubbard model proposed by Anderson to describe the cuprates, there is one state per copper ion. The Hamiltonian reads as follows,

$$H = -t \sum_{<i,j>,\sigma} (d_{i\sigma}^\dagger d_{j\sigma} + h.c.) + U \sum_i n_{i\uparrow} n_{i\downarrow} \qquad (6.1)$$

where, assuming a hole-doped material,

$d_{i\sigma}^\dagger \equiv$ Operator which creates a hole in the d-orbital at the ith-ion with spin σ. They satisfy the CAR (Cannonical Anticommutation Relations):

$$\{d_{i\sigma}, d_{j\sigma'}^\dagger\} = \delta_{ij}\delta_{\sigma\sigma'} \qquad (6.2)$$

$n_{i\sigma} = d_{i\sigma}^\dagger d_{i\sigma}$ Counts the number of holes at the ith-ion with spin σ.

t Is the hopping parameter. The holes may hoppe from one site in the lattice of the CuO_2 planes to one of its nearest-neighbours (that is what $<i,j>$ means). This parameter measures the kinectic energy of the holes.

U Is the onsite repulsion among holes. It mimics the Coulomb repulsion and only happens when two holes coincide at the same site, thus Pauli principle forces them to have opposite spins as shown in (6.1).

$|vacuum> = |0>$ Is the vacuum state which means all the d-orbitals in the copper atoms are full of electrons.

A particular state of the system made up of N holes is $[|\Psi_0> = \prod_{i=1}^N d_{i\sigma_i}^\dagger |0>]$

Thus far we have motivated the connection of the one-band Hubbard model to the cuprate compounds. By all means, this candidate for the microscopic hamilatonian is not fully substantiated in several respects. For instance, the fact that the interaction are restricted to onsite interactions maight be quite restrictive or, the fact that the oxygens are not fully taken into account could be an oversimplication (see [4]).

Even if we restrict to this simple model to hopefully catch up the main features of the superconductivity in the cuprates, the task is pretty formidable. Despite its harmless appeareance, the model is terribly difficult to handle even with the aid of computers. Couriously enough, it is harder to study than QCD on the lattice. Only in one dimension the model is integrable and that will be the subject of chapter 10 of these notes. In any other dimensions the hope is not to integrate the model (no interacting theory in $2 + 1$ dimensions is known to be integrable), but to be able to recognize the ground state of the system and some of its potentially interesting properties to high-T_c superconductivity like wether there is spin-charge separation or not. This would be enough to retrieve the low-energy / long-distance properties of the model.

6.2.2 Three Band Hubbard Model

This is a Hubbard-like hamiltonian more refined for it takes into account explicitly the effect of the oxygen p-orbital which carry a hole upon doping, assuming we are dealing with a Lanthanum cuprate as in example 1. It was proposed by Emery in [5]. We have now 2 types of hole creation operators,

$d_{i\sigma}^{\dagger}$ Creates a hole in the orbital $d_{x^2-y^2}$ of the copper atom at site i.

$p_{i\sigma}^{\dagger}$ Creates a hole in te orbitals p_x, p_y of the oxygen atom at site j. See figure 6.11.

They both satisfy CAR relations and commute between them.
Thus, the three-band Hubbard hamiltonian reads as follows,

$H = -t_{pd} \sum_{<i,j>} (p_{j\sigma}^{\dagger} d_{i\sigma} + h.c.) \equiv$ Hybridation of the $d_{x^2-y^2} - p_{\sigma}$ orbitals.

$-t_{pp} \sum_{<j,j'>} (p_{j\sigma}^{\dagger} p_{j'\sigma} + h.c.) \equiv$ Hopping among the p-orbitals.

$+U_d \sum_i n_{i\uparrow}^d n_{i\downarrow}^d \equiv$ Coulomb repulsion among d-orbitals.

$+U_p \sum_j n_{j\uparrow}^p n_{j\downarrow}^p \equiv$ Coulomb repulsion among p-orbitals.

$+U_{dp} \sum_{<i,j>} n_i^d n_j^p \equiv$ Coulomb repulsion between p-d-orbitals.

$+\epsilon_d \sum_i n_i^d \equiv$ Energy needed to create a hole in the d-orbitals.

$+\epsilon_p \sum_i n_i^p \equiv$ Energy needed to create a hole in the p-orbitals.

Hybersten et al. ([6]) have carried out band structure calculations to roughly estimate the actual values in the three-band Hubbard hamiltonian which yield the values (in eV),

$\Delta = \epsilon_p - \epsilon_d$	~ 3	U_d	~ 10
t_{pd}	$\sim 1-2$	U_p	~ 4
t_{pp}	~ 1	U_{pd}	~ 1

We see that these values fall into the strong coupling regime. Let us point out that positive value of $\Delta = \epsilon_p - \epsilon_d$ means that the first hole added to the system will energetically prefer to occupy the d-orbital of the copper ions, as it is observed in the undoped cuprates compounds which have a hole per unit cell. Adding more holes and being in the strong coupling limit where $U_d \gg \Delta$, the new holes will tend to occupy the p-orbitals as it is suppoused to be the case in the doping mechanism of the hole-doped cuprates.

The question as to wether the three-band model reduces to a simpler model like the one-band Hubbard model or similar remains controversial. Zhang and Rice ([7]) have worked out this program based upon the following scenario. Consider a copper ion surrounded by say four oxygen atoms. A hole in a p-orbital can have its spin parallel (singlet) or antiparallel (triplet) with respect to the hole at the copper ion. These authors performed a second order perturbative calculation in the strong coupling limit showing that the spin singlet has the lowest energy. They further assumed that working in the singlet subspace does not change the physics of the problem. Then, the hole originally located at the oxygen is replaced by a spin singlet centered at the copper. This is equivalent to removing one Cu spin-1/2 from the square lattice of copper spins, and thus the effective model corresponds to spins and holes (absence of spin) on a two dimensional square lattice. In this fashion, the oxygen ions have been effectively renormalized away. Their calculations yield an effective hamiltonian which replaces the three-band Hubbard model and it happens to be the so-called $t - J$ model previously introduced by Anderson ([1]),

$$H = -t \sum_{<ij>\sigma} [c_{i\sigma}^\dagger (1 - n_{i-\sigma}) (1 - n_{j-\sigma})c_{i\sigma} + h.c.] + J \sum_{<ij>} (\mathbf{S}_i \cdot \mathbf{S}_j - \frac{1}{4}n_i n_j) \quad (6.3)$$

where $\mathbf{S}_i$ are spin-1/2 operators at the sites i of a two dimensional square lattice and j is the antiferroamgnetic coupling between nearest-neighbours sites. The hopping term allows the movement of the electrons without changing their spin and the double occupancy has been *explicitly* excluded due to the presence of the projector operators $1 - n_{j-\sigma}$. As a consequence the number of possible states per site of the lattice has been reduced to 3, namely, hole, electron with spin up or electron with spin down.

This reduction of the three-band Hubbard model to the $t - J$ model is not supported by all the researcher in the field. For instance, Emery and Reiter ([8]) have argued that the resulting quasiparticles of the three-band model have both charge and spin, in contradiction to the $Cu - O$ singlet that form the effective one-band $t - J$ model.

References

[1]Anderson, P. W. *1987, Science* **235**, *1196*

[2]Dagotto, E., *Correlated Electrons in High Temperature Superconductors* Florida State University preprint, 1993. (submitted to Rev. Mod. Phys.)

[3]Haldane, F. D. M. *1981, J. Phys.* C**14**, *2585*

[4]Anderson, P. W. and Schrieffer, J. R. *1991, Phys. Today* **44**, *55*

[5]Emery, V. J. *1987, Phys. Rev. Lett.* **58**, *2794*

[6]Hybertsen, M., M. Schluter, N. E. Christensen, *1989, Phys. Rev. B* **39**, *9028*

[7]Zhang, F. C., and T. M. Rice, *1988, Phys. Rev. B* **37**, *3759*

[8]Emery, V. J., and G. Reiter, *1988, Phys. Rev. B* **38**, *11938*

7. The Mott Transition and the Hubbard Model

7.1 Mott Theory of the Metal-Insulator Transition

In the previous chapter we have adopted the framework in which the Hubbard model is the microscopic model describing the conduction electrons relevant to the high-T_c superconductivity in the CuO_2 planes. Although the issue of how many bands are necesary to account for the observed properties of the cuprate compounds is not yet solved, nevertheless the key ingredient of the model and its relatives is that it takes into account the *correlation among electrons*. Thus we have the following identification,

$$\{Hubbard\ Model\} \equiv \{Model\ for\ Correlated\ Electrons\ on\ a\ Lattice\}$$

Hubbard proposed his model in a famous paper in 1963 [1] and subsequently he further developed it in two other papers [2], [3]. Altogether they are known as Hubbard I, Hubbard II and Hubbard III.

The importance of the Hubbard model is twofold whether depending we are considering the early days of the model or nowdays,

Nowdays It is a model potentially describing the high-T_c superconductors.

Early days It is a model describing the Mott transition of a metal to an insulator.

As a matter of fact, both aspect of the model are intimately related for as we have seen in the previous chapter, the onset of the high-T_c superconductivity occurs in the vecinity of a Mott transition induced by the doping mechanism of the cuprate compounds.

In order to gain some insight into the Mott transition, let us briefly recall what the standard band theory (Bloch-Wilson) tells us about the metal or insulator behaviour of the solid state.

The standard band model resides on the one-electron approximation. The Bloch functions extend over the entire lattice so that every electron is delocalized. Actually, the three basic assumptions made by the band theory are:

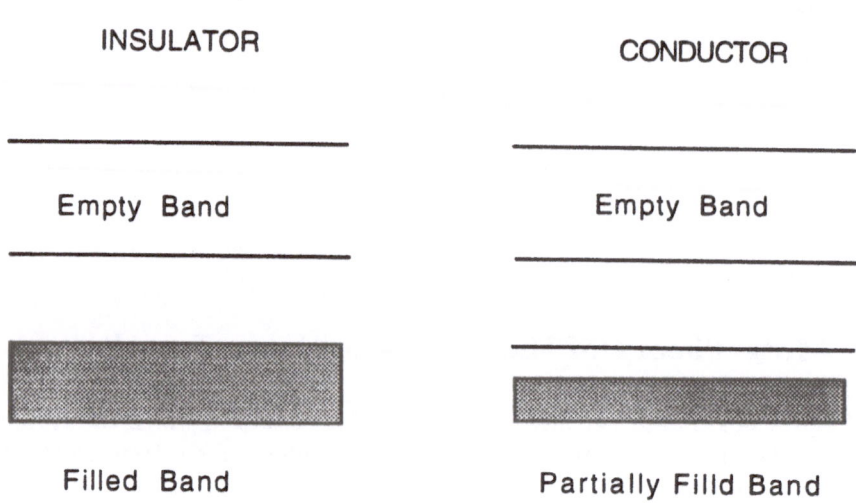

Fig. 7.1. Band structure of metals and insulators according to standard band theory (Bloch-Wilson).

- Strict periodicity of the lattice.

- Replacement of the actual electron-electron interaction with the mean interaction of the Hartree-Fock approximation.

- Neglect of the lattice vibrations, with the assumption that these could be taken into account later by perturbative calculation (weak electron-phonon coupling).

The restriction to *periodic lattices* is needed to be able to introduce a momentum k-vector as a good quantum number, and from it the delocalized states of the band model. In figure 7.1 we recall that when the conduction band is full of states, there is an energy gap preventing the electrons to freely move upon the application of an external electric field. This situation corresponds to an *insulator*. If the conducting band is not filled, say half-filled, then there is no gap and the conduction electrons are able to move. This situation corresponds to a *metal*. The band model loses validity when we consider alloys or amorphous phases, where disorder plays a major role (Anderson localization) in the description of the metal-insulator phase transition.

The introduction of a *mean electron-electron interaction* is needed to allow us, in the one-electron approximation, to regard the possible states of the electron observed as being totally independent of the occupation of the other states by electrons. The behaviour of the Bloch electron is determined solely by the periodic potential in which it moves. This implies the *neglection of correlations* between the valence electrons in a crystal. In the following section we shall use the Hubbard model to investigate the role of correlations.

There are plenty of applications of the band model to many solid compounds. However, Mott, in 1949 [4], pointed out that there are experimental indications of failures of the band model which are more apparent in solids whose band structures contain narrow d-bands like Niquel oxides. According to the band model all these compounds should be metals (as long as their d-bands are not completely filled). In practice, however, one finds both metals and insulators among them, with differences in conductivity of the order of 10^{20}. Some of them change from insulator to metals when increasing the temperature.

A second aspect of these Mott insulators is that they exhibit magnetic properties. Experimentally it is found that *magnetism is strongly related to the fact that these materials are insulators.*

Remark. Notice the emphasis made upon the magnetic properties of these insulators which was also apparent in the high-T_c materials.

At this point it is convenient to introduce the following definitions which will be useful in what follows.

Mott Insulator $\equiv$ Insulating material whose properties are not explained by the standard band model of Bloch-Wilson and which exhibits magnetic structure.

Mott Transition $\equiv$ Transition in a metal to a Mott insulator.

Mott's address of the problem was to blame the Coulomb repulsion for these anomalous behaviours in many transition metal compounds. Before studying his theory, let us mention a predecessor of the Mott insulator, namely, the *Wygner crystal* which was introduced by E. P. Wigner in 1938.

Wigner considered an interacting electron gas in a homogeneously distributed positive charge background (jellium model or two-component plasma model or Coulomb gas). The energy of the electron is computed in the Hartree-Fock approximation. It is often given as a function of the mean electron separation in the electron gas r_0 defined by

$$(4\pi/3)r_0^3 = 1/n \qquad (7.1)$$

where n = electron concentration], or by the dimensionless quantity r_s,

$$r_s = r_0/a_0 \qquad (7.2)$$

where a_0 is the Bohr radius. In most metals r_s lies between 2 and 6. The larger r_s means low electron concentration, and viceversa.

It can be proved that the mean energy of a Hartree-Fock electron of effective mass m^* as a function of r_s is

$$\bar{E}_{HF} = \left(\frac{2.21}{r_s^2} \left(\frac{m}{m^*} \right) - \frac{0.916}{r_s} \right) \; ryd. \tag{7.3}$$

The second term is the contribution from the exchange interaction taken into account in the Hartree-Fock approximation.

As the electron concentration decreases (increasing mean electron separation r_s), the kinetic energy contribution in (7.3) (first term in the RHS) decreases relative to the potential energy contribution (second term on the RHS). When the second term is much more significant, one finds that in the state of lowest energy the electrons are organized in a crystal-like array, i.e., the electrons arrange themselves in such a way as to be as far away from each other as possible. This also means a localization of the electrons through correlation effects, which is called *Wigner crystallization.*

Now let us consider the description of a *Mott insulator* from a qualitative point of view according to Mott's line of reasoning. For simplicity, consider a monovalent metal. Then there is one valence electron (*s*-electron) associated to each lattice atom. The valence band of the metal is then half-filled so that occupied and empty states are adjacent in the *s*-band. As a result, the valence electrons are delocalized and move freely about the crystal. As far as the standard band model is concerned, this is all we need to know to explain the metallic properties of the crystal.

Let us now increase the lattice spacing, while still maintaining the relative arrangement of the lattice ions. This need not be a real experiment but just think of different transition metals with different lattice constants. In figure 7.2 we see that the consequence is a reduction of the width of the *s*-band. If we keep increasing the lattice spacing to such an extent that there is practically no interaction between lattice ions, the band reduces to the discrete *s*-orbital of the isolated atom. However, in the limit of isolated atoms, each atom is certainly neutral, i.e., it has one of the valence electrons localized in its neighbourhood. Consequently, metallic conduction is not possible any more, although according to the band approximation we still have a half-filled band!.

Moreover, Mott also argues that this mechanism allows an abrupt rise in the conductivity at the metal-insulator phase transition as shown in figure 7.3 where 2ϵ, the activation energy, is plotted against the lattice spacing d. It is the energy necessary to create a pair of charge carriers in the insulating phase ($d > d_c$). A sharp transition (gap) between the metallic and insulating character like this is known as a *Mott transition.*

The band model approximation thus breaks down for narrow bands. The localization of the electron on the lattice atoms means a correlation among electrons. For narrow bands such correlations have to be taken into account.

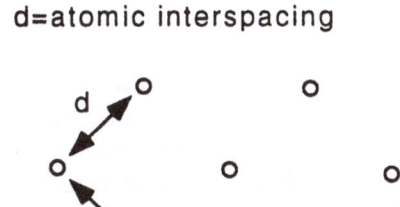

Fig. 7.2. The atomic interspacing as parameter controlling the Mott transition.

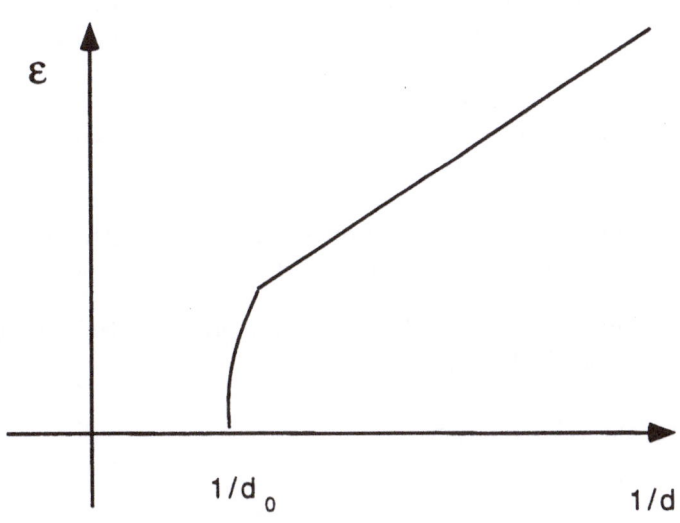

Fig. 7.3. Activation energy plotted as a function of the spacing d showing an abrupt rise at the metal-insulator phase transition.

7.2 The Hubbard Approximation

Mott's description of the metal-insulator transition in the metal transition compounds considered above is qualitative. It reveals what influence correlation can have on the statements derived from the one-electron approximation. For a quantitative formulation we now outline an approach which, by taking correlation effects into account in the band model, leads to a localized description. This approach is precisely the Hubbard model which he originally introduced as a microscopic model to explain the Mott transition.

The basic ingredient of the model is a collection of atomic orbitals $\phi_\mu(x)$, $\mu = 1,\ldots,m$ which can be of three types,

Non-degenerated $\mu = 1, 2$ There is not angular orbital degeneracy but only spin degeneracy. A typical example is a s-orbital. These one-band models are the subject of the paper Hubbard I [1].

Degenerated $\mu = 1, 2, \ldots, m$, $m = 2(2l + 1)$ where l is the orbital angular momentum number. In this case there are orbitals either of type p or d, or etc. These degenerated-band models are the subject of the paper Hubbard II [2].

Several bands This is the most general case which is a combination of the previous cases: there are several bands which are degenerated. As an example we may cite the 3-band Emery model described in chapter 1.

The Hubbard model is described within the second quantization formalism of many-body theory, also called the occupation number formalism. We introduce fermionic operators,

$c_{i\mu}$ Operator which destroys one electron on the orbital $\phi_\mu(\mathbf{x} - \mathbf{R}_i)$ centered at the ith ion on the lattice, see figure 7.4.

$c_{i\mu}^\dagger$ Operator which creates one electron on the orbital $\phi_\mu(\mathbf{x} - \mathbf{R}_i)$ centered at the ith ion on the lattice.

They satisfies the CAR relations,

$$\{c_{i\mu}, c_{j\nu}^\dagger\} = \delta_{i,j}\delta_{\mu,\nu} \tag{7.4}$$

The simplest case corresponds to have only spin up, spin down degeneracy, $c_{i\uparrow}$, $c_{i\downarrow} \longrightarrow$ one-band Hubbard model.

In the case of degenerated-band Hubbard model we need to dress the fermionic operators with an extra index m_l taking values on the angular momentum orbilats. That is,

$$c_{i,m_l,m_s}, \quad m_l = -l, \ldots, l. \quad m_s = \uparrow, \downarrow.$$

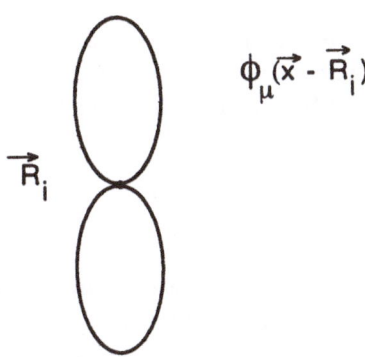

$\phi_\mu(\vec{x} - \vec{R}_i)$

$\vec{R}_i$

Fig. 7.4. Electron orbital centeres at the $i - th$ ion of the lattice.

The first approximation that Hubbard makes is the *tight binding approxima-tion* in which one electron can hoppe only one lattice spacing at each time. Whithin this approximation, the hamiltonian H for the electrons described by the orbitals ϕ_μ splits into a "free" kinetic term H_0 and an interacting term H_I accounting for correlations among the electrons,

$$H = H_0 + H_I \tag{7.5}$$

where H_0 is

$$H_0 = \sum_{<i,j>} \sum_{\mu,\nu} T_{ij}^{\mu\nu} c_{i\mu}^\dagger c_{i\nu} \tag{7.6}$$

and the hopping amplitude is given by

$$T_{ij}^{\mu\nu} = \int d^3x \, \phi_\mu^*(\mathbf{x} - \mathbf{R}_i) \left(-\frac{\hbar^2}{2m} \nabla_x^2 + V(|\mathbf{x}|) \right) \phi_\nu(\mathbf{x} - \mathbf{R}_j) \tag{7.7}$$

$V(\mathbf{x}) \equiv$ Crystal ion potential seen by a single electron.
As for the interacting hamiltonian we have,

$$H_I = \frac{1}{2} \sum_{i,j,k,l} \sum_{\mu,\nu,\sigma,\tau} < i\mu, j\nu | \frac{1}{r} | k\sigma, l\tau > c_{i\mu}^\dagger c_{j\nu}^\dagger c_{k\sigma} c_{l\tau} \tag{7.8}$$

where the matrix elements describing Coulomb interactions between electrons associated with different lattice ions have the explicit form

$$< i\mu, j\nu | \frac{1}{r} | k\sigma, l\tau > = e^2 \int d^3x \, d^3x' \frac{1}{|\mathbf{x} - \mathbf{x}'|} \times$$

$$\phi_\mu^*(\mathbf{x} - \mathbf{R}_i) \phi_\sigma(\mathbf{x} - \mathbf{R}_k) \phi_\nu^*(\mathbf{x}' - \mathbf{R}_j) \phi_\tau(\mathbf{x}' - \mathbf{R}_l) \tag{7.9}$$

7.2.1 Hubbard approximation.

The interaction between the electrons at the same ion will of course play the major role. As a first step in considering correlation Hubbard makes the approximation, bearing his name, of taking only matrix elements in which $i = j = k = l$. Thus,

$$H_I \simeq \frac{1}{2} \sum_i \sum_{\mu,\nu,\sigma,\tau} < i\mu, j\nu|\frac{1}{r}|k\sigma, l\tau > c_{i\mu}^\dagger c_{i\nu}^\dagger c_{i\sigma} c_{i\tau} \qquad (7.10)$$

In the simplest case of having only one orbital, the interacting piece of the hamiltonian takes the familiar Hubbard form

$$H_I = U \sum_i n_{i\uparrow} n_{i\downarrow} \qquad (7.11)$$

where the remaining matrix element has been called

$$U = < ii|\frac{1}{r}|ii > \qquad (7.12)$$

Interestingly enough, all the Colulomb repulsion, i.e. the correlation, has been encoded in only one parameter $U!!$.
For the orbitals of type $3d$ in the transition metals the U parameter is of the order

$$U_{3d} \sim 20 \ eV$$

However this is just its bare value for the renormalized or effective value after taking into account a combination of many effects is

$$U_{eff} \sim 10 - 15 \ eV$$

Coming back to the kinetic term, let us split the hopping amplitude into two pieces,

$$T_{ij}^{\mu\nu} = T_0 \delta^{\mu\nu} \delta_{ij} + t_{ij}^{\mu\nu} \qquad (7.13)$$

with $t_{ij}^{\mu\nu} \neq 0 \Longleftrightarrow < ij >$ are nearest-neighbours, according to the tight binding approximation.

$T_0 \equiv$ Energy of the unperturbed μ-orbital.

$t_{ij}^{\mu\nu} \equiv$ Transition amplitude for one electron in the μ-orbital at the ion i to hoppe to the ν-orbital at the ion j.

Sometimes we adopt the convention setting $T_0 - 0$ for it is just a global constant in the hamiltonian. In this case and for the one-band Hubbard model we arrive at the well-known expression

$$H = -t \sum_{<i,j>\sigma} \left(c_{i\sigma}^\dagger c_{j\sigma} + c_{j\sigma}^\dagger c_{i\sigma} \right) + U \sum_i n_{i\uparrow} n_{i\downarrow} \tag{7.14}$$

In the particular case of one spatial dimension the kinetic term can be written as

$$H_0 = -t \sum_{i,\sigma} \left(c_{i+1,\sigma}^\dagger c_{i,\sigma} + c_{i,\sigma}^\dagger c_{i+1,\sigma} \right) + U \sum_i n_{i\uparrow} n_{i\downarrow} \tag{7.15}$$

7.2.2 Hubbard parameters

In the simplest Hubbard model with only one band all the information is encoded in two parameters, t and U. Actually, what really matters is the ratio of them: U/t.

However apart from these parameters which show up in the Hubbard hamiltonian there is another which is implicit, namely, the density n of electrons. As it happens, the Hubbard hamiltonian H commutes with the operator $N_e = \sum_{i,\sigma} n_{i,\sigma}$ which is the total number of electrons,

$$[H, N_e] = 0 \tag{7.16}$$

Thus, the hamiltonian H can be solved (diagonalized) by restricting to sectors with a fixed number of electrons. It is costumary to introduce the electronic density as

$$n = \frac{N_e}{N_{\text{lattice sites}}}$$

The range of values for n depends on the number of bands,

1 band $\qquad 0 \leq n \leq 2$

$\mu = 1, \ldots, m \qquad 0 \leq n \leq 2^m$

Definition 7.1 Half-filling *Half-filling $\Longleftrightarrow n = 1$ (for 1 band model). That is, one electron per site, regardless of the spin.*

Then, altogether we see that the spectrum of the Hubbard model depens on the parameters U/t and n. The phase diagram in thus tridimensional with the temperature as the extra axis, see figure 7.5.

Remark. Where is the connection between the parameter d introduced in the qualitative description of the Mott transition and the parameters appearing in

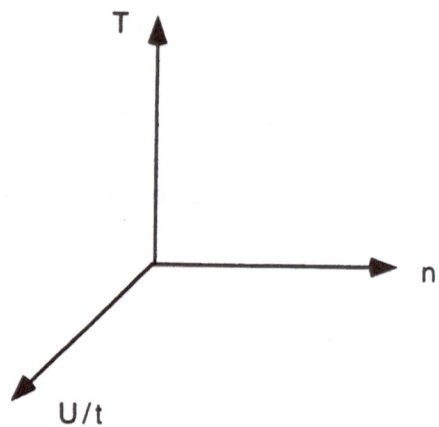

Fig. 7.5. The three parameters defining the phase space of the Hubbard model.

the Hubbard model? It is clear from the above definition of $t_{ij}^{\mu\nu}$ that it measures the overlapping of the orbitals $\phi_\mu(\mathbf{x} - \mathbf{R}_i)$ and $\phi_\nu(\mathbf{x} - \mathbf{R}_j)$. From (7.7) we see that the lattice spacing d is encoded in the position of the ions $\mathbf{R}_i, \mathbf{R}_j$. The bigger is t the bigger the movility of the electrons, assuming we keep U is fixed. Then the parameter d corresponds to U/t.

7.2.3 Solvable limits of the Hubbard model.

The Hubbard model can be solved exactly in two limits each one corresponding to extreme physical situations. This is true for any number of spatial dimensions, but we will assume that we are at *half-filling* ($n = 1$).

$U = 0$ **Model of Free Electrons.** In this limit the interaction between electrons is turned off. As a result, the system becomes a Fermi gas so that the electrons are capable of moving freely all over the lattice. This situation corresponds to a conductor and the system is a METAL.

The solution of the problem is carried out introducing a set of planes waves in the fermionic operator formalism,

$$c_{i\sigma} = \frac{1}{\sqrt{N_0}} \sum_{\mathbf{k}} e^{\mathbf{k} \cdot \mathbf{R}_i} c_{\mathbf{k}\sigma} \qquad (7.17)$$

where the sum $\sum_{\mathbf{k}}$ extends over the 1st Brillouin zone.

The free hamiltonian takes the following form in Fourier representation, after substituting (7.17) in $H_{U=0}$,

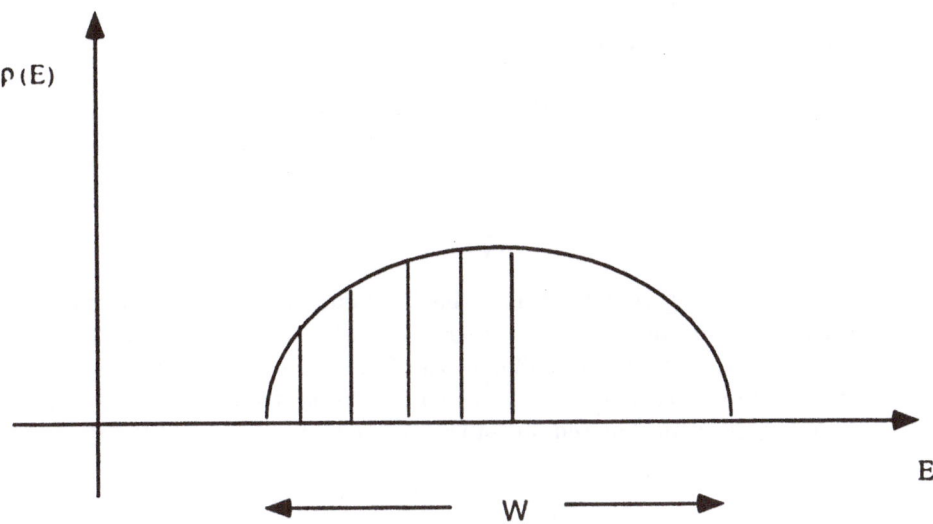

Fig. 7.6. Schematic plot of the density of states in the free electron approximation of the Hubbard model.

$$H_{U=0} = \sum_{\mathbf{k},\sigma} \epsilon(\mathbf{k}) c^{\dagger}_{\mathbf{k}\sigma} c_{\mathbf{k}\sigma} \qquad (7.18)$$

where, by traslational invariance, the purely kinetic energy of the electrons depends only on the nearest-neighbours hopping amplitudes to a certain site of the lattice, say the origin,

$$\epsilon(\mathbf{k}) = \sum_{<i,0>} t_{i0} e^{\mathbf{k}\cdot\mathbf{R}_i} \qquad (7.19)$$

Example. In a rectangular two dimensional lattice the energy of the conduction band of electros is given by

$$\epsilon = -2t[\cos(k_x a) + \cos(k_y a)] \qquad (7.20)$$

$$|k_{x,y}| \leq \frac{\pi}{a}$$

The width w of the band is proportional to the hopping amplitude t which in turn is a function of the filling factor and the type of the lattice. In figure 7.6 it is squematically shown the density of states $\rho(E)$ against the energy.

$t = 0$ **Model of Confined Electrons to the Lattice Ions** This limit recives several denominations,

- Zero band-width limit ($w = 0 \sim t = 0$).
- Atomic limit.
- Strong coupling limit ($U \gg t$).

In this extreme situation the Hubbard hamiltonian contains only interaction terms,

$$H_{t=0} = U \sum_i n_{i\uparrow} n_{i\downarrow} + T_0 \sum_{i,\sigma} n_{i\sigma} \qquad (7.21)$$

We are assuming that to mimic the Coulomb repulsion the coupling is positive $U > 0$ and also that we are dealing with a half-filling situation $n = < \sum_i (n_{i\uparrow} + n_{i\downarrow}) > = 1$. Then, the fundamental state of the system consists of one electron per site. If we have N_0 electrons on the lattice, the ground state is constructed out of the creation operators and the vacuum,

$$|\{\sigma_i\} >_{GS} = \prod_{i=1}^{N_0} c_{i,\sigma_i}^\dagger |0>$$

where the vacuum state is defined as usual with respect to the fermionic destruction operators,

$$|0>: c_{i,\sigma_i}|0> = 0 \ , \ \forall i, \sigma$$

It is apparent that this ground state is highly degenerated for the ground state energy E_0 is insensitive to the spin of the electrons. Actually, there are 2^{N_0} configurations with the same energy.

Once the ground state is known, we can also construct easily the excited states. The first excited state corresponds to having one site of the lattice doubly occupied by two electrons, in the second excited state there are two sites with double occupancy, and so on and so forth. See figure 7.7.

$$
\begin{array}{lll}
E_0 & = & N_0 T_0 \qquad \equiv \text{One electron per site} \\
E_1 & = & N_0 T_0 + U \quad \equiv \text{1 lattice site double occupied} \\
E_2 & = & N_0 T_0 + 2U \ \equiv \text{2 lattice site double occupied} \\
\vdots & & \vdots \qquad\qquad\quad \vdots
\end{array}
$$

- T_0 is therefore the energy needed to bind an electron on an isolated atom.
- $T_0 + U$ is the energy needed to attach a second electron, of opposite spin. Hence U is the Coulomb interaction energy of two electrons located at the same atom.

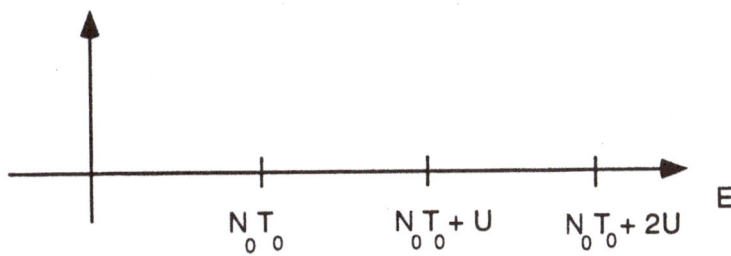

Fig. 7.7. Excited states in the strong coupling approximation of the Hubbard model.

7.2.4 Hubbard's Results.

In his original work, Hubbard studied the Mott transition in the atomic limit. It is not a casuality that his paper was entitled *"Electron Correlation s in Narrow Energy Bands"*. It is not our purpose to present a full detailed account of his computations. They are too lengthy to be reproduced here and the reader is referred to Hubbard's original work cited in the bibliography. We shall concentrated instead on his results which are very remarkable for he succeeded in describing quantitatively metal-insulator transition of Mott type.

To study the Mott transition in the atomic limit Hubbard employs Greenfunctions techniques. In a discrete model, this amounts to computing expressions of the form

$$G_{jk}^\sigma(E) \sim \int < c_{j\sigma}(E) c_{k\sigma}^\dagger(E) > \ e^{iEt} dt \qquad (7.22)$$

where we are using units in which the energy $E \sim \omega$ (frecuency).

In the atomic limit and for a model with only one band, Hubbard finds

$$G_{ij}^\sigma(E) = \frac{1}{2\pi} \ \delta_{ij} \left\{ \frac{1 - \frac{n}{2}}{E - T_0} + \frac{\frac{n}{2}}{E - T_0 - U} \right\} \qquad (7.23)$$

where it is assumed that the occupation number for each species of spin is

$$< n_{i\sigma} >= \frac{1}{2} n \ \ n = 1 \text{ for half-filling}$$

From the one-particle Green function it is possible to retrieve the density of pseudo-particle states which turns out to be,

$$\rho_\sigma(E) = \frac{i}{N_0} \lim_{\epsilon \to 0+} \sum_j \left[G_{jj}^\sigma(E + i\epsilon) - G_{jj}^\sigma(E - i\epsilon) \right] \qquad (7.24)$$

$$= \left(1 - \frac{n}{2}\right) \ \delta(E - T_0) + \frac{n}{2} \ \delta(E - T_0 - U) \qquad (7.25)$$

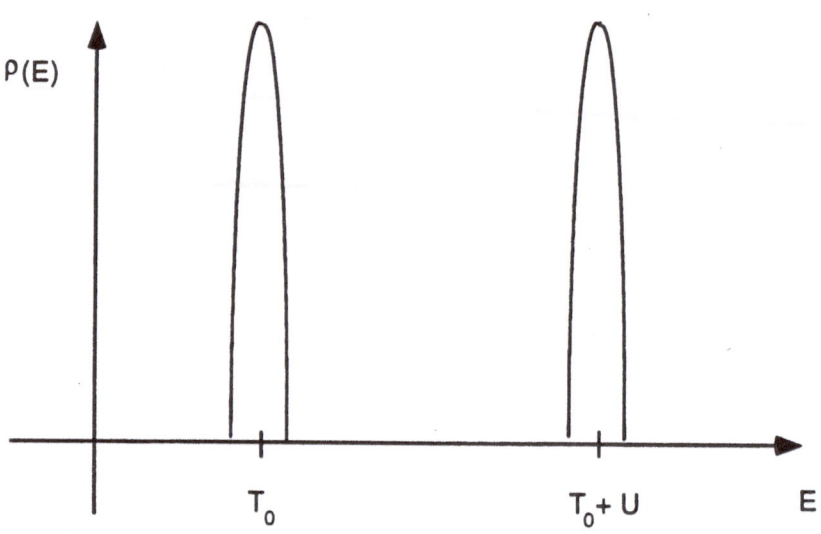

Fig. 7.8. The Lower Hubbard band (LHB) and the Upper Hubbard band (UHB).

From this result we see that the system behaves as if only there would be two energy levels with energy T_0 and $T_0 + U$ and with $1 - n/2$ and $n/2$ states per atom, respectively. See figure 7.8.

Upon starting to add electrons to the system, they begin occupaying the level with energy $E_0 = T_0$ which is called *"Lower Hubbard Band"* LHB. This band is filled when

$$2 \text{ spin} \times \left(1 - \frac{1}{2}\right) \text{(LHB states)} = n \text{ (number of electrons)} \implies n = 1 \text{ (half-filling)}$$

From this situation on ($n > 1$) the band with energy $E = T_0 + U$ starts to be filled. This band is called *"Upper Hubbard Band"* UHB.

Remark. We see that precisely at half-filling there is an energy gap of value U $\implies$ The system is an *insulator*. In fact, the presence of an electric field in the electronic system is not enogh to produce an electric current for it is needed an energy of order U to create a hole in the filled conduction band and to promote the corresponding electron to the Upper Hubbard Band.

Summary of Exact Solutions.

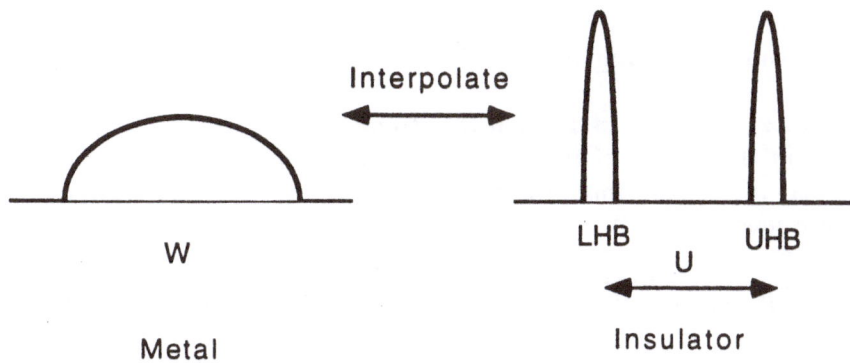

Fig. 7.9. Interpolation between the weak and strong coupling regimes as proposed by Hubbard.

$U = 0$	$t = 0$
Free Electrons	Localize Atomic Electrons
1 Conduction Band	2 Conduction Bands (LHB, UHB)
band width $w \sim t$	band width $w = 0$
metal	*insulator*

Remark. The big issue now is how to manage to interpolate between these two extreme situations. See figure 7.9.

We are facing the problem of the Mott transition in which d (Mott) $\sim U/t$ (Hubbard). The key question is to find a critical value of the parameter $(U/t)_c$ for which the metal-insulator transition occurs. Except for the particular case of $D = 1$ dimensions, it is not known the answer to this crucial question (despite the zillion of papers about this issue.)
Thus far, there are known two approaches to address the determination of this phase transition.

7.2.5 Weak Coupling Approach.

The strategy is to start from the region $U \sim 0$: $(u/t \ll 1)$. Thus this amounts to a *fermi gas* + perturbations. From the general theory of Landau Fermi liquid one is tempted to think that the outcome of this calculation should be a Fermi liquid. This may be valid for small U/t and low electronic densities ($n \ll 1$). But in the strong coupling limit and $d = 2$, which suppousedly is the relevant in the high-T_c superconductivity, it might be possible that the electron liquid is not a Fermi liquid according to Anderson's ideas and experimental data showing deviations from conventional Landau theory.

7.2.6 Strong Coupling Approach.

The strategy now is to start from the strong coupling regime $U/t \gg 1$. The analysis in this region of the parameter space is facilitated by the following important result concerning the relation of the Hubbard model to other very famouses models as well,

$$H_{\text{Hubbard}} \xrightarrow{U/t \gg 1} \text{t-J Model} \xrightarrow{n=1} \text{Heisenber Model } (J = 4t^2/U)$$

We shall deal with this chain of connections in the next chapter.

Coming back to Hubbard's results, he was the first studying the strong coupling regime using a series of approximations,

" ... only spatially homogeneous solutions in which all atoms are equivalent will be investigated. Such solutions can describe non-magnetic and ferromagnetic states of the system, but not antiferromagnetic states, spiral spins or other sort of states with more complicated structures." *Hubbard-II*

The result of his investigations can be summarized by means of the following equation for the one-particle Green function,

$$G^\sigma(\mathbf{q}, E) = \frac{1}{2\pi N_0} \left\{ \frac{A_{\mathbf{q},\sigma}^{(1)}}{E - E^{(1)}{}_{\mathbf{q}}} + \frac{A_{\mathbf{q},\sigma}^{(2)}}{E - E^{(2)}{}_{\mathbf{q}}} \right\} \qquad (7.26)$$

$$A_{\mathbf{q},\sigma}^{(1)} + A_{\mathbf{q},\sigma}^{(2)} = 1 \qquad (7.27)$$

The density of quasiparticles associated to this one-particle Green function is,

$$\rho_\sigma(E) = \frac{1}{N_0} \sum_{\mathbf{q}} \left(A_{\mathbf{q}\sigma}^{(1)} \, \delta(E - E_{\mathbf{q}\sigma}^{(1)}) + A_{\mathbf{q}\sigma}^{(2)} \, \delta(E - E_{\mathbf{q}\sigma}^{(2)}) \right) \qquad (7.28)$$

The form of the two possible particle energies $E_{\mathbf{q}\sigma}^{(1)}$, $E_{\mathbf{q}\sigma}^{(2)}$ is shown in figure 7.10,
It is very illustrative to see what happens with these two energy levels when the two extreme situations are considered.

- In the free electron limit $U \to 0$ the two energy levels fuse together forming a unique level. This corresponds to having one energy band in the model of electrons as was previously explained when the analysis of solvable situations were considered. See figure 7.11.

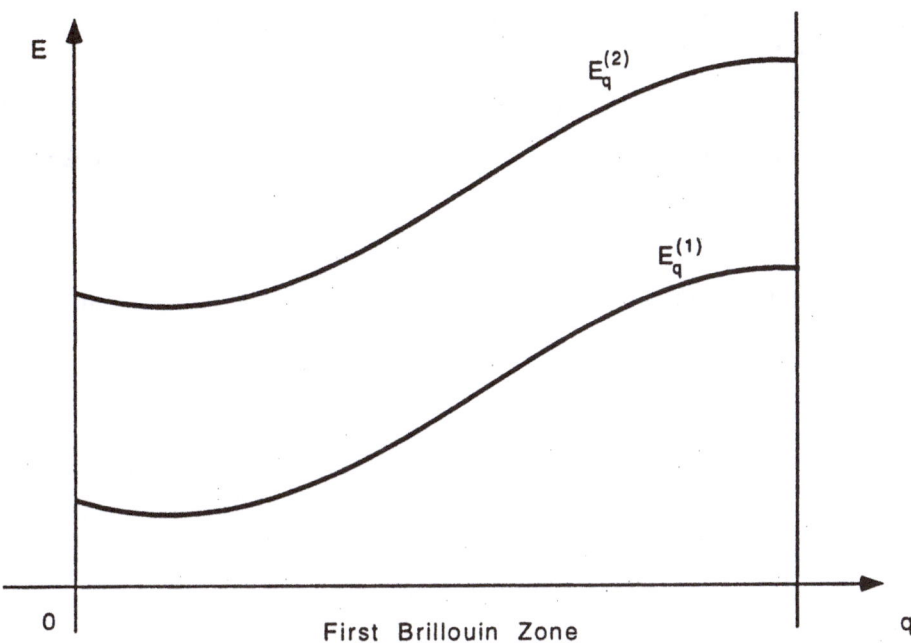

Fig. 7.10. The two possible particle energies in the strong coupling Hubbard approximation.

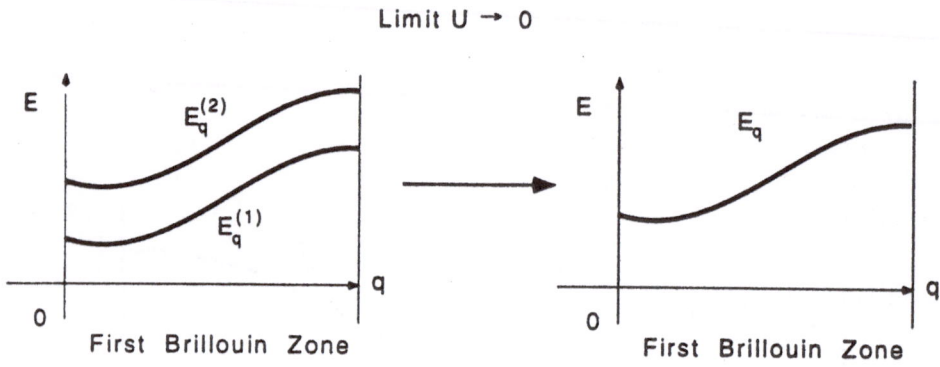

Fig. 7.11. The fusion of the two energy levels in the free electron limit.

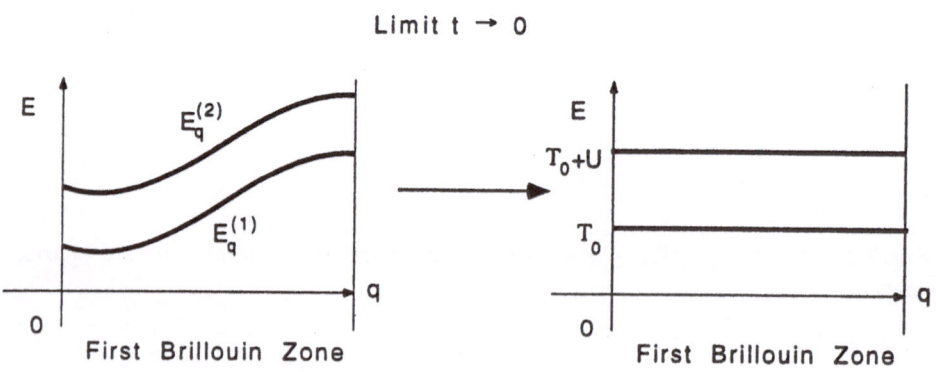

Fig. 7.12. The two energy levels in the atomic limit.

- In atomic limit $t \to 0$ they keep a distance apart which ultimately turns out to be U. At that moment, the energy levels are constant in energy (see figure 7.12). This situation is reminescent of the two energy bands that show up when this limit is considered in the Hubbard model.

To see that it is possible to interpolate between the above two situations, Hubbard considered the following simplified situation,

1. Electronic density at half-filling.

2. Non-Ferromagnetic System with $< n_{i\uparrow} >=< n_{i\downarrow} >= 1/2$.

3. The density of states of the unperturbed band has the simple form,

$$\rho(E) = \frac{1}{N} \sum_{\mathbf{k}} \delta(E - E_{\mathbf{k}})$$

$$\rho(E) = \begin{cases} \frac{4}{\pi\Delta} \sqrt{1 - \left(\frac{E-T_0}{2\Delta}\right)^2} & |E - T_0| < \frac{1}{2}\Delta \\ 0 & |E - T_0| > \frac{1}{2}\Delta \end{cases} \qquad (7.29)$$

With this assumptions Hubbard finds the behaviour shown in figure 7.13. There the density of quasiparticles $\rho(E)$ is plotted against energy. Depending on the ratio Δ/U, where Δ is the width of the band in the band model without correlations and U is the effective Coulomb repulsion of an electron pair at an ion, the s-band splits into two separate bands. In the ground state, the lower half of the states is occupied. As long as the widths of the bands are smaller than their separation $(T_0 + U) - T_0 = U$, there will be a gap between the two bands. At a critical amount of splitting (determined by the lattice constant) the gap will disappear. We have thus in the above been able to show the transition from metallic to insulating behaviour in a solid.

Another interesting graphic results when plotting the borders of the two bands (x-axis) as a function of the ratio Δ/U, see figure 7.14. We see that there is a critical value for this ratio for which the two Hubbard bands merge. This corresponds precisely to the Mott transition which according to Hubbard's computations occurs at,

$$\Delta/U|_{\text{critical}} = 1.15 \qquad (7.30)$$

This figure can potentially explain an important phenomenon which happens in the group of transition metal compounds: upon increasing the temperature a jump in the conductivity is observed which can amount to many orders of magnitude. A possible explanation is that with increase of the temperature the lattice constant crosses the threshold value (7.30) at which the localized electrons become delocalized.

Despite the remarkable success of Hubbard's approximation in quantitatively describing the metal-insulator phase transition, it is far from being fully satisfactory. Among some criticisms that we can pose to his solution, let us mention the following,

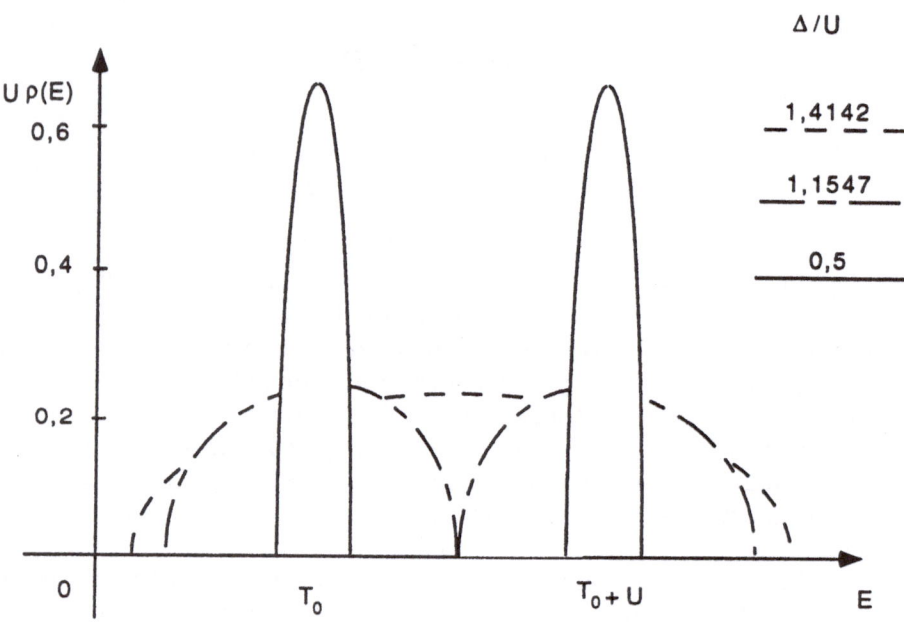

Fig. 7.13. Interpolation between the free electron limit and the two bands of the atomic limit as studied by Hubbard.

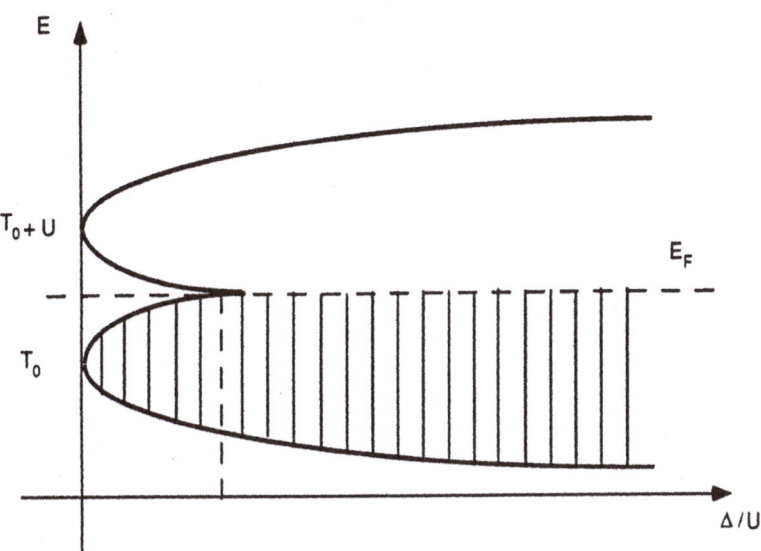

Fig. 7.14. The bordes of the two Hubbard bands as a function of the ratio Δ/U.

- Hubbard's solution does not take into account the possible existence of antiferromagnetic solutions. We have already mention that antiferromagnetism plays a major role in the high-T_c superconductors.

- In the region $\Delta/U > (\Delta/U)_c$ the system should correspond to a metal state, however it has been shown that there is *not* exits a well defined *fermi surface*, i.e., there is not a discontinuity in the density of states. This is explained by the authors because the life-time of the quasiparticles is finite at the Fermi level, namely,

$$\tau^{-1} = U^2/3\Delta$$

References

[1]Hubbard, J., *1963, Proc. R. Soc. London* **A276**, *238*
[2]Hubbard, J., *1964, Proc. R. Soc. London* **A277**, *237*
[3]Hubbard, J., *1964, Proc. R. Soc. London* **A281**, *401*
[4]Mott, N. F., *1949, Proc. R. Soc. London* **A62**, *416*

8. Strong Coupling Limit and Some Exact Results

8.1 The Strong Coupling Limit

The purpose of this chapter is to explain one of the most interesting limits of the Hubbard model, that is, the strong coupling limit or atomic limit as was introduced in chapter 7 following the early days terminology. This limit is supposed to be of great relevance in the description of the high-T_c superconductivity, for it is apparent that to take into account strong correlation effects between the electrons in the CuO_2 conducting planes it is necessary that the Coulomb repulsion U be large as compared to the kinetic energy t-term.

We shall be dealing with the description of the following chain of models related by the strong coupling limit,

$$H_{\text{Hubbard}} \xrightarrow{U/t \gg 1} \text{t-J Model} \xrightarrow{n=1} \text{Heisenberg Model } (J = 4t^2/U) \qquad (8.1)$$

Hubbard's idea for solving his model was to treat in an exact fashion the zero-band-width limit, i.e., $U/t = \infty$ and henceforth to carry out approximations exclusively in those quantities which vanish when U/t goes to infinity. As we saw in the previous chapter, within this approximation we are belitling correlations of magnetic type which eventually can lead to a sensible modification of the "picture" of the model. However, as we have mentioned several times by now, the magnetic properties of the model play a paramount role in the $U/t \longrightarrow \infty$ limit.

In the strong coupling limit $U/t \longrightarrow \infty$, the main effect is to prevent two electrons with opposite spins from occupying the same lattice site. Thus, it is appropiate to decompose the Hilbert space of states H of the Hubbard model in the following fashion,

$$\mathcal{H} = \oplus_{l=0}^{M} \mathcal{H}_l \qquad (8.2)$$

where

$\mathcal{H}_l \equiv$ Hilbert subspace in which there are l sites occupied by a couple of electrons with opposite spins.

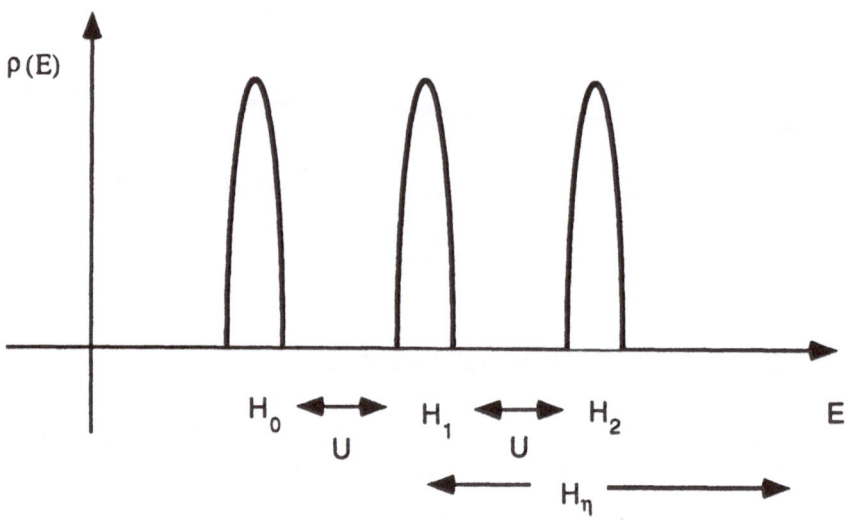

Fig. 8.1. Qualiative picture underlying the Hilbert space decomposition for the Hubbard model in the strong coupling limit.

and M is the number of doble occupied lattice sites.

Moreover, when dealing with the strong coupling limit, energy considerations impose a further decomposition

$$\mathcal{H} = \mathcal{H}_0 \oplus \mathcal{H}_\eta \tag{8.3}$$

where

$\mathcal{H}_0 \equiv$ Hilbert space corresponding to having no doubly occupied lattice site.

$\mathcal{H}_\eta \equiv$ Hilbert space corresponding to having at least one doubly occupied lattice site.

The qualitative picture underlying this decomposition is depicted in the figure 8.1.

Thus, it would be of great interest to find an effective hamiltonian that would allow us to restrict ourselves to the Hilbert subspace $\mathcal{H}_0$. In the languaje of Hubbard's works this amounts to working within the Lower Hubbard Band LHB.

The idea underlying this procedure is to devise a canonical transformation which partially "diagonalizes" the Hubbard hamiltonian. That is to say,

$$H = \begin{pmatrix} H_1 & H_\eta \\ H_\eta & H_2 \end{pmatrix} \longrightarrow H_{eff} = \begin{pmatrix} ? & 0 \\ 0 & ? \end{pmatrix}$$

The notation we are employing deserves some explanation. The Hubbard hamiltonian is conveniently split into two terms, the kinetic term and the interacting term,

$$H = H_0 + V \tag{8.4}$$

Let us introduce projector operators Π_0 and Π_η onto the Hilbert suspaces $\mathcal{H}_0$ and $\mathcal{H}_\eta$, respectively. Then, the entries in the matrix hamiltonian are defined to be ,

$$H_1 = \Pi_0 H \Pi_0, \quad H_2 = \Pi_\eta H \Pi_\eta, \quad H_\eta = \Pi_0 H \Pi_\eta = \Pi_\eta H \Pi_0 \tag{8.5}$$

Notice as well that the potential part of the Hubbard hamiltonian only has projection onto the subspace of doubly occupied lattice sites,

$$V = \Pi_\eta V \Pi_\eta \tag{8.6}$$

Now, let us decompose the Hubbard hamiltonian into diagonal and off-diagonal parts according to,

$$H = \tilde{H}_0 + \tilde{H}_\eta \tag{8.7}$$

where

$$\tilde{H}_0 = \Pi_0 H_0 \Pi_0 + \Pi_\eta H_0 \Pi_\eta + V \tag{8.8}$$

$$\tilde{H}_\eta = \Pi_0 H_0 \Pi_\eta + \Pi_\eta H_0 \Pi_0 \tag{8.9}$$

The strategy now is somehow to "integrate" the coupling between the subspaces $\mathcal{H}_0$ and $\mathcal{H}_\eta$ such that the physics thereafter will be described by an effective hamiltonian H_{eff} wich in order to be acceptable, it has to be diagonal with respect to the decomposition (8.3), i.e.,

$$\Pi_0 H_{eff} \Pi_\eta = 0. \tag{8.10}$$

To this aim, let us introduce a one parameter family of Hamiltonians and unitary transformations as follows,

$$H(\epsilon) \equiv \tilde{H}_0 + \epsilon \tilde{H}_\eta \tag{8.11}$$

$$\mathcal{U}(\epsilon) \equiv e^{i\epsilon S}; \quad S^\dagger = S \tag{8.12}$$

where ϵ is a parameter in $[0,1]$. In order to acomplish the requirement (8.10) for the effective hamiltonian, we search for a transformation S satisfying,

$$H_{eff} \equiv e^{i\epsilon S} H(\epsilon) e^{-i\epsilon S} = \tilde{H}_0 + O(\epsilon^2) \tag{8.13}$$

Notice that we are treating the parameter ϵ as if it were a "small" parameter even though equation (8.13) is not a real Taylor expansion. The idea is to work with this parameter to eliminate crossed terms in $\mathcal{H}_0$ and $\mathcal{H}_\eta$ to lowest order in ϵ. The real small parameter is t/U will appear below.

Expanding equation (8.13) to second order in ϵ,

$$H_{eff} = \tilde{H}_0 + \epsilon(H_\eta + i[S, \tilde{H}_0]) + \epsilon^2(i[S, \tilde{H}_\eta] + \frac{1}{2}[S, [\tilde{H}_0, S]]) + O(\epsilon^3) \quad (8.14)$$

The vanishing of the ϵ-term yields the condition determining the generator S of the transformation diagonalizing H_{eff},

$$[\tilde{H}_0, S] + i\tilde{H}_\eta = 0. \quad (8.15)$$

where $\tilde{H}_0$ and $\tilde{H}_\eta$ are given. Once S is known, as will be further pursue, the effective hamiltonian is obtained upon substituting equation (8.15) in (8.14) and setting $\epsilon = 1$,

$$H_{eff} = \tilde{H}_0 + \frac{i}{2}[S, \tilde{H}_\eta] \quad (8.16)$$

Our two unknown quantities are H_{eff} and S. We will determine them by giving their components with respect to the decomposition (8.3), that is , $\Pi_0 H_{eff} \Pi_0$, ..., $\Pi_0 S \Pi_0$, ... In this fashion, equation (8.15) yields the following components for the effective hamiltonian,

$$\Pi_0 H_{eff} \Pi_0 = \Pi_0 H \Pi_0 + \frac{i}{2}[(\Pi_0 S \Pi_\eta)(\Pi_\eta H \Pi_0) - (\Pi_0 H \Pi_\eta)(\Pi_\eta S \Pi_0)] \quad (8.17)$$

$$\Pi_\eta H_{eff} \Pi_\eta = \Pi_\eta H \Pi_\eta + \frac{i}{2}[(\Pi_\eta S \Pi_0)(\Pi_0 H \Pi_\eta) - (\Pi_\eta H \Pi_0)(\Pi_0 S \Pi_\eta)] \quad (8.18)$$

$$\Pi_\eta H_{eff} \Pi_0 = \frac{i}{2}[(\Pi_\eta S \Pi_\eta)(\Pi_\eta H \Pi_0) - (\Pi_\eta H \Pi_0)(\Pi_0 S \Pi_0)] \quad (8.19)$$

According to conditions (8.10), equation (8.19) must vanish identically. The other two equations above express the fact that the diagonal parts of the effective hamiltonian are determined by the off-digaonal parts of the S transformation. From the condition (8.15) we can likewise determine the components of the S generator. To shorten the discussion we will skip some details of the computation which can be found in [1] and state the outcome of it, namely, the conditions on the diagonal parts of S can be fulfilled by demanding the alternative conditions

$$\Pi_0 S \Pi_0 = \lambda \Pi_0; \quad \Pi_\eta S \Pi_\eta = \lambda \Pi_\eta \quad (8.20)$$

where λ is an arbitrary real number, while the off-diagonal part is determined by the condition,

$$(\Pi_0 S \Pi_\eta)(\Pi_\eta H \Pi_\eta) - (\Pi_0 H \Pi_0)(\Pi_0 S \Pi_\eta) = i \Pi_0 H \Pi_\eta \qquad (8.21)$$

In what follows we shall restrict the value of the otherwise arbitrary parameter λ to the simplest case, $\lambda = 0$. Thus, equation (8.21) is the only condition that we need to obtain the diagonal parts of the effective hamiltonian. We will make a somewhat formal analysis of this condition. Our unknown is $\Pi_0 S \Pi_\eta$ so that we need the inverse of the operator $\Pi_\eta H \Pi_\eta$. In the full Hilbert space of states (8.3) this inverse does not exist, but when the strong coupling limit comes about it is apparent that the dominant part of the hamiltonian H is the potential term associated to the subspace $\mathcal{H}_\eta$. In this subspace the operator $\Pi_\eta H \Pi_\eta$ is definite positive and invertible, so we can write,

$$\Pi_0 S \Pi_\eta = i(\Pi_0 H \Pi_\eta) \cdot (\Pi_\eta H \Pi_\eta)^{-1} + (\Pi_0 H \Pi_0) \cdot (\Pi_0 S \Pi_\eta) \cdot (\Pi_\eta H \Pi_\eta)^{-1} \quad (8.22)$$

The solution to this equation is found perturbatively by the recursion method,

$$\Pi_0 S \Pi_\eta = i \sum_{n=0}^{\infty} (\Pi_0 H \Pi_0)^n \cdot (\Pi_0 H \Pi_\eta) \cdot (\Pi_\eta H \Pi_\eta)^{-n-1} \qquad (8.23)$$

Notice that the order of magnitude of $(\Pi_0 H \Pi_0) \cdot (\Pi_\eta H \Pi_\eta)^{-1}$ is t/U and therefore equation (8.23) is the real strong coupling limit expansion which we were after. Making further approximations like $(\Pi_0 S \Pi_\eta) \sim U \cdot \Pi_\eta$, we eventually obtain the following expression for the off-diagonal component of the generator S, to lowest order,

$$\Pi_0 S \Pi_\eta = i(\Pi_0 H \Pi_\eta) \cdot (\Pi_\eta H \Pi_\eta)^{-1} \sim \frac{i}{U} \Pi_0 H \Pi_\eta \qquad (8.24)$$

As we explained above, the off-diagonal terms of S fully determines the diagonal parts of the effective hamiltonian, which from (8.24) turn out to be,

$$\Pi_0 H_{eff} \Pi_0 = \Pi_0 H \Pi_0 - \frac{1}{U} \Pi_0 H \Pi_\eta \Pi_\eta H \Pi_0 \qquad (8.25)$$

$$\Pi_\eta H_{eff} \Pi_\eta = \Pi_\eta H \Pi_\eta + \frac{1}{U} \Pi_\eta H \Pi_0 \Pi_0 H \Pi_\eta \qquad (8.26)$$

As we are only interested in the low energy sector, spanned by $\mathcal{H}_0$, of the current approximation, we may restrict ourselves to the equation (8.25). Moreover, it can be check that the only projector which is relevant in this equation is only Π_1, the one corresponding to the subspace of one doubly occupied site.

In the low energy sector, the first term in (8.25) is a kinetic term T_h which transfers one electron from a singly occupied site to an empty one and it can be given the following form,

$$T_h = - \sum_{<ij>\sigma} t_{ij} (1 - n_{i,-\sigma}) c_{i\sigma}^\dagger c_{i\sigma} (1 - n_{i,-\sigma}) \qquad (8.27)$$

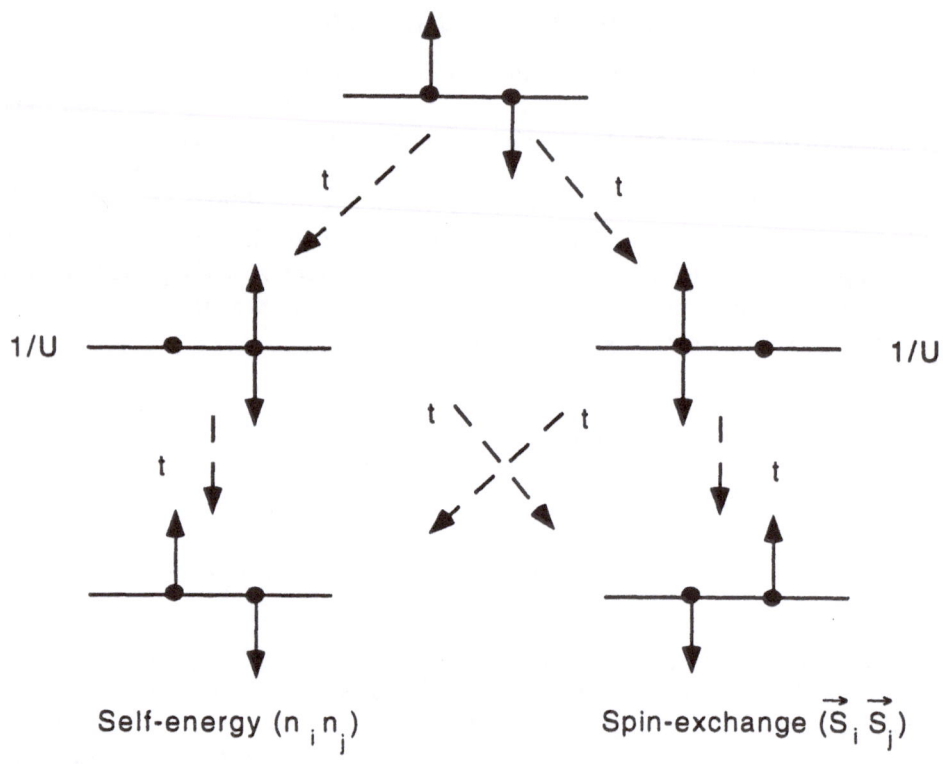

Fig. 8.2. Qualitative description of the origin of the exchange interaction in the $t - J$ model.

As for the second term $-\frac{1}{U}\Pi_0 H \Pi_\eta H \Pi_0$, it can be decomposed into two parts [1],

$$\Pi_0 H \Pi_\eta H \Pi_0 = \Pi_0 (H^{(1)} + H^{(2)}) \Pi_0 \tag{8.28}$$

where

- $H^{(1)}$: This is a two-site term corresponding to an electron virtually hopping from a site i to a site j and backwards to site i. In figure 8.2 we describe this procces and it is shown that the order of this term is t^2/U. The explicit form of this term can be seen to be,

$$H^{(1)} = \sum_{<ij>} \sum_{\sigma\tau} |t_{ij}|^2 (1 - n_{i,-\sigma}) c^\dagger_{i\sigma} c_{j\sigma} n_{j,-\sigma} n_{j,-\tau} c^\dagger_{j\tau} c_{j\tau} (1 - n_{j,-\tau}) \tag{8.29}$$

As a matter of fact, this hamiltonian can be given a further illuminating expression,

$$H^{(1)} = \sum_{<ij>} 2|t_{ij}|^2 \left\{ \mathbf{S}_i \cdot \mathbf{S}_j - \frac{1}{4} \sum_{\sigma\sigma'} n_{i\sigma}(1 - n_{i,-\sigma})n_{i\sigma'}(1 - n_{i,-\sigma'}) \right\}$$

(8.30)

where $\mathbf{S}_i$ are the spin operators associated to the electrons,

$$\mathbf{S}_i \equiv \frac{1}{2} \sum_{\sigma\sigma'} c_{i\sigma}^\dagger \boldsymbol{\sigma}_{\sigma\sigma'} c_{j\sigma'}$$

(8.31)

• $H^{(2)}$: This is a three-site term corresponding to an electron virtually hopping from a site i to a site j and then to a site k different from the initial one. The order of this term is the same as the previous one and its expression is,

$$H^{(2)} = \sum_{<ijk>} \sum_{\sigma\tau} t_{ij}t_{jk}(1 - n_{i,-\sigma})c_{i\sigma}^\dagger c_{j\sigma} n_{j,-\sigma} n_{j,-\tau} c_{j\tau}^\dagger c_{k\tau}(1 - n_{k,-\tau})$$ (8.32)

Altogether the effective hamiltonian in the low-energy sector takes the form (we drop the Gutzwiller projectors to lighten the notation),

$$H_{eff} = T_h + \sum_{<ij>} J_{ij} \left\{ \mathbf{S}_i \cdot \mathbf{S}_j - \frac{1}{4}n_i n_j \right\} - \frac{1}{U} \cdot H^{(2)}$$

(8.33)

The term proportional to $J_{ij} = 4|t_{ij}|^2/U$ is known as Anderson's exchange energy term and its origin is the virtually hopping processes described in figure 8.2. Notice that the factor 4 instead of 2 in J_{ij} comes about because the sum in (8.33) is over unordered pairs sites.

It is worthwhile to pointing out that the strong coupling limit of the Hubbard model is not exactly the t-J model, as usually stated,

$$H_{t-J} = -t \sum_{<ij>,\sigma} (c_{i\sigma}^\dagger c_{j\sigma} + h.c.) + J \sum_{<ij>} (\mathbf{S}_i \cdot \mathbf{S}_j - \frac{1}{4}n_i n_j)$$

(8.34)

for there is an extra term due to $H^{(2)}$ which being of the same order t^2/U, cannot be strictly neglected. Little is known about the estructure and implications of this term which is currently under study [1].

Interestingly enough, there is an important situation in which this extra term can be eliminated. This situation occurs when the system is at half-filling. This is so because of the presence of the Gutzwiller projector Π_0 in the low energy sector of the effective hamiltonian, preventing any final event with doubly occupied sites to happen. In the same fashion, the kinetic term is neither relevant at half-filling. Therefore,

Table 8.1. Solvability of the Heisenberg and Hubbard models depending on the spa-tial dimension.

	Heisenberg	Hubbard
	Heisenberg	Hubbard
$d = 1$	solvable	solvable
$d = 2, 3$	non-solvable	non-solvable

$$H_{eff}^{h-f} = \sum_{<ij>} J_{ij}(\mathbf{S}_i \cdot \mathbf{S}_j - \frac{1}{4}n_i n_j) \tag{8.35}$$

the Hubbard model is equivalent up to second order in t/U to an antiferrromag-net Heisenberg model supporting the idea that the magnetic properties are of relevant to this model, even though they were originally discarded for simplicity by Hubbard in his first studies.

We have thus obtained a hierarchy of models in which the Hubbard model represents the one with highest complexity. We can express this fact by the following relation of models and the dimension of the space of its states for a given system with N electrons,

$$\text{Heisenberg Model} \quad \subset \quad \text{t-J Model} \quad \subset \quad \text{Hubbard Model}$$
$$2^N \qquad\qquad 3^N \qquad\qquad 4^N$$

In the Heisenberg model there are two possible states per site: up and down. In the t-J model there is the additional possibility of having one site empty, while in the Hubbard model it is possible, in addition, to have a doubly occupied site.

8.2 Exact Results for the Hubbard Model and its Strong Coupling Limit Relatives

The outcome of the strong coupling limit studied in the previous section is that the solution of the Hubbard model in this regime goes through the solution of the Heisenberg model. Unfortunately, in the dimensions of physical interest $d = 2, 3$ the Heisenberg model is not known to be exactly solvable. The situation, as far as the solvability is concerned, is ilustrated in the following table,

It looks as if the fate of this to models is intimately related. Further in this section we shall see more examples of this issue.

Despite that the panorama stated in the previous table may look grief, there are a number of exact results which are known even in dimensions higher than one.

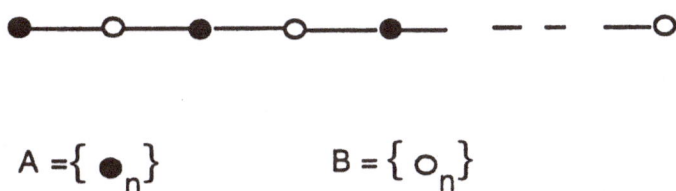

$$A = \{ \bullet_n \} \qquad\qquad B = \{ \circ_n \}$$

Fig. 8.3. One dimensional open bipartite lattice.

The theorems to be stated below can be formulated on very general grounds. To start with, we introduce following Lieb the concept of *graph* denoted by Λ as a collection of sites and links which connect the vertices. The reason to introduce the notion of graph instead of working with the usual lattice is to emphasize the fact that most of the results are independent of whether the system exhibits translational invariance or not.

Given the graph we can associate to it the *hopping matrix* T whose elements t_{xy} are defined according to the convention,

$$t_{xy} = 0 \ \text{ if x,y are sites not connected by a link,} \quad t_{xx} \equiv 0 \qquad (8.36)$$

The matrix T is always self-adjoint for it is actually the discrete version of the Laplacian. However, in the presence of an external magnetic field taking values on the links of the graph, the hopping matrix may have complex entries.

Among the different types of graphs there is a class of them which is worthy to single out,

Bipartite Graph The bipartite graph Λ is the disjoint union of two subsets $\Lambda = A \cup B$ there is no link between x, y if $x \in A$ and $y \in A$ or if $x \in B$ and $y \in B$.

Some examples of bipartite graphs are given below.

- In $d = 1$.

  The linear open chain depicted in figure 8.3 is always bipartite.

  As for the closed linear chain, it is bipartite whenever the number of sites is even, as it is exemplified in figure 8.4.

- In $d = 2$.

  Examples of bipartite graphs are the square lattice and the honeycomb lattice (see figure 8.5), while the triangular lattice is not bipartite.

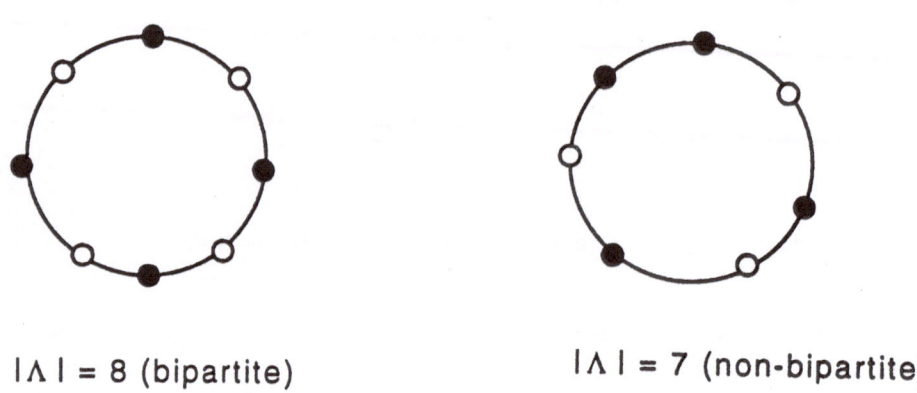

|∧| = 8 (bipartite) |∧| = 7 (non-bipartite)

Fig. 8.4. Examples the closed one-dimensional bipartite and non-bipartite lattices.

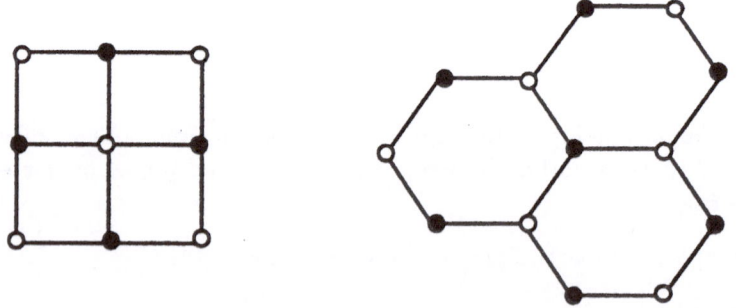

Fig. 8.5. Examples of two-dimensional bipartite lattices: square and hexagonal lattices.

Theorem The necessary and sufficient condition for a graph to be bipartite is that every plaquette of the graphs has an even number of links.

For the sake of completeness, let us write down the Hubbard model hamiltonian on a general graph Λ:

$$H = - \sum_{x,y \in \Lambda, \sigma} t_{x,y} c_{x\sigma}^\dagger c_{y\sigma} + \sum_{x \in \Lambda} U_x (n_{x\uparrow} - \frac{1}{2})(n_{x\downarrow} - \frac{1}{2}) \qquad (8.37)$$

where $t_{xy} = t_{yx}$ $(H^\dagger = H)$.
The factor $(n_x - \frac{1}{2})$ instead of n_x is introduced to facilitate the study of the half-filling case which is defined by the condition,

$$N_e = < \sum_x (n_{x\uparrow} + n_{x\downarrow}) > = |\Lambda| \qquad (8.38)$$

where $|\Lambda|$ is the number of sites in the graph.
At half-filling we have seen in the previous section that the strong coupling limit of the generic Hubbard model is the Heisenberg model which takes the following generic form on a graph Λ

$$H_{eff} = \sum_{x,y \in \Lambda} J_{xy} (\mathbf{S}_x \cdot \mathbf{S}_y - \frac{1}{4}) \qquad (8.39)$$

where

$$J_{xy} = |t_{xy}|^2 (\frac{1}{U_x} + \frac{1}{U_y}) \qquad (8.40)$$

Therefore if the potential is repulsive $U_x > 0 \forall x$ we obtain an antiferromagnetic Heisenberg model $J_{xy} > 0 \forall x, y$.
As far as the ground state of the Heisenberg model is concerned it is possible to state the following theorem [3],

Theorem 1 (Lieb, Lieb-Mattis)

The total spin S for the ground state of a Heisenberg antiferromagnet of spin 1/2 defined on a bipartite graph is given by

$$S = \frac{1}{2} ||A| - |B|| \qquad (8.41)$$

where $|A|, |B|$ are the number of sites in the subgraphs A and B, respectively.

Notice that it is very remarkable that the range of applicablity of this theorem is *every bipartite graph* no matter which is the spatial dimension where the model is defined.

As it happens, the normal situation is $|A| = |B|$ and therefore the ground state of a Heisenberg antiferromagnet is a *spin singlet*.

Along with this theorem there exist a parallel version regarding the nature of the ground state of the Hubbard model [3], [4].

Theorem 2 (Lieb)

1. For a repulsive Hubbard model ($U_x > 0$ and $t_{xy}real$) defined on a bipartite graph, at half-filling and with an even number of electrons, the ground state is unique and its spin S is

$$S = \frac{1}{2} ||A| - |B||$$ (8.42)

2. For an atractive Hubbard model ($U_x < 0$, $\forall x$ and $t_{xy}real$) the ground state on any connected graph and for any even number of electrons (not just at half-filling) is unique and has spin

$$S = 0$$ (8.43)

Another relevant aspect regarding the nature of the ground state of the Heisenberg and Hubbard models is the existence or not of long range order (LRO). Such ordering can be genererically of two types: ferromagnetic or antiferromagnetic, but it is precisely the latter which is of interest in these models.

From now on, and unless otherwise stated, we shall restrict ourselves to homogeneous models

$$t_{xy} = t, \quad U_x = U \quad \forall x, y$$ (8.44)

and to hypercubic lattices in the corresponding dimension.

Theorem 3 For an antiferromagnetic Heisenberg model of s-spins in d-diemensions, there exist LRO in the ground state for the following cases:

1. $d \geq 3$, $s \geq 1/2$
2. $d = 2$, $s \geq 1$

In $d = 2$ and $s = 1/2$ which is the most interesting case from the point of view of its applications to high-T_c superconductivity, the general belief is that there exist LRO but its existence is yet to be proven.

In $d = 1$ and $s = 1/2$ the ground state is the Bethe solution in terms of magnons to be described in chapter 10. This state does not show LRO, quite the contrary it is closer to the concept of resonating valence bond states introduced in the next chapter.

All the theorems stated so far are referred to zero temperature. If the temperature is taken into account then the Hohenberg-Mermin-Wagner theorem

Table 8.2. Known exact results for the AF-Heisenberg model and open problems.

AF-Heisenberg	Ground State	$T > 0$
$d = 1$	$s = 1/2$,	$\not\exists$ LRO
$d = 2$	$s = 1/2$, LRO (?)	$\not\exists$ LRO
$d = 2$	$s \geq 1$, LRO	$\not\exists$ LRO
$d \geq 3$	$s = 1/2$, LRO	$s = 1/2$, LRO (?)
$d \geq 3$	$s \geq 1$, LRO	$s \geq 1$, LRO

Table 8.3. Known exact results for the Hubbard model and open problems.

Hubbard ($U > 0$, h-f)	Ground State	$T > 0$
$d = 1$	$\not\exists$ LRO	$\not\exists$ LRO
$d = 2$	LRO (?)	$\not\exists$ LRO
$d \geq 3$	LRO	LRO (?)

asserts that in $d = 1$ or 2 there not exist LRO. In $d = 3$ and $s \geq 1$ it can be shown that there exist LRO.

In the following table we summarize the known exact results described thus far as well as the open problems along with the corresponding expectations,

Likewise we have relaxed the condition of zero temperature, it is also very interesting to consider situations away from half-filling in the Hubbard model. Maybe the most remarkable result in this regard is due to Nagaoka [5].

Theorem 4 (Nagaoka)

The ground state of the Hubbard model in the limit $U = +\infty$ with one hole ($N_e = |\Lambda| - 1$) is a ferromagnetic state with total spin

$$S = \frac{N}{2} \tag{8.45}$$

For this theorem to hold we have to assume that the graph under consideration has loops supporting non-trivial permutations of the particles. This condition holds for every graph in dimension greater than 1.

When more than a hole is present it has been proven that Nagaoka's result is not longer true, so that the ferromagnetism cited above in the very strong limit is a subtle and controversial matter.

The question as to whether there exist or not LRO in the two-dimensional Hubbard model has been given rise to very fruitful developments.

If the existence of LRO is assumed then it is justified to use the Neel state as the starting point to improve the approximations to the ground state of the system. Let us recall that the Neel state is determined on a bipartite graph by

$$\text{Neel State} = \begin{cases} \uparrow & \text{subgraph } A \\ \downarrow & \text{subgraph } B \end{cases}$$

This state has total spin $\frac{1}{2}||A| - |B||$, and it exhibits LRO.

In this case the ground state, at half-filling, is characterized by being anti-ferrromagnetico:

$$< n_{i\uparrow} >= \frac{1}{2} + (-1)^i m \qquad\qquad (8.46)$$

$$< n_{i\downarrow} >= \frac{1}{2} - (-1)^i m \qquad\qquad (8.47)$$

where $(-1)^i = 1$ if $i \in A$ and $(-1)^i = -1$ if $i \in B$ for the bipartite lattice, and m is the antiferromagnetic order parameter which can be computed within the Hartree-Fock approximation.

In the opposite side there exist another possibility in which the ground state is not assumed to exhibit LRO, as this is the case in 1 dimension. Along this line of study the most atractive proposal has been made by Anderson [6] who establishes a picture of resonating valence bonds (RVB) for the ground state of the AF-Heisenberg model which is the subject of the next chapter.

References

[1] Balachandran, A.P., Ercolessi, E. , Morandi, G., Srivastava, A.M., *Hubbard Model and Anyon Superconductivity* Lecture Notes in Physics. Vol. 38. World Scientific.

[2] Lieb, E.H., Mattis, D.C., *1965, J. Math. Phys. 6, 304.*

[3] Lieb, E.H., *The Hubbard Model: Some Rigorous Results and Open Problems* Proceedings of the Conference "Advances in Dynamical Systems and Quantum Physics" Capri, May, 1993. World Scientific.

[4] Lieb, E.H., *1989, Phys. Rev. Lett.* **62**, *1201* [Errata **62**, 1927 (1989)]

[5] Nagaoka, Y., *1966, Phys. Rev.* **147**, *392*

[6] Anderson, P. W. *1987, Science* **235**, *1196*

9. Resonating Valence Bond States and High-T_c Superconductivity

9.1 Resonating Valence Bond States (RVBS)

The main result stated in the previous chapter is that in the strong coupling limit $(U/t \longrightarrow \infty)$ and at half-filling the Hubbard model can be mapped onto an antiferromagnetic Heisenber model with an exchange coupling constant given by $J = 4t^2/U$. This shows once more the major role played by the magnetic properties of the states in the study of the physics of the Hubbard model at strong coupling which happens to be the appropiate regime in the description of high-T_c superconductors (see chapters 6, 7). Thus, let us explicitly establish the relationship

$$\begin{array}{ccc} \text{Hubbard Ground State} & & \text{AF-Heisenberg} \\ \text{at Half-Filling } n_\uparrow + n_\downarrow = 1/2 & \leftrightarrow & \text{Ground State} \end{array}$$

Unfortunately, the exact structure of the AF-Heisenberg ground state in $d = 2$ dimensions is unknown, though recent rumerical results [1] point out towards a state exhibiting long range order (LRO) of antiferromagnetic type.

In 1987 Anderson revived an old idea proposed early by himself in 1973 and suggests that the ground state of the Heisenberg antiferromagnet in 2 dimensions could have a structure similar to what happens in 1 dimensions where the state is a linear superposition of nearest-neigbours singlet states.

To be more precise and advancing some ideas to be fully developed in the next chapter, the solution of the Heisenberg model in 1 dimensions was found by Bethe in 1931 [2] and can be expressed as a superposition of spin waves called "magnons",

$$|\psi_0> = \sum_P A_{k_{P_1},\dots,k_{P_{N/2}}} e^{k_{P_1}x_1 + \dots + k_{P_{N/2}}x_{N/2}} |x_1,\dots,x_{N/2}> \tag{9.1}$$

where the sum extends over all the permutations of the group $S_{N/2}$, k_i are the moments of the magnons,

$$A_{k_{P_1},\dots,k_{P_{N/2}}} \equiv \text{ is the scattering amplitude of } N/2 \text{ magnons}$$

and

$$|x_1, \ldots, x_{N/2}> = | \uparrow \ldots \overset{x_1}{\downarrow} \ldots \overset{x_{N/2}}{\downarrow} \ldots \uparrow>$$

are the states spanning the Hilbert space of states in which there are $N/2$ overturned spins.

Notice that in this ground state we may find all the possible combinations of $N/2$ spins up and $N/2$ spins down,

$$\uparrow \downarrow \uparrow \downarrow \downarrow \uparrow \downarrow \uparrow \ldots, \quad \text{etc.}$$

Therefore, the picture emerging from this one-dimensional ground state has nothing to do with the concept of long range order, at least in an obvious way. As a matter of fact, the Bethe state is the quite the opposite to the concept of perfectly ordered Neel state:

$$|\text{Neel } 1 > = | \uparrow \downarrow \uparrow \downarrow \ldots \uparrow \downarrow>$$

$$|\text{Neel } 2 > = | \downarrow \uparrow \downarrow \uparrow \ldots \downarrow \uparrow>$$

Notice that strictly speaking there exist two possible Neel states related one another by one lattice site shift.

These Neel states are exact ground states of the Ising model hamiltonian in 1 dimensions,

$$H_{\text{Ising}} = J \sum_{i=1}^{N} \sigma_i^z \sigma_{i+1}^z, \quad J > 0 \tag{9.2}$$

and the classical energy of the Neel ground state is

$$E_0^{\text{Neel}} = -JN \tag{9.3}$$

When we go to the Heisenberg hamiltonian, what we are doing is to turn on "quantum fluctuations" which perturbate the Neel state causing the spins to fluctuate in such fashion as to totally disappear the perfect one-dimensional antiferromagnetic order.

$$H_{\text{Heisenberg}} = J \sum_i (\sigma_i^z \sigma_{i+1}^z + 2(\sigma_i^+ \sigma_{i+1}^- + \sigma_i^- \sigma_{i+1}^+)) \tag{9.4}$$

The second term in brackets in the RHS undiagonalizes the Neel state.

Model	Ising	$\longrightarrow$	Heisenberg
Ground State	Neel	$\longrightarrow$	Bethe

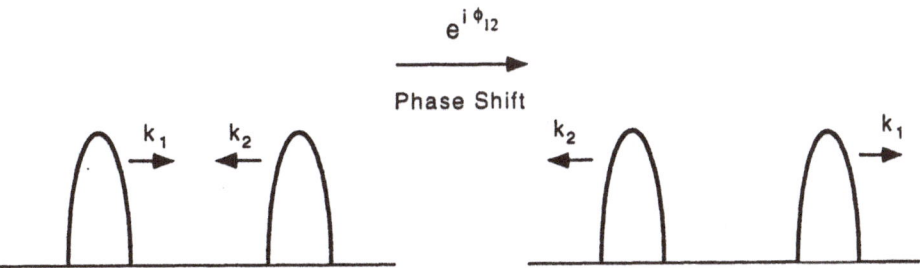

Fig. 9.1. Elastic scattering of two magnons in one dimension.

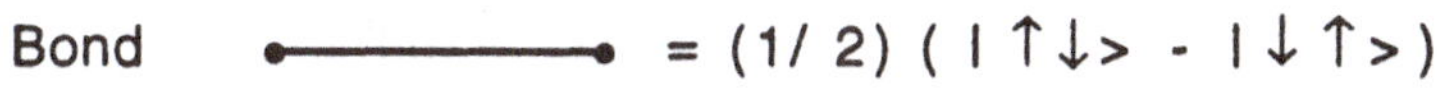

Bond •————————• $= (1/2) (| \uparrow\downarrow> - | \downarrow\uparrow>)$

Fig. 9.2. A valence bond state constructed out of two spin singlets.

In $d = 2$ we do not know the exact form and nature of the ground state $|\psi_0 >$ for the Heisenberg model. It is not an easy task trying to generalize the formula for the one-dimensional Bethe state. The reason for this to happen is because the one-dimensional magnons associated to the Heisenberg model undergo elastic scattering (see figure 9.1),

This property of elastic scattering makes the model integrable in $d = 1$ dimensions. Now, the would-be "magnons" in $d = 2$ dimensions do not scatter elastically [1] so that the equivalent to the Bethe ansatz in two dimensions is doomed to failure.

Confronting these difficulties, Anderson proposes to represent the ground state not like a sea of magnons but instead a sea, or better to say a "liquid", of spin singlets resonating among different configurations. The resonating valence bond states (RVBS) construction rely on two main ideas:

Valence Bond The spin singlet or valence bond is grafically represented in figure 9.2.

---

[1] The Heisenberg model in $d = 2$ dimensions is thought to be non-integrable as is it the case with the majority of bidimensional hamiltonians.

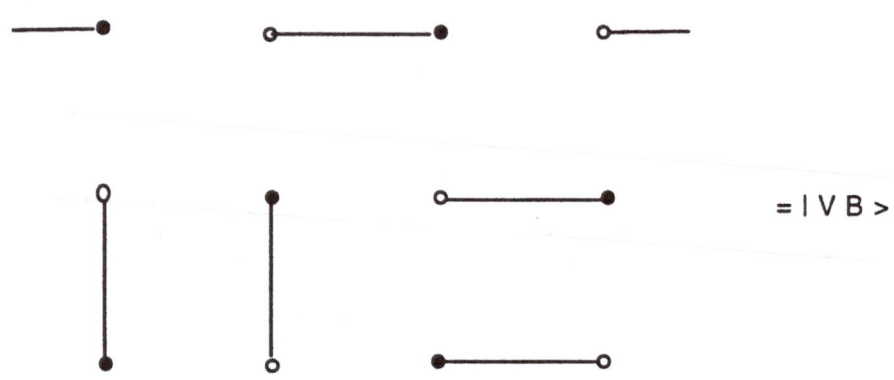

Fig. 9.3. A piece of lattice showing several valence bond states.

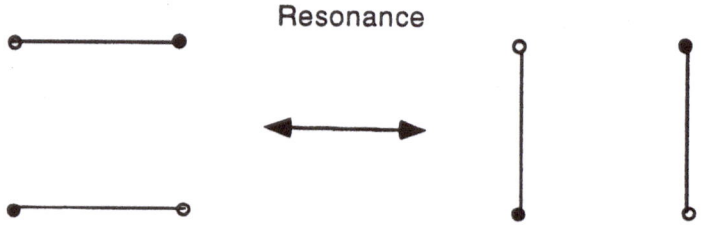

Fig. 9.4. The resonating mechanism in a couple of valence bond states.

Mathematically, the *valence bond* state is identified with the *antisymmetric tensor* $\epsilon_{\sigma_1\sigma_2}$ with $\sigma_i = \pm$ for the valence bond state is expressed as

$$|VB> \equiv \sum_{\sigma_1\sigma_2} \epsilon_{\sigma_1\sigma_2}|\sigma_1\sigma_2 > \qquad (9.5)$$

An example of valence bond state on a square lattice is depicted in figure 9.3.

Notice that each spin on every vertice of the lattice is forming a singlet state with some of its nearest-neighbour vertices.

Resonance   Two nearest-neighbour bonds are allowed to resonate among themselves, causing the energy of the configuration to decrease. This resonance phenomenon is illustrated in figure 9.4.

Altogether, a RVB state is constructed out the superposition of VB states resonating among themselves, namely,

$$|RVB> \equiv |VB> +|\tilde{V}\tilde{B}> +\ldots \tag{9.6}$$

This somehow is reminescent of the old Pauling resonace of chemical bonds like
in the benzene.
The description so far outlined deserves some remarks.

Remark.
    The VB states form an overcomplete basis of the Hilbert subspace of sin-
glet states in the Heisenberg model and therefore the ground state is a linear
combination of this type of states.

Remark.
    Using RVB states an estimation of the ground state energy can be carried
out using the variational method. In $d = 1$ the resulting energy is off the exact
energy by a 10 % (see below).

There exist a concrete example served as the testing ground where the RVB
ideas where applied in an explicit fashion. This is the planar triangular lattice
on which an anisotropic Heisenberg model is defined [3]. The hamiltonian is,

$$H_{\text{Heisenberg}} = J \sum_i (\sigma_i^z \sigma_{i+1}^z + 2\alpha(\sigma_i^+ \sigma_{i+1}^- + \sigma_i^- \sigma_{i+1}^+)) \tag{9.7}$$

where $0 \leq \alpha \leq 1$ is the anisotropic parameter. The nature of the model crucially
depends on this parameter. The two extreme cases are,

$$\begin{cases} \alpha = 0 & \longrightarrow \quad \text{Ising} \\ \alpha = 1 & \longrightarrow \quad \text{Heisenberg} \end{cases} \tag{9.8}$$

This model is also known as XXZ model (XXX $\equiv$ isotropic Heisenberg) or
also Heisenberg-Ising model.
Let us point out that the triangular lattice is a "tripartite lattice". This means
that the lattice can be decomposed into 3 sublattices A,B and C in such way
that every point in the lattice is only connected to points of the remaining two
lattices, as can be seen in figure 9.5.
Assuming firstly that the spins behave classically, we can search for configurations as the ones depicted in figure 9.6, which is characterized by an angle say
θ. The value which produces the minimum energy is easily found to be

$$\cos\theta = \frac{1}{1+\alpha} \tag{9.9}$$

The two extreme values corresponding to (9.8) are given by

$$\begin{cases} \alpha = 0 & \Longrightarrow \quad \theta = 0 \\ \alpha = 1 & \Longrightarrow \quad \theta = \frac{\pi}{3} \end{cases} \tag{9.10}$$

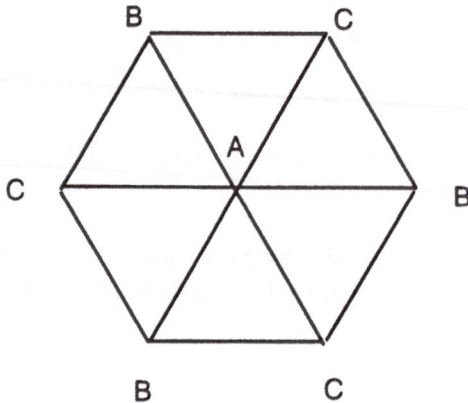

Fig. 9.5. The triangular lattice considered as a tripartite lattice.

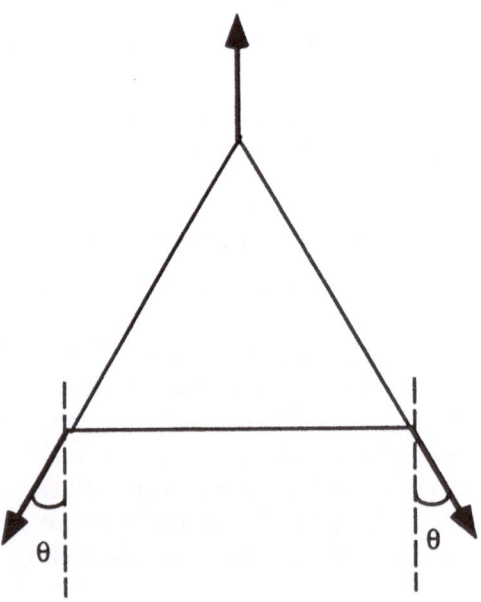

Fig. 9.6. Mean field theory treatment of the Heisenberg antiferromagnet in the triangular lattice.

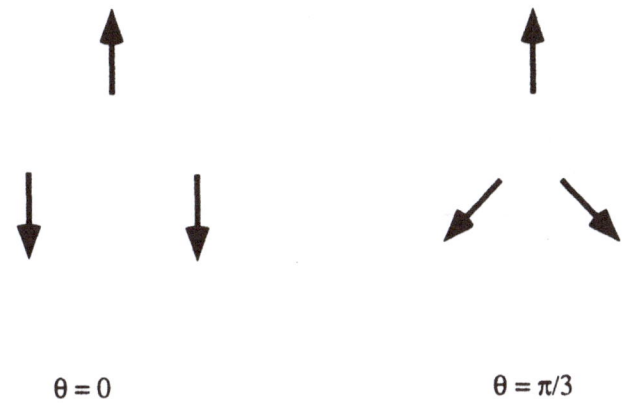

$\theta = 0$ $\theta = \pi/3$

Fig. 9.7. The two extreme classical unit cell configurations of the Heisenberg antiferromagnet in the triangular lattice.

These values correspond to the planar configurations shown in figure 9.7. The configuration associated to $\theta = \frac{\pi}{3}$ may be considered as the Neel state in a triangular lattice for it corresponds to a situation in which the spins are maximally opposed to each other.

The classical energy per site for an anisotropic Heisenberg model of spins with magnitude s is given by

$$E^{cl}/N = -Js^2\frac{1 + \alpha + \alpha^2}{1 + \alpha} \tag{9.11}$$

We are interested in studying the disturbances on the Ising model caused by the anisotropic part

$$H^{\pm} = \sum_i 2\alpha(\sigma_i^+ \sigma_{i+1}^- + \sigma_i^- \sigma_{i+1}^+) \tag{9.12}$$

of the hamiltonian. Thus, we shall be treating the anisotropic parameter α as a small parameter and henceforth perform a perturbative expansion,

$$E^{cl}/N \approx -Js^2 - Js^2\alpha^2 \tag{9.13}$$

where the first term corresponds to the classical Ising energy and the second is the classical anisotropic correction.

We may wonder whether it is possible somehow to lower the classical correction which is order $O(\alpha^2)$. The answer is positive for a triangular lattice. To see this, let us notice that taking the classical unit cell configuration of spins depicted in figure 9.7 ($\theta = 0$ case), we have to generate the whole configuration on the entire lattice by translating this unit cell all over the plane. In doing so we realize that the triangular lattice exhibits frustration as shown by the two

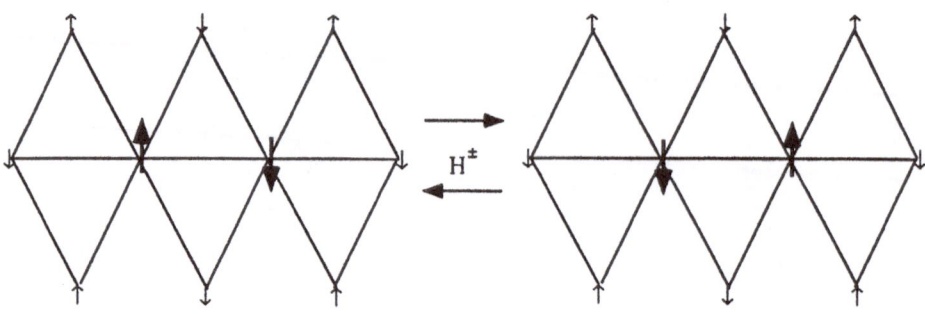

Fig. 9.8. Resonating valence bond states of the triangular lattice.

possible singlet spin configurations which are different but energetically equivalent. These two states resonate among one another and they are connected by the quantum fluctuation hamiltonian $H^{\pm}$. These are the RVB states of the triangular lattice. In addition to the singlet states it is also possible to have triplet spin states just by placing the inner spins of figure 9.8 parallel to each other.

If we now restrict the hamiltonian to the subspace of singlet and triplet states it takes the form,

$$H = \begin{pmatrix} E_0 & \alpha \\ \alpha & E_0 \end{pmatrix} \tag{9.14}$$

and α is again a small parameter. Diagonalizing (9.14) we inmediately find the triplet and singlet energies,

$$E = \begin{cases} E_0 + \alpha & \implies \textit{Triplet} \\ E_0 - \alpha & \implies \textit{Singlet} \end{cases} \tag{9.15}$$

We see in this way that it is more likely to form a singlet spin state than a triplet. Moreover, the RVB states are more favorable to be formed than the classical situation (9.13) for the former is order α in perturbation theory while the latter is order α^2.

The outcome of this discussion for the triangular lattice is that it is possible to lower the energy of the ground state when combining different valence bond states, i.e., by resonance.

Notice that so far the parameter α has been restricted to be small while the case of interest is the isotropic Heisenberg model $\alpha = 1$. However, Anderson and Fazekas [3] argued that the same conclusions remain valid in the limit $\alpha \longrightarrow 1$. They proposed the following identification,

Ground State of   Resonating Valence
AF-Heisenberg Model $\equiv$ Bond States
on Triangular Lattice   (Quantum Spin Liquid)

Remark.

The existence of RVB states on a given lattice depends crucially on the type of lattice at hand, namely its coordination number. In fact, we have seen that such states are likely to happen on a triagular lattice for it allows frustration to be formed, while on a square lattice we already mentioned that numerical analysis show evidence in favor of antiferromagnetic correlations, i.e., LRO.

9.2 Anderson's RVB Ground States in High-T_c Superconductors

In his famous 1987 paper [4] Anderson proposed the resonating valence bond states as the most correct construction to unveil the high-T_c superconductivity mechanism, which in turn revived the interest on this class of states.

Anderson's Hypothesis This may be stated in the following equivalence,

$$La_2CuO_4\text{-}Ground\ State\ (undoped) \equiv Resonating\ Valence\ Bond\ State$$

Moreover, he also establishes an analogy between the RVB states and the BCS states in the ordinary superconductivity, which sounds reasonable for after all the main point is to explain the origin of a new type of superconductivity (however, this analogy is just a technicality at this moment).

RVB / BCS Relationship The precise statement of this relationship is,

$$\Psi_{RVB} = \Pi_0 \Pi_{N/2} \Psi_{BCS} \qquad (9.16)$$

where Π_0 is the Gutzwiller projector introduced in chapter 8 which prevents any lattice site from being doubly occupied, and $\Pi_{N/2}$ is a projector over states with N electrons defined as (see [4], [5]),

$$\Pi_{N/2}\Psi_{BCS} = \frac{1}{n!}(b^\dagger)^{N/2}|0> \qquad (9.17)$$

In order to construct the resonating valence bond state Ψ_{RVB} it is convenient to introduce an operator $b^\dagger_{ij}$ which creates bonds out of two lattice sites i and j,

$$b_{ij} \, | \, 0 > = \quad \text{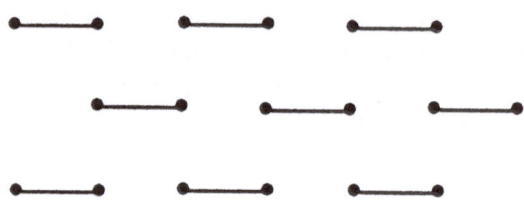}$$

Fig. 9.9. The operator that creates a valence bond state.

Fig. 9.10. The spin-Peierls state.

$$b^\dagger_{ij} = \frac{1}{\sqrt{2}}(c^\dagger_{i\uparrow}c^\dagger_{j\downarrow} - c^\dagger_{i\downarrow}c^\dagger_{j\uparrow}) \tag{9.18}$$

The action of this operator is illustrated in figure 9.9. Notice as well that $b^\dagger_{ji} = b^\dagger_{ij}$.

An important property about the bond is its mobility. There exists a particular case in which the bonds are fixed giving rise to a type of state known as "spin-Peierls" state depicted in figure 9.10. It consists of a crystal lattice of ordered singlets.

Observe that in this extreme case there are not resonating states for these require the existence of pairs of neighbour bonds (see figure 9.10) which are absent in the "spin-Peierls" state .

When the bonds are moving a mean field theory description can be undertaken based on the operator $b^\dagger$ which is the creator of moving bonds. Is is defined as,

$$b^\dagger = \sum_{i,j} g_{ij} b^\dagger_{ij} \tag{9.19}$$

where g_{ij} is an undertermined function by now, which by translational invariance can only depend on the distance between two sites,

$$g_{ij} = g_{ji} = g(|\mathbf{R}_i - \mathbf{R}_j|) \tag{9.20}$$

and in order to avoid double occupancy we also demand

$$g_{ii} = g(0) = 0 \tag{9.21}$$

It is more convenient to work in Fourier space, so in momentum space representation the moving-bonds creation operator takes the form,

$$b^\dagger = \sum_{\mathbf{k}} a(\mathbf{k}) c^\dagger_{\mathbf{k}\uparrow} c^\dagger_{-\mathbf{k}\downarrow} \tag{9.22}$$

where the Fourier coefficient is given by

$$a(\mathbf{k}) \sim \sum_{\mathbf{R}_i} e^{-i\mathbf{k}\cdot\mathbf{R}_i} g(\mathbf{R}_i) \tag{9.23}$$

It has to comply with the equivalent condition to (9.21), namely,

$$\sum_{\mathbf{k}} a(\mathbf{k}) = 0 \tag{9.24}$$

As it happens to be the case, a valence bond state is a boson and then it is possible to form a condensate of $N/2$ bonds ($\approx N$ electrons) in the following way,

$$\Psi = (b^\dagger)^{N/2} |0> \tag{9.25}$$

where $|0>$ is the vacuum state in the Fock space generated by the operators $c^\dagger_{i\uparrow}, c^\dagger_{j\downarrow}$. The problem now has to do with the condition $\sum_{\mathbf{k}} a(\mathbf{k}) = 0$ for it is not enough as to eliminate the double occupancy situations when multiplying more than one $b^\dagger$ operator. Thus, it is mandatory to carry out a Gutzwiller projection leading to,

$$\Psi_{RVB} = \Pi_0 (b^\dagger)^{N/2} |0> \tag{9.26}$$

Alternatively, one could have employed the quasifermionic $d^\dagger_{i\sigma}$ operators which were introduced in the description of the $t - J$ model in chapter 8, namely,

$$d^\dagger_{i\sigma} = \Pi_0 c^\dagger_{i\sigma} \Pi_0 = c^\dagger_{i\sigma}(1 - n_{i,-\sigma}) \tag{9.27}$$

and then to write $b^\dagger_{ij}$ in terms of the $d^\dagger$'s in such fashion that the Gutzwiller projector would have been automatically implemented.

The relationship between the Ψ_{RVB} state and the Ψ_{BCS} can be obtained from the following observation; as the moving-bond-states creation operator $b^\dagger$ takes the form (9.22)

$$b^\dagger = \sum_{\mathbf{k}} a(\mathbf{k}) b^\dagger_{\mathbf{k}} \tag{9.28}$$

with $c^\dagger_{\mathbf{k}\uparrow} c^\dagger_{-\mathbf{k}\downarrow}$ being precisely a Cooper pair creation operator [2]. Given the fermionic character of the creation operator $(b^\dagger_{\mathbf{k}})^2 = 0$, it is possible to express the BCS state as

---

[2]This operator creates a pair of electrons with opposite spins and momenta.

$$\Psi_{BCS} = e^{\sum_k a(k) b_k^\dagger} \psi_0 \tag{9.29}$$

From this eventually we obtain,

$$\Pi_{N/2} \Psi_{BCS} = \left(\sum_k a(k) b_k^\dagger \right)^{N/2} \psi_0 \tag{9.30}$$

$$\Psi_{RVB} = \Pi_0 \Pi_{N/2} \Psi_{BCS} \tag{9.31}$$

All these ideas can be most suitably implemented within a mean field theory approximation, for in particular the Hubbard hamiltonian at half-filling can be entirely expressed by means of the moving-bonds creation operators:

$$H_{h-f} = J \sum_{<ij>} (\mathbf{S}_i \cdot \mathbf{S}_j - \frac{1}{4}) = -J \sum_{<ij>} b_{ij}^\dagger b_{ij} \tag{9.32}$$

This identity can be directly checked using the Pauli matrices relationship $\sigma_{\alpha\beta} \cdot \sigma_{\gamma\delta} = 2\delta_{\alpha\delta}\delta_{\beta\gamma} - 2\delta_{\alpha\delta}\delta_{\gamma\delta}$. However, it is more illuminating to introduce the spin singlet operators which precisely appear in the effective hamiltonian at half-filling,

$$e_{ij} = -(\mathbf{S}_i \cdot \mathbf{S}_j - \frac{1}{4}) \tag{9.33}$$

Explicitly the action of these projection operators is

$$e_{ij} \left\{ \begin{array}{l} |\uparrow\uparrow> \\ |\downarrow\downarrow> \\ (|\uparrow\downarrow> + |\downarrow\uparrow>) \end{array} \right\} = 0 \tag{9.34}$$

$$e_{ij}(|\uparrow\downarrow> - |\downarrow\uparrow>) = |\uparrow\downarrow> - |\downarrow\uparrow> \tag{9.35}$$

where i, j denote nearest-neigbour sites. Thus, the effective hamiltonian is precisely the superposition of all these singlet-state projectors,

$$H_{h-f} = -J \sum_{ij} e_{ij} \tag{9.36}$$

and in this fashion the identity (9.32) is manifest for $e_{ij} = b_{ij}^\dagger b_{ij}$.
In passing, it is worthy to point out that these projector operators $e_{i,i+1} \equiv e_i$ satisfy the Temperly-Lieb-Jones algebra [6] ,

$$e_i^2 = \beta e_i$$

$$e_i e_{i+1} e_i = e_i \tag{9.37}$$

These relations are graphically proved in figure 9.11.
Identity (9.32) may look misleading for it is quadratic in creation/destruction operators and we may be tempted to conclude that the model is free and hence integrable. Obviously this is not the case as have stated several times by now.

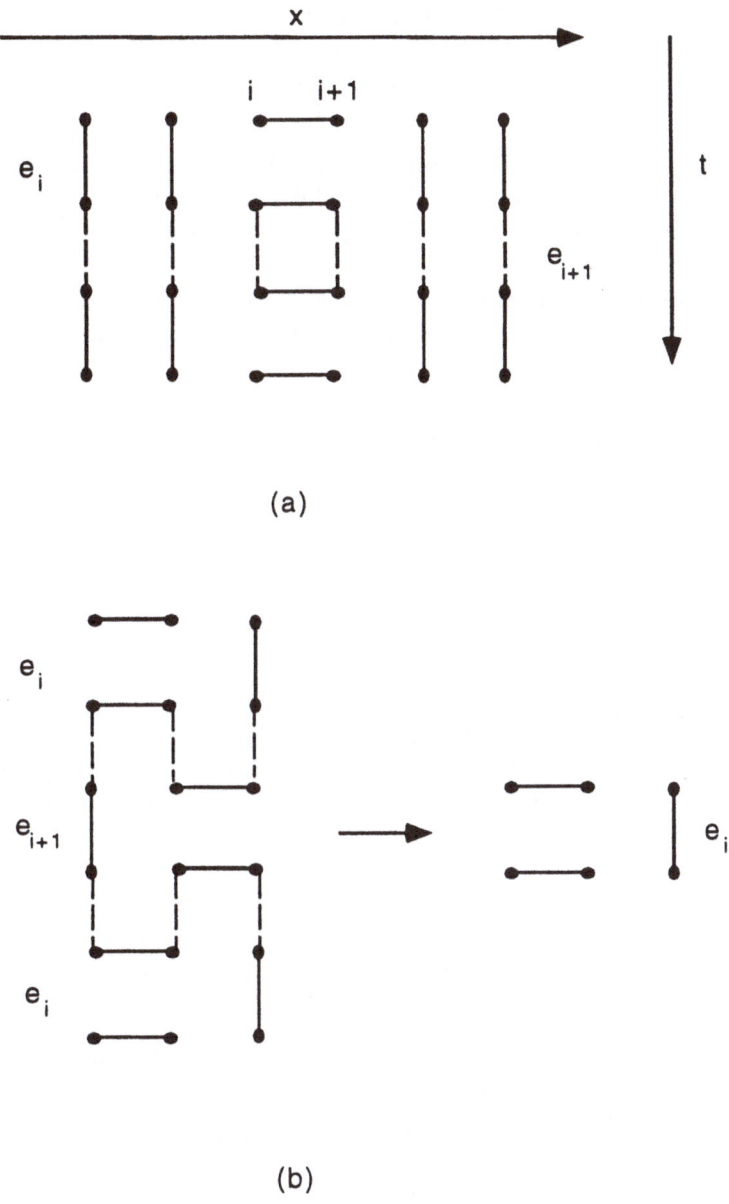

Fig. 9.11. The graphical description of the generators of the Temperly-Lieb-Jones algebra.

The key point to notice is that the bond operators do not satisfy canonical commutation relations (CCR), but instead they obey

$$[b_{ij}, b_{jk}^\dagger] = \delta_{ik}(1 - \frac{n_j}{2}) - \frac{1}{2}\sum_\sigma c_{k\sigma}^\dagger c_{i\sigma} \tag{9.38}$$

for i, j and j, k nearest-neigbours, while all the others commutators vanish. At half-filling the commutators (9.38) simplifly for they vanish, but nevertheless they are not canonical. This fact is due to the nontrivial dynamics of the bond operators in this system which is not free.

Away from half-filling we should have to include a hopping term to account for the holes present in the system thereby the effective hamiltonian within this approximation would be

$$H_{eff} = -t\delta \sum_{<ij>,\sigma} (c_{i\sigma}^\dagger c_{j\sigma} + h.c.) - J \sum_{<ij>} b_{ij}^\dagger b_{ij} \tag{9.39}$$

The negative sign in the J-term of the hamiltonian suggests that effectively the bonds can condensate in the a ground state, such as we have previously described. The mean field theory approximation relies on the assumption of a non-vanising average value for the bond state, namely,

$$\Delta_{ij} = \sqrt{2} < b_{ij} > \tag{9.40}$$

This strongly resembles the BCS treatment of ordinary superconductivity. Following this analogy, Anderson and coworkers [7] went on to develope a Hartree-Fock theory which is not our purpose to present here but some important conclusions are:

1. There exists a pseudo Fermi surface at half-filling. The system behaves as an insulator because even though there are moving bonds, these do not transport charge. There are also spin excitation which happen to be gapless.

2. Away from half-filling, upon doping, the system becomes metallic (that is, there is charge transportation). Then, a gap opens on the pseudo Fermi surface so that the system becomes a superconductor.

9.3 Excitation Spectrum of RVB States

One of the reasons why the knowledge of the structure and nature of the ground state for any system is of paramount importance is to determine the type of elementary excitations that this ground state can support. The elementary excitations are eventually the states to be confronted someway with experimental results if the model is going to be successful explaining the physics under consideration. Let us point out that the elementary excitations can be collective excitations whose nature may have nothing to do with the elementary particles entering in the formulation of the corresponding hamiltonian.

We shall be describing now some relevant features of the excited states that a RVB ground state can originate even though this is a highly controversial issue which has produced lots of results which many times are contradictory.

It was Anderson who firstly proposed [4] that the excitation spectrum of high-T_c superconductors should exhibit the phenomenon of spin-charge separation, of which there is some experimental evidence [4], [8]. In the spectrum of the one-dimensional Hubbard model this separation of spin and charge degrees of freedom is manifest as shall be described in chapter 10. This separation due to the different velocities of these two types of excitations is at the basis of the Luttinger liquid behaviour present in the low energy physics of the Hubbard model.

The two types of excitations associated to the spin and charge degrees of freedom are known as spinons and holons.

Definition 9.1 Spinon   *Elementary excitation carrying spin 1/2 but zero charge.*

Definition 9.2 Holon
   Elementary excitation carrying charge but zero spin.

Let us illustrate pictorically how these excitations look like on a one-dimensional lattice in the strong coupling limit. In this regime the double occupancy is unlikely and due to the strong short-range antiferromagnetic interactions a local configuration may look as follows,

$$\cdots \uparrow\downarrow\uparrow\downarrow\uparrow\downarrow\uparrow\downarrow\uparrow\downarrow\uparrow\downarrow \cdots$$

Upon doping with a hole the configurations becomes,

$$\cdots \uparrow\downarrow\uparrow\downarrow\uparrow \; 0 \; \uparrow\downarrow\uparrow\downarrow\uparrow\downarrow \cdots$$

Now a crucial point here is that because the kinetic term of the hamiltonian does not flip spins, it is possible to move the hole independently of the spins in the following way,

$$\cdots \uparrow\downarrow \; 0 \; \uparrow\downarrow \; \uparrow\uparrow \; \downarrow\uparrow\downarrow\downarrow \cdots$$

Likewise it is possible to perform spin exchange without moving the hole so that eventually we arrive to the following configuration where the spin-charge separation is manifest,

$$\cdots \uparrow\downarrow \; 0 \; \uparrow\downarrow\uparrow\downarrow\uparrow\downarrow\uparrow\uparrow\downarrow \cdots$$

Thus, the original configuration of one added hole has evolved into this last configuration in which there is a hole surrounded by one up and one down spins which is the *holon* of positive charge and zero spin, and on the right side there is one overturned spin on a Neel configuration of zero spin, and this is precisely the *spinon* with spin 1/2 and zero charge.

Two remarks are in order regarding this simple picture just described. Firstly, this construction is based on an antiferromagnetic initial configuration and not a RVB state as we want in our present case. Secondly, if we want to extend this picture to a two-dimensional Neel state we drive into difficulties for, as the reader may check as an exercise, the simple spin-charge separation described above fails when it is embedded into 2 dimensions: the hole appears to have spin and the would-be holon and spinon cannot be disentangled into separate entities[3]

Now let us come back to the problem as to how a RVB ground state can support spinon and holon excitations. Kivelson et al. [9] have described qualitatively this phenomenon. The spinon arises when breaking one valence bond thereby creating two dangling bonds with ±1/2 spins which will be described as white and black sites respectively. These are spinons that can move freely over the entiere lattice of valence bond states as independent excitations resembling neutral solitons (see figure 9.12).They do not carry charge for breaking a valence bond does not add any charge to the system. As they happen to be topological in nature, it is convenient to introduce a topological charge ±1 associated to the ±1/2 spinons. For a given configuration the topological charge is a conserved number which can be computed by counting the number of white spinons minus the number of black spinons enclosed in a contour which does not intersect any bond, as in figures 9.12-13,

$$\text{Total Charge} = \sum_{\text{contour}} (\# \bullet - \# \circ) = \sum_{\text{spinons}} (\bullet - \circ)$$

Likewise solitons and antisolitons, white and black spinons can only be annihilated in pairs. The presence of these spinons implies the existence of a long range topological order in a valence bond state.

As for the holons, they appear when doping holes. An added hole amounts to a charged particle with spin 1/2. This is not a holon but it can be combined with an already present spinon into a singlet state which does have charge (see figure 9.13).

When coming to the issue of the statistics of these spinon-holon excitations the subject becomes very controversial and there is not agreement in the literature. Kivelson et al. [9] conclude after an analysis based on adiabatical exchange (braiding) of excitations that spinons are fermions while holons are bosons. However, many authors disagree on this conclusion [10], [11], [12], [13], [14].

Another prediction of Kivelson et al. is the existence of a gap in the spinon spectrum, but this again has been questioned by other authors [4], [7]. This

---

[3]They behave as the two extremes of a string-like dislocation over the antiferromagnetic sea whose length increases as the two extremes move all over.

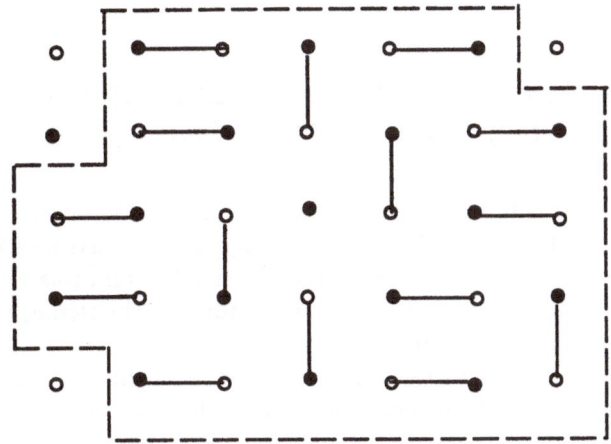

Fig. 9.12. A spinon state from a ground state of valence bond states.

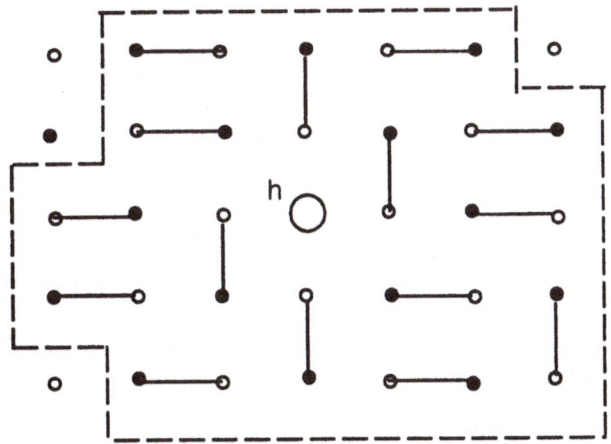

Fig. 9.13. A holon state from a ground state of valence bond states.

would be a remarkable result for in that case one is tempted to think of a condensation of holons as the mechanism of high-T_c superconductivity. However, the charge of the order parameter predicted in this scenario is one eletron charge which is half of the observed charge .

9.4 Other Applications of RSV States: The Majumdar-Ghosh and AKLT Constructions

As it has been already explained, the valence bond states were revived by Anderson [4] in an attempt to describe the magnetic properties of the Heisenberg antiferromagnet ground state in two dimensions. The aim was to explain the insulating normal state of high-T_c superconductors via the strong coupling regime of the Hubbard model at half-filling.

Even though these expectations have not been confirmed to date [5], the ideas of RVB states have evolved independently of these potential applications as a whole subject of interest in itself.

After the attempts described so far in order to construct a RVB ground state for the Heisenberg model in two dimensions, the natural questions arises as to whether is possible to propose any hamiltonian whose ground state is a valence bond state. The answer is positive and twofold depending the approach pursued to achieve such goal.

There exists one approach called the Majumdar-Ghosh hamiltonian [15] in which the building blocks are spin 1/2 matrices and the original idea is to add to the Heisenberg hamiltonian a second term involving next-to-nearest-neigbour interaction, namely,

$$H_{M-G} = \sum_i (\mathbf{S}_i \cdot \mathbf{S}_{i+1} + \frac{1}{2}\mathbf{S}_i \cdot \mathbf{S}_{i+2}) \tag{9.41}$$

It can be shown that this one-dimensional hamiltonian has a valence bond ground state of the form,

$$| > = \epsilon_{\alpha_1 \alpha_2} \epsilon_{\alpha_3 \alpha_4} \cdots \epsilon_{\alpha_{N-1} \alpha_N} |\alpha_1 \ldots \alpha_N > \tag{9.42}$$

where $\alpha_i = \pm 1$, and ϵ_{ij} is the antisymmetric tensor associated to every consequtive valenc bond. This is an exac result.

This result motivates the introduction of a variant of the t-J model that can be denoted by t-J-J' model as,

$$H_{t-J-J'} = -t \sum_i (c_i^\dagger c_{i+1} + h.c.) + J \sum_i (\mathbf{S}_i \cdot \mathbf{S}_{i+1} - \frac{1}{4}n_i n_{i+1}) +$$

$$+ J' \sum_i (\mathbf{S}_i \cdot \mathbf{S}_{i+2} - \frac{1}{4}n_i n_{i+2}) \tag{9.43}$$

This type of hamiltonian may have potential applications in the physics of strongly correlated electrons. Let us mention that as far as the existence of a gap in its spectrum is concerned, this model exhibits a gap depending on the values of the coupling constants: when $J'/J > 0.25$, the gap opens in the spectrum.

There is another alternative to obtain exact VB ground states known as the AKLT construction after Affleck-Kennedy-Lieb-Tasaki which is remarkable not only for the previous feature but also because it allows to construct exact ground states for dimension higher than 1.

Unlike the Majumdar-Ghosh construction, the AKLT construction works within the framework of nearest-neighbour interactions, but considers spin matrices higher than 1/2 as the building blocks. In this fashion this treatment connects with Haldande's earlier work on the existence of a gap in the spectrum of isotropic Heisenberg spin chain depending on the integer or half-integer nature of the spin.

Using non-linear sigma model techniques Haldane predicted [16], [17], [18] the following result:

Haldane's Conjeture In $d = 1$ dimensions, the isotropic AF-Heisenberg model presents the following gap phase structure,

$$s \text{ Half-Integer} \implies \text{NO GAP}$$
$$s \text{ Integer} \implies \text{GAP ("Haldane" gap)}$$

The authors of the AKLT construction were able to present for the first time an isotropic model exhibiting the gap phase predicted by Haldane in one dimension. The model is constructed in two steps. Firstly, the spin-1 building blocks are constructed out of two spin-1/2 variables by symmetrization. Let us denote $\psi_{\alpha\beta}$ the orthogonal basis obtained by taking the symmetric part of the tensor product of two spin 1/2 spaces with variables ψ_α, ψ_β:

$$\psi_{\alpha\beta} = \psi_\alpha \otimes \psi_\beta + \psi_\beta \otimes \psi_\alpha = \psi_{\beta\alpha} \tag{9.44}$$

This building block is graphically shown in figure 9.14. where each lattice site has a sort of internal structure of two points corresponding to the two symmetrized underlying one-half spins.

The second step consists in joining together these type of lattice sites to reproduce the whole lattice. This process is acomplished by establishing valence bonds between consecutive sites. Mathematically, this amounts to the following state

$$\Psi_{VBS} = \psi_{\alpha_1\beta_1}\epsilon^{\beta_1\alpha_2}\psi_{\alpha_2\beta_2}\epsilon^{\beta_2\alpha_3}\psi_{\alpha_3\beta_3}\epsilon^{\beta_3\alpha_4}\psi_{\alpha_4\beta_4} \tag{9.45}$$

This state is graphically depicted in figure 9.14 where it is shown that all bonds are singlet states.

Fig. 9.14. Construction of a valence bond state out of sites occupied by spin-1 states.

The important observation now is that the projection operator over spin-2 states associated to two consequtive sites vanish identically:

$$\Pi^{s=2}\left[\ldots\psi_{\alpha_i\beta_i}\epsilon^{\beta_i\alpha_{i+1}}\psi_{\alpha_{i+1}\beta_{i+1}}\ldots\right]=0 \tag{9.46}$$

The reason for this is because the bond being a singlet contributes zero while the remaining sites on the extremes having spin one-half give spin 0 or 1, but never 2. Thus, the sum of all the local spin-2 projectors acting on the VB state is identically zero,

$$\left\{\sum_i \Pi_{i,i+1}^{s=2}\right\}|VBS>=0 \tag{9.47}$$

Now a clever choice of local hamiltonian $H_{i,i+1}$ is precisely the spin-2 projection operator,

$$H_{AKLT}=\sum_i H_{i,i+1}=\sum_i \Pi_{i,i+1}^{s=2} \tag{9.48}$$

for in this fashion the hamiltonian is obviously positive semidefinite and from (9.47), the valence bond state is the exact ground state. What is left is to express the projector operator in terms of spin matrices as it is customary. This is easily achieved because that operator can only be constructed out ot 1, $\mathbf{S}_i \cdot \mathbf{S}_{i+1}$ and $(\mathbf{S}_i \cdot \mathbf{S}_{i+1})^2$. The result is

$$\Pi_{i,i+1}^{s=2}=\frac{1}{3}+\frac{1}{2}\mathbf{S}_i \cdot \mathbf{S}_{i+1}+\frac{1}{6}(\mathbf{S}_i \cdot \mathbf{S}_{i+1})^2 \tag{9.49}$$

The AKLT hamiltonian thus takes the following form, after droping constant terms,

$$H_{AKLT}=\sum_i \mathbf{S}_i \cdot \mathbf{S}_{i+1}+\frac{1}{3}(\mathbf{S}_i \cdot \mathbf{S}_{i+1})^2 \tag{9.50}$$

The authors of this construction went on to prove the following results [19],

- The hamiltonian has an exact $SO(3)$ symmetry and is translational invariant.

- The model has a unique ground state.

- A gap in the spectrum of the hamiltonian opens inmediately above the ground state.

- The correlation functions in the ground state present exponential decay.

$$< VBS|\mathbf{S}_0 \cdot \mathbf{S}_n|VBS >= (-3)^{-n} \quad \text{(exact)} \quad (9.51)$$

We can think of a family of AKLT hamiltonians depending on a parameter β:

$$H_\beta = \sum_i \mathbf{S}_i \cdot \mathbf{S}_{i+1} - \beta(\mathbf{S}_i \cdot \mathbf{S}_{i+1})^2 \quad (9.52)$$

The case $\beta = 0$ corresponds to the Heisenberg model which is not integrable by Bethe ansatz but according to Haldane it is believe to have a unique massive state. After the AKLT theory the case $\beta = -1/3$ has a unique massless ground sate. However, the model for $\beta = 1$ is solvable by Bethe ansatz [19]. It has a unique ground state with no energy gap and appears to have power law decay of the correlation functions.

Let us finally mention that the AKLT construction presented above for spin-1 matrices can be easily generalized for higher spins just by considering the symmetrized tensor product of more than two spin-1/2 spaces at each lattice site. For instance, a spin-3/2 can be obtained out of three spin-1/2 variables at every lattice site. Then, when connecting these type of sites altogether by means of valence bonds we arrive at a two dimensional honeycomb lattice. This is a quite remarkable result for it means that we can construct exact ground states for interacting hamiltonians in dimensions higher than 1, even though the entire spectrum is unknown.

References

[1]Fradkin, E., *1991 Field Theories of Condensed Matter Systems.* Addison-Wesley, Redwood City. 1991.

[2]Bethe, H.A., *1931, Z. Phys.* **71**, *205*

[3]Fazekas, P., Anderson, P. W., *1974, Phil. Mag.* **30**, *432*

[4]Anderson, P. W. *1987, Science* **235**, *1196*

[5]Balachandran, A.P., Ercolessi, E. , Morandi, G., Srivastava, A.M., *Hubbard Model and Anyon Superconductivity* Lecture Notes in Physics. Vol. 38. World Scientific.

[6]Gomez, C., Ruiz-Altaba, M., Sierra , G., *Quantum Groups in 2-dimensional Physics* Cambridge University Press (in press)

[7]Anderson, P.W., Baskaran, G., Zou, Z., Hsu, T., *1987, Phys. Rev. Lett.* **58**, *2790*

[8]Dagotto, E., *Correlated Electrons in High Temperature Superconductors* Florida State University preprint, 1993. (submitted to Rev. Mod. Phys.)

[9]Kivelson, S.A., Rokhsar, D.S., Sethna, J.P., *1987, Phys. Rev. B* **35**, *8865*

[10]Chen, Y.H., Wilczek, F., Witten, E., Halperin , B., *1989, Int. J. Mod. Phys. B* **3**, *1001*

[11]Fetter , A.L., Hanna ,C.B., Laughlin , R.B., *1989, Phys. Rev. B* **35**, *9679*

[12]Greiter, M., Wilczek, F., Witten, E., *1989, Mod. Phys. Lett. B* **3**, *903*

[13]Laughlin , R.B., *1988, Science* **242**, *525*

[14]Laughlin , R.B., *1988, Phys. Rev. Lett.* **60**, *2677*

[15]Majumdar, C.K., Ghosh, D.K., *1969, J. Math. Phys.* **10**, *1388*

[16]Haldane, F.D.M. *1983, Phys. Lett. A* **93**, *464*

[17]Haldane, F.D.M. *1983, Phys. Rev. Lett.* **50**, *1153*

[18]Haldane, F.D.M. *1985, J. Appl. Phys.* **557**, *3359*

[19]Affleck, Kennedy, T., Lieb, E.L., Tasaki, H., *1988, Comm. Math. Phys.* **115**, *477*

10. The Hubbard Model at $D = 1$

10.1 The Bethe Ansatz

This chapter is devoted to the exact solution of the Hubbard model for the particular case of 1 dimension. In this simplified case most of the open questions reviewed in previous chapters are able to be answered. This is so because the particular kinematical properties of 1 dimensions. For integrable models, the particle scattering is elastic, i.e., not only is the total momentum of the particles conserved, but the individual momenta of the particle remain unchanged after the collision with other particles. This strong requirement makes the problem solvable and the mathematical framework in which this properties are formulated is known as the *Bethe ansatz*.

In 1931 H. Bethe [1] solved the eigenvalue problem associated to the Heisenberg hamiltonian in 1 dimension (with periodic boundary conditions),

$$H = J \sum_{j=1}^{L} \sigma_j \cdot \sigma_{j+1} \tag{10.1}$$

The solution of this problem is the starting point for the theory of integrable models. A generalization of this model is the anisotropic Heisenberg model or also called XXZ model,

$$H_{XXZ} = J \sum_{j=1}^{L} \left(\sigma_j^x \cdot \sigma_{j+1}^x + \sigma_j^y \cdot \sigma_{j+1}^y + \Delta(\sigma_j^z \cdot \sigma_{j+1}^z - 1) \right) \tag{10.2}$$

There are plenty of one-dimensional integrable models which are related to this model one way or the other. We may assert that the XXZ model is ubiquitous in one dimension. For instance, it can be related to a model of spinless interacting fermions,

$$H = -t \sum_{j=1}^{L} \left(c_j^\dagger c_{j+1} + c_{j+1}^\dagger c_j \right) + U \sum_{j=1}^{L} n_j n_{j+1} \tag{10.3}$$

where $n_j = c_j^\dagger c_j$.

The mapping between both models is carried out by means of the standard Jordan-Wigner transformation which relates spin operators (satisfying commutation relations) with fermionic operators (satisfying anticommutaion relations),

$$\begin{cases} \sigma_i^+ = c_i e^{i\pi \sum_{j=1}^{i-1} n_j} \\ \sigma_i^- = (\sigma_i^+)^+ \\ \sigma_i^z = 2c_i^\dagger c_i - 1 \end{cases} \qquad (10.4)$$

After little algebra we arrive at the relation between the spin parameters (J, Δ) and the fermionic parameters (t, U),

$$\begin{cases} J = -t \\ J\Delta = U \implies \Delta = -U/t \end{cases} \qquad (10.5)$$

These two models are models defined on the lattice. From the form of the spinless fermion hamiltonian above we see that it is the lattice version of the continous g-ology models studied in the second part of these notes.

Another interesting model, for it is also related to the previous ones, is the problem of N particles interacting through a two-body delta potential. It was exactly soved by Yang [2] in 1967, and it served as the basis for the resolution of the Hubbard model by Lieb and Wu [3] in 1968.

It is illustrative to solve both the anisotropic Heisenberg model and the N-particle delta model in parallel in order to see their analogies and differences. The Heisenberg model introduces the ordinary Bethe ansatz while the many-body δ-model introduces the notion of *nested Bethe ansatz* necessary to solve the Hubbard model in the next section.

10.1.1 XXZ MODEL. 1 and 2 magnon sectors.

The starting point is to choose the state with all spins up as the reference state:

$$|0> = |\uparrow\uparrow \dots \uparrow> \qquad (10.6)$$

This state has zero energy $H|0> = 0$ as it is seen upon acting with (10.2). We have substracted $(\sigma^z \sigma^z - 1)$ in order for this to be true.

As the XXZ hamiltonian commutes with the total spin number operator,

$$[H_{XXZ}, \frac{1}{2} \sum_j \sigma_j^z] = 0 \qquad (10.7)$$

it is possible to diagonalize the hamiltonian for sectors of well-defined number of total spin S_z. That is, the number of spins up (or down) is conserved. Hence, the previous state has $S_z = \frac{1}{2}L$. The following spin sector corresponds to $S_z = \frac{1}{2}L - 1$ and one spin has been fliped down. A basis for the subspace $\mathcal{H}_1$ of one-spin-down states is generated by the states of the form,

$$|x> = |\uparrow\uparrow \dots \overset{x}{\downarrow} \dots \uparrow\uparrow> \qquad (10.8)$$

Every state in the $\mathcal{H}_1$ subspace can be expanded in the form,

$$|\psi> = \sum_{x=1}^{L} f(x) \, |x> \tag{10.9}$$

Let us introduce the following definition,
Application of the hamiltonian operator to the state $|x>$ yields,

$$H|x> = 2J \left(|x+1> + |x-1> -2\Delta|x> \right) \tag{10.10}$$

Then, the Schrodinger equation $H|\psi> = E|\psi>$ amounts to the condition,

$$2J \left(f(x+1) + f(x-1) - 2\Delta f(x) \right) = Ef(x) \tag{10.11}$$

A solution for this equation is given by the plane wave solution,

$$f(x) = e^{ikx} \tag{10.12}$$

with energy,

$$E(k) = 4J(\cos k - \Delta) \tag{10.13}$$

Magnon $\equiv$ It is an excited state of the Heisenberg model belonging to the subspace of one-spin-down states.

The name magnon used to designate the state (10.9) calls for the spin-wave picture associated to this states (See figure 10.1). Equation (10.13) is the dispersion relation fo the magnon of quasimomentum k.

The periodicity of the lattice reduces the allowed momentum vectors,

$$f(x+L) = f(x) \Longrightarrow e^{ikx} = 1 \tag{10.14}$$

$$k = \frac{2\pi I}{L}, \; I = 0, 1, \ldots, L-1. \tag{10.15}$$

Notice that in the case $\Delta = 0$ (also called XX model) the dispersion relation (10.13) corresponds precisely to the dispersion relation of a free fermion. This is something that makes sense by virtue of the mapping to the spinless fermions (10.5).

Let us proceed now to the 2-magnons sector $S_z = \frac{1}{2}L - 2$ where interactions between magnons takes place for the first time. The most general wave function corresponding to this subspace is,

$$|\psi> = \sum_{1 \leq x_1 < x_2 \leq L} f(x_1, x_2) \, |x_1, x_2> \tag{10.16}$$

where the basis vectors are of the form,

$$|x_1, x_2> = \sigma_{x_1}^{-} \sigma_{x_2}^{-} |0> = |\uparrow \ldots \overset{x_1}{\downarrow} \ldots \overset{x_2}{\downarrow} \ldots \uparrow> \tag{10.17}$$

and the ordering $1 \leq x_1 < x_2 \leq L$ is always assumed.
Now the eigenvalue equation reduces to the condition, when $x_1 \leq x_2 - 2$,

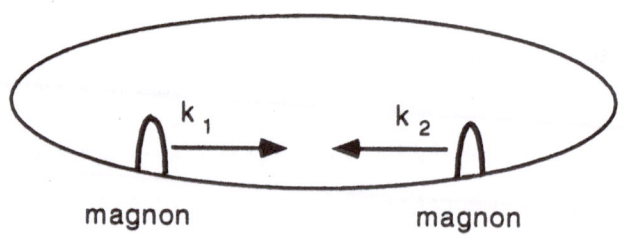

Fig. 10.1. Interaction of magnons in the one dimensional Heisengberg model.

$$2J \left(f(x_1 - 1, x_2) + f(x_1 + 1, x_2) + \right.$$

$$\left. + f(x_1, x_2 - 1) + f(x_1, x_2 + 1) - 4\Delta f(x_1, x_2) \right) = E \; f(x_1, x_2) \qquad (10.18)$$

and when $x_1 = x_2 - 1$, the condition is

$$2J \left(f(x - 1, x + 1) + f(x, x + 2) + f(x, x - 1) - 2\Delta f(x, x + 1) \right) =$$

$$= E \; f(x, x + 1) \qquad (10.19)$$

If, as a first attempt to solve these conditions, we try a naive extension of the plane wave solution (10.12),

$$f(x_1, x_2) = e^{i(k_1 x_1 + k_2 x_2)} \qquad (10.20)$$

then the total energy is $E = E(k_1) + E(k_2)$ with the individual magnon energies as in the one-magnon sector,

$$E(k_j) = 4J(\cos k_j - \Delta) \qquad (10.21)$$

However, the ansatz (10.20) is not admisible for it represents a non-periodic solution, and we do not have to forget that the spin system is in a ring (see figure 10.1).

A clever ansatz (Bethe ansatz) corresponds to make a linear superposition of plane-wave like solutions (see figure 10.2),

$$f(x_1, x_2) = A_{12} \; e^{i(k_1 x_1 + k_2 x_2)} + A_{21} \; e^{i(k_2 x_1 + k_1 x_2)} \qquad (10.22)$$

As it happens, the eigenvalue condition (10.18) is automatically satisfied by the Bethe ansatz (10.22). Upon imposing the condition (10.19) we deduce the value of the amplitude ratio A_{21}/A_{12}. Noticing that the Bethe ansatz (10.22) satisfies the relation

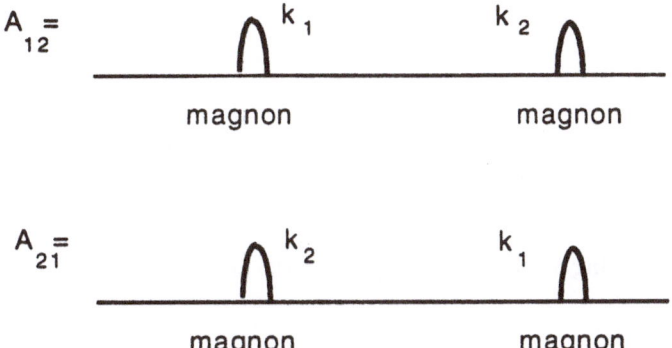

Fig. 10.2. Scattering amplitudes for magnons in the one dimensional Heisengberg model.

$$2J(f(x-1, x+1) + f(x+1, x+1) + f(x, x) + f(x, x+2)-$$

$$-4\Delta f(x, x+1)) = Ef(x, x+1) \tag{10.23}$$

the eigenvalue condition (10.19) turns out to be equivalent to a simpler relation,

$$f(x, x) + f(x+1, x+1) = 2\Delta f(x, x+1) \tag{10.24}$$

From this equation we can easily obtain the ratio,

$$\frac{A_{21}}{A_{12}} = -\frac{1 - 2\Delta e^{k_2} + e^{i(k_1+k_2)}}{1 - 2\Delta e^{k_1} + e^{i(k_1+k_2)}} \tag{10.25}$$

Notice that in the particular case $\Delta = 0 \implies \frac{A_{21}}{A_{12}} = -1$, independently of k_1, k_2. Thus, $f(x_1, x_2) = -f(x_2, x_1)$ and we have again a free fermion.

10.1.2 Physical meaning of A_{21}/A_{12}.

It corresponds to the scattering amplitude for the process in which two states interact. That is, it is the S-matrix (in this particular case, a number) for the scattering of magnons. See figure 10.3.

$$S_{12} = \frac{A_{21}}{A_{12}} = \frac{\text{Amplitude (21)}}{\text{Amplitude (12)}} = \text{Scattering Amplitude} \tag{10.26}$$

Finally, let us impose the *periodicity condition* which for the case of two overturned spins reads as follows,

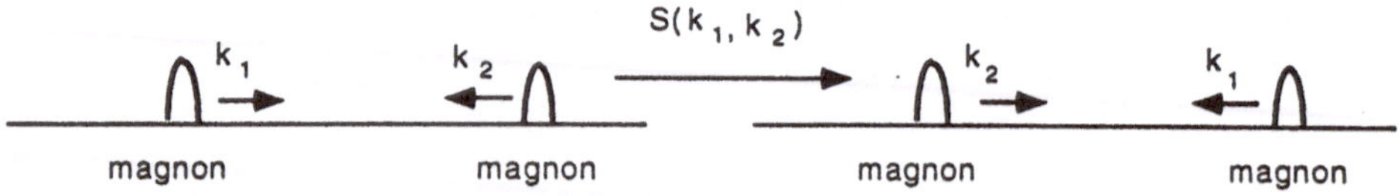

Fig. 10.3. Scattering S-matrix for magnons in the one dimensional Heisengberg model.

$$f(x_1, x_2) = f(x_2, x_1 + L)$$
$$x_1 < x_2 \qquad x_2 < x_1 + L$$

Then, the two-spin amplitudes satisfy,

$$A_{12} = A_{21}\, e^{ik_1 L}$$
$$A_{21} = A_{12}\, e^{ik_2 L}$$

Recall that from translational invariance the following relation for the momenta holds

$$e^{i(k_1+k_2)L} = 1 \tag{10.27}$$

These two conditions (10.1.2) have a direct physical significance if we write them in the following fashion, see figure 10.4,

$$\left(\frac{A_{21}}{A_{12}}\right) e^{ik_1 L} = 1 \tag{10.28}$$

The first factor of the LHS corresponds to the scattering of the momentum 1 by the momentum 2. The second factor is a kinematical phase factor. Altogether this condition represents the single-valuedness of the wave function.
Inserting equation (10.26) into equations (10.1.2) we arrive at a system of equations for the allowed momenta of the two-magnon state,

$$e^{ik_1 L} = -\frac{1 - 2\Delta e^{k_1} + e^{i(k_1+k_2)}}{1 - 2\Delta e^{k_2} + e^{i(k_1+k_2)}} \tag{10.29}$$

$$e^{ik_2 L} = -\frac{1 - 2\Delta e^{k_2} + e^{i(k_1+k_2)}}{1 - 2\Delta e^{k_1} + e^{i(k_1+k_2)}} \tag{10.30}$$

These are the Bethe equations for $M = 2$.
Before proceeding with the general case of multi-magnon solutions, let us turn to the model of N particles with an interacting delta potential.

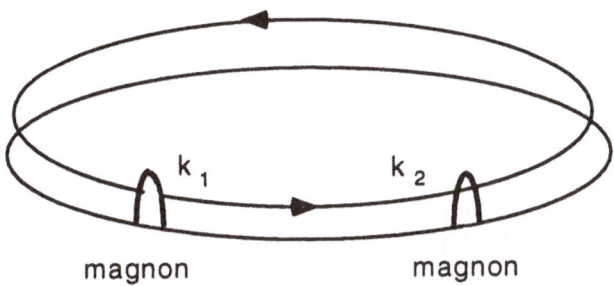

Fig. 10.4. Translational invariance for the magnons in the Heisenberg model.

10.1.3 δ-function many-body system. 1 and 2 particle solutions

We consider a one-dimensional system of N spinless particles interacting between them through a δ-function potential. Thus the hamiltonian reads as follows,

$$H = -\sum_{i=1}^{N} \frac{\partial^2}{\partial x_i^2} + 2c \sum_{i<j} \delta(x_i - x_j) \tag{10.31}$$

where $x_i \in [0, L]$ denotes the position of the particles on the line and periodic boundary conditions are also assumed. The particles only interact, with strength c, when they happen to be at the same position. This somehow resembles the Hubbard model.

If the number of particles is $N = 1$, the problem is trivial to solve as in the Heisenberg hamiltonian, for the solution is a plane wave,

$$\psi(x) = e^{ikx} \tag{10.32}$$

with energy,

$$E = k^2 \tag{10.33}$$

Let us consider the case of two particles $N = 2$ where the interaction is no longer trivial. We are going to follow closely Bethe's approach.

First important observation We are not going to impose any a priori symmetry on the wave function, i.e., it could not only be fully symmetric (bosons) or fully antisymmetric (fermions), but shall seek for a more general situation in which the symmetry of the wave function can be symmetric with respect to the exchange of two certain particles and antisymmetric with respect to others.

To be more precise, the fully symmetric case means that the wave functions transforms under a representation $[N]$ of the permutation group of N elements S_N, where $[N]$ stands for a Young tablaux with N horizontal boxes. On the other extreme, the fully antisymmetric case corresponds to the representation $[1^N]$ of S_N group, where $[1^N]$ stands for Young tablaux with N vertical boxes.

Now, the more general case of symmetry considered by C. N. Yang is somehow a mixture between these two extreme cases in which the symmetry of the wave function corresponds to the representation $[2^M 1^{N-2M}]$, where it consists of M horizontal lines with 2 boxes each one and a vertical line with $N - 2M$ boxes underneath the first set of boxes. In this fashion, whenever two labels in two vertical boxes are interchanged, the wave function picks up a minus sign while it remains unchanged when two labels in two horizontal boxes are exchanged. It must be remarked that the physical situation underlying this general symmetry of the system does *not* corresponds to having a system of bosons and fermions together.

From the above exposition it must be apparent that the most general Bethe ansatz for the problem at hand is given by

$$\psi(x_1, x_2) \equiv \psi_I(x_1, x_2) = A_{12}\, e^{i(k_1 x_1 + k_2 x_2)} + A_{21}\, e^{i(k_2 x_1 + k_1 x_2)} \qquad (10.34)$$

when $x_1 < x_2$,
and by

$$\psi(x_1, x_2) \equiv \psi_{II}(x_1, x_2) = B_{12}\, e^{i(k_1 x_2 + k_2 x_1)} + B_{21}\, e^{i(k_2 x_2 + k_1 x_1)} \qquad (10.35)$$

when $x_1 > x_2$.

Notice that we are distinguishing between the different possible sortings of the particles. There are $N!$ orderings corresponding to the different permutations of the particles. This is a "degree of freedom" which was not present in Heisenberg spin problem, for there the spins fill the whole lattice and are not able to move, they can only flip.

As we see, now there are needed 4 two-particle amplitudes $A_{12}, A_{21}, B_{12}, B_{21}$ instead of only two as in the spin model.

The wave functions ψ_I, ψ_{II} are subject to both continuity and discontinuity conditions,

- Continuity condition,

$$\psi_I(x, x) = \psi_{II}(x, x) \qquad (10.36)$$

- Discontinuity of the derivative of the wave function as required from the form of the hamiltonian (10.31)

  In order to set up this condition we have to integrate the wave function on a strip around $x_1 = x_2$, see figure 10.5,

$$\int_0^L dx_2 \int_{x_2-\epsilon}^{x_2+\epsilon} dx_1 = \int_0^L dx_1 \int_{x_1-\epsilon}^{x_1+\epsilon} dx_2 \qquad (10.37)$$

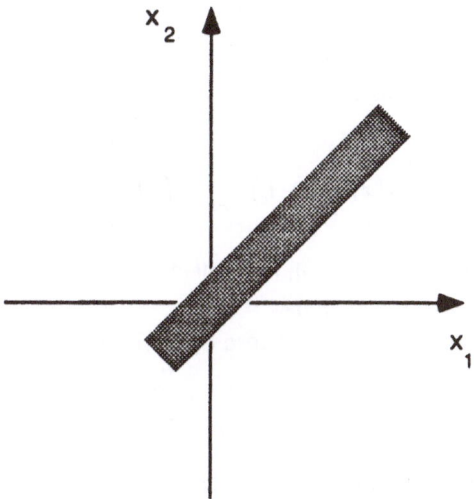

Fig. 10.5. Integration strip of the two-body wave function in the Yang model.

where ϵ is a small quantity. The discontinuity equation turns out to be,

$$\left(\frac{\partial \psi_{II}}{\partial x_1} - \frac{\partial \psi_I}{\partial x_1} + \frac{\partial \psi_I}{\partial x_2} - \frac{\partial \psi_{II}}{\partial x_2}\right)\bigg|_{x_1=x_2=x} = 2c\psi_I(x,x) \qquad (10.38)$$

These two conditions in turn lead to the system of equations,

$$\begin{cases} A_{12} + A_{21} = B_{12} + B_{21} \\ ik_{12}(-A_{12} + A_{21} - B_{12} + B_{21}) = 2c(A_{12} + A_{21}) \end{cases} \qquad (10.39)$$

From these two equations we can eliminate two amplitudes. The most convenient choice is

$$\begin{pmatrix} A_{21} \\ B_{21} \end{pmatrix} = \begin{pmatrix} \frac{-ic}{k_{12}-ic} & \frac{k_{12}}{k_{12}-ic} \\ \frac{k_{12}}{k_{12}-ic} & \frac{-ic}{k_{12}-ic} \end{pmatrix} \begin{pmatrix} A_{12} \\ B_{12} \end{pmatrix} \qquad (10.40)$$

At this point we introduce a matrix notation to make the analysis easier, namely,

$$\xi_{21} = R_{12}(k_1,k_2)\xi_{12} \qquad (10.41)$$

where the R-matrix is given by the expression,

$$R_{12}(k_1,k_2) = \frac{-ic}{k_{12}-ic}\,1 + \frac{k_{12}}{k_{12}-ic}\,P_{12} \qquad (10.42)$$

Now it is in this model when we obtain a real scattering matrix when relating the amplitudes for the process $(1,2) \longrightarrow (2,1)$.

Again we have to consider periodicity conditions for the δ-many-body system, which have to be implemented in the Bethe ansatz. Now the periodicity conditions turn out to be,

$$\begin{matrix} \psi_I(x_1, x_2) = \psi_{II}(x_1 + L, x_2) \\ x_1 < x_2 \end{matrix} \implies \left\{ \begin{matrix} A_{12} = B_{21}\ e^{ik_1 L} \\ A_{21} = B_{12}\ e^{ik_1 L} \end{matrix} \right. \tag{10.43}$$

$$\begin{matrix} \psi_{II}(x_1, x_2) = \psi_I(x_1, x_2 + L) \\ x_1 > x_2 \end{matrix} \implies \left\{ \begin{matrix} B_{12} = A_{21}\ e^{ik_2 L} \\ B_{21} = A_{12}\ e^{ik_2 L} \end{matrix} \right. \tag{10.44}$$

These conditions also yield the relation $e^{i(k_1+k_2)L} = 1$ which expresses the translational invariance of the solutions.

Again it is convenient to organize these conditions in a matrix form,

$$e^{ik_1 L} \begin{pmatrix} A_{21} \\ B_{21} \end{pmatrix} = \begin{pmatrix} B_{12} \\ A_{12} \end{pmatrix} \tag{10.45}$$

which in turn, with the aid of the R-matrix (10.41), can be written as

$$e^{ik_1 L} R_{12} \xi_{12} = P_{12} \xi_{12} \tag{10.46}$$

where P_{12} is a permutation given by $\begin{pmatrix} 0 & 1 \\ 1 & 0 \end{pmatrix}$.

Multiplying equation (10.46) by P_{12} and introducing the actual S-matrix

$$S_{12} \equiv P_{12} R_{12} = \frac{1}{k_{12} - ic} \begin{pmatrix} k_{12} & -ic \\ -ic & k_{12} \end{pmatrix} \tag{10.47}$$

we arrive at the eigenvalue problem to be solved,

$$e^{ik_1 L} S_{12} \xi_{12} = \xi_{12} \tag{10.48}$$

This equation tells us that the vector of two-particle amplitudes ξ_{12} has to be an eigenvector of the scattering S-matrix with eigenvalue $e^{-ik_1 L}$. The meaning of this equation is apparent and, as in the case of the Heisenberg model, it amounts to the single-valuedness of the wave function when a particle undergoes a complete lap around the lattice.

In this case of two particles it is easy to find the eigenvalues of the S_{12}-matrix which turn out to be 1 and $\frac{k_{12}+ic}{k_{12}-ic}$. Then the momenta of the two particle system must satisfy one of the following two conditions,

$$\left\{ \begin{matrix} e^{ik_1 L} = 1 & \text{free case} \\ e^{ik_1 L} = \frac{k_{12}+ic}{k_{12}-ic} & \text{interacting case} \end{matrix} \right. \tag{10.49}$$

10.1.4 XXZ Model. Multi-magnon solutions.

The general case of having M overturned spins, $S_z = \frac{L}{2}$ sector, can be handled straightforwardly once the two spin down sector has been worked out.

Let us write the wave vector in the subspace of M spins down as

$$|\psi> = \sum_{1 \leq x_1 < \dots < x_M \leq L} f(x_1, \dots, x_M) \, |x_1, \dots, x_M > \qquad (10.50)$$

$$S_z = \frac{L}{2} - M \qquad (10.51)$$

The Bethe ansatz amounts to expressing the wave function as a linear superposition of one-particle wave functions, namely,

$$f(x_1, \dots, x_M) = \sum_P A_P e^{i(k_{P_1} x_1 + \dots + k_{P_M} x_M)} \qquad (10.52)$$

with $1 \leq x_1 < \dots < x_M \leq L$. The sum $\sum_P$ extends over all the possible $M!$ permutations of the overturned spins $1, \dots, M$. The variables $k_1, k_2, \dots, k_M$ are M distinct numbers among themselves (not necessarily reals.)
To relate the notation that we are using now in the general case with the $M = 2$ case, just notice the following correspondence,

$$A_{12} = A_{P=e} \qquad e \equiv \text{identity permutation.}$$
$$A_{21} = A_{P=\tau} \qquad \tau \equiv \text{transposition } 12 \to 21.$$

To proceed further in the analysis of the general case the key point is to assuming the factorization of the amplitudes A_P of 3 or more magnons into two-magnons amplitudes. That this is indeed possible is due to the integrability of the model. With the help of the two-magnons scattering matris (10.22) it is possible to relate the 3-magnons amplitude A_{321} in the configuration (321) to the amplitude in the reference configuration 123 as,

$$A_{321} = S_{12} \, A_{312} = S_{12} S_{13} \, A_{132} = S_{12} S_{13} S_{23} \, A_{123} \qquad (10.53)$$

This factorization is of course independent of the ordering in which the S-matrices are applied. Later we shall see this condition yields the Yang-Baxter equation. This analysis leads to the following set of Bethe equations which determine the allowed momenta for the magnons in the multimagnon state,

$$e^{ik_j L} \prod_{l=1, l \neq j}^{M} S(k_j, k_l) = 1. \quad j = 1, \dots, M. \qquad (10.54)$$

where S is the scattering S-matrix introduced in the solution of the two-magnons sector.

10.1.5 δ -function many-body system. Multiparticle solutions.

As we have at our disposal a problem with N particles, it is possible to form $N!$ wave functions each corresponding to every spatial ordering of the particles. This extra freedom was not present in the Heisenberg spin model and makes necessary to consider two sets of permutations. One set serves to sort the positions $x's$ of the particles and are denoted by $Q's$. The second set serves to order the

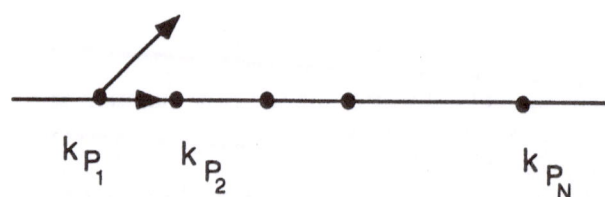

Fig. 10.6. The ordering of momenta for used to enumerate the particle-state vectors in the Yang model.

momenta $k's$ and are denoted by $P's$. Therefore, the most general Bethe ansatz takes the form,

$$\psi(x_1,\ldots,x_N) = \sum_P A_{Q,P} e^{i(k_{P_1}x_{Q_1}+\ldots+k_{P_N}x_{Q_N})} \qquad (10.55)$$

where the permutations Q are such that the values of the positions $x_1,\ldots,x_N$ satisfy $0 \le x_{Q_1} \le x_{Q_2} \le \ldots \le x_{Q_N} \le L$.

The relation of this equation with the particular case of $N = 2$ considered above is given by the matrix,

$$A_{Q,P} = \begin{pmatrix} A_{e,e} & A_{e,\tau} \\ A_{\tau,e} & A_{\tau,\tau} \end{pmatrix} = \begin{pmatrix} A_{12} & A_{21} \\ B_{12} & B_{21} \end{pmatrix} \qquad (10.56)$$

In the general multiparticle case, $A_{Q,P}$ is a $N! \times N!$ matrix. Let us call ξ_P the column vector of the matrix $A_{Q,P}$,

$$\xi_P = \begin{pmatrix} A_{e,P} \\ A_{\tau_{12},P} \\ \vdots \end{pmatrix} \qquad (10.57)$$

This vector represents all the possible amplitudes in which the momenta $k_{P_1},\ldots,k_{P_N}$ are ordered from left to right as shown in figure 10.6.

We would like to generalize the relation

$$\xi_{21} = R_{12}(k_1, k_2)\, \xi_{12}$$

to the case of N particles. Let us suppose that we have two permutations P and P' which only differ in the momenta of the particles located at i and $i+1$, see figure 10.7.

Then the relation of the two vectors $\xi_{P'}$ and ξ_P is:

$$\xi_{P'} = R_{i,i+1}(k_m, k_n)\, \xi_P \qquad (10.58)$$

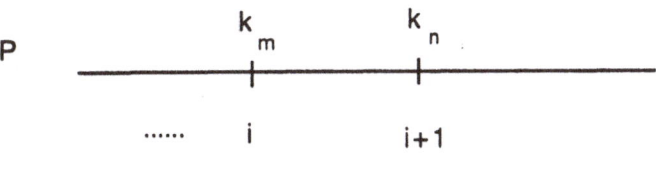

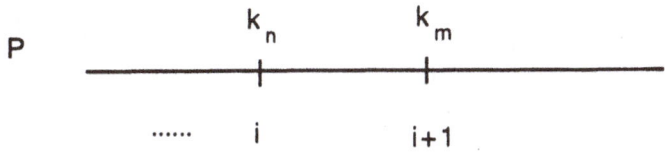

Fig. 10.7. The action of the permutation operator on the momenta labels of certain state.

where the R matrix is given by

$$R_{i,i+1}(k_m, k_n) = \frac{-ic}{k_{mn} - ic} \, 1 + \frac{k_{mn}}{k_{mn} - ic} \, P_{i,i+1} \tag{10.59}$$

and the permutation $P_{i,i+1}$ acts on the amplitude vectors in an obvious way,

$$P_{i,i+1} \begin{pmatrix} \vdots \\ A_{...a,b...;P} \\ A_{...b,a...;P} \\ \vdots \end{pmatrix} = \begin{pmatrix} \vdots \\ A_{...b,a...;P} \\ A_{...a,b...;P} \\ \vdots \end{pmatrix} \tag{10.60}$$

A graphical representation of this equation in shown in figure 10.8. In the horizontal direction we represent the spatial ordering of the particle momenta and the vertical lines represent the world lines of the particles in the discrete time evolution.

Let us recall that conditions of the form (10.58) represent the conditions that the vector amplitudes ξ_P have to fulfill in order for the continuity and discontinuity conditions to be satisfied in the multiparticle solution case. However, this conditions pose a consistency problem which have to be solved before we carry on. Namely, it is clear that there are $N! \times (N-1)$ conditions of the type (10.58) for $N!$ vectors of the form ξ_P.

The check that this redundant conditions are in fact compatible resides on the Yang-Baxter property which the R-matrices satisfy. To make the discussion clear, let us consider two vector amplitudes which differ in the position of 3 particles, say

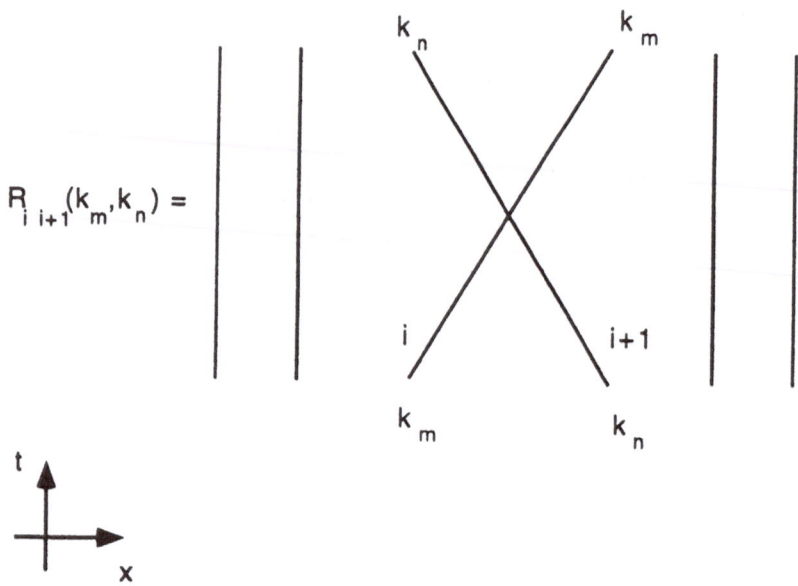

$$R_{i\,i+1}(k_m,k_n) =$$

Fig. 10.8. The action of the R-matrix on the momenta of a certain state.

$$\xi_{P_i} = \begin{pmatrix} 1 \\ 2 \\ 3 \end{pmatrix}, \quad \xi_{P_f} = \begin{pmatrix} 3 \\ 2 \\ 1 \end{pmatrix}$$

There are 2 possible ways to bring the vector ampltitude ξ_{P_f} to the initial form ξ_{P_i} and this is the origin of the mentioned redundancy. These two ways are depicted squematically in figueres 10.9 and 10.10. The compatibility of both ways is encoded in the Yang-Baxter property expressed graphically in figure 10.10 and analitically it means that the R matrices must comply with the condition,

$$R_i(k_{23})\ R_{i+1}(k_{13})\ R_i(k_{12})\ =\ R_{i+1}(k_{12})\ R_i(k_{13})\ R_{i+1}(k_{23}) \qquad (10.61)$$

where we employ the notation $R_i \equiv R_{i+1}$. This condition can be seen to be satisfied by the R matrices of our problem simply by sustituting (10.59) in (10.61) and using the property which the transpositions obey,

$$P_i\ P_{i+1}\ P_i\ =\ P_{i+1}\ P_i\ P_{i+1} \qquad (10.62)$$

Once the compatibility check has been performed for the simpler case of 3 particles, the extension to any number of different particle positions with respect to the initial vector amplitude can be checked straightfowardly.
Thus the outcome of the continuity and discontinuity conditions is that given a vector amplitude ξ_P for any given permutation of the momenta, then it can be

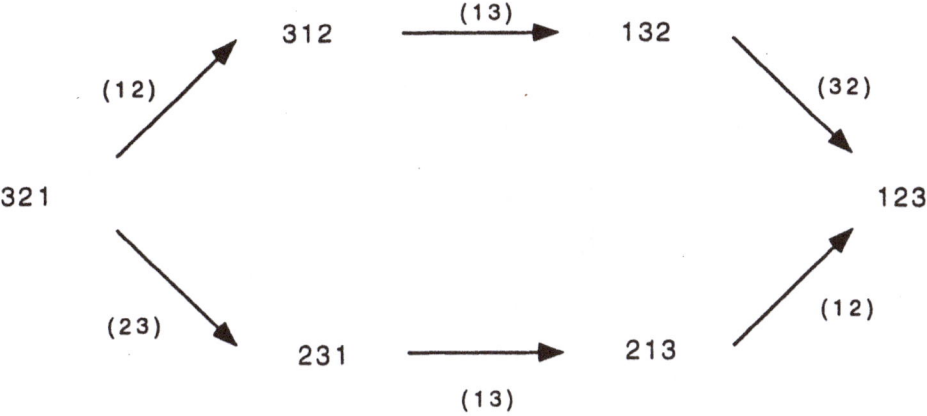

Fig. 10.9. The two possible paths of braidings leading to the Yang-Baxter equation for a three-particle state.

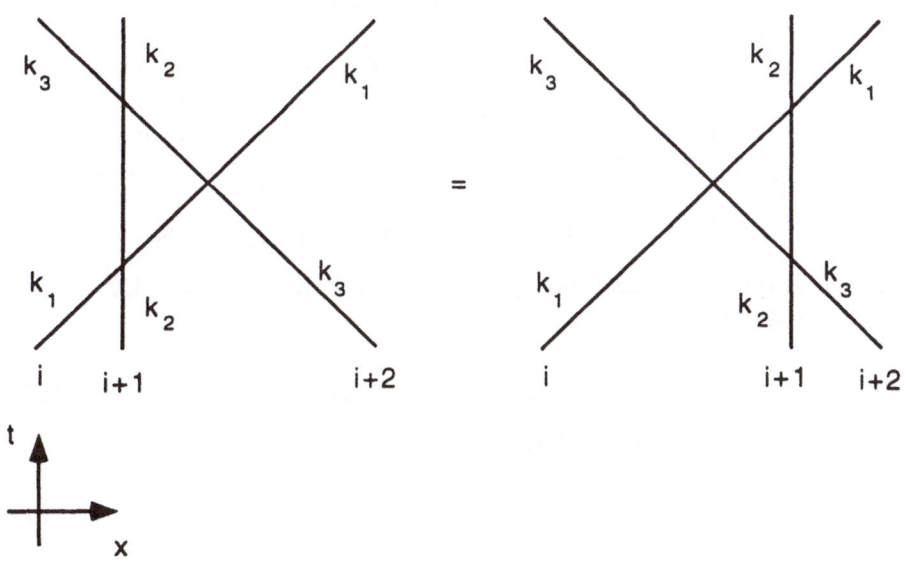

Fig. 10.10. Schematic description of Yang-Baxter relation.

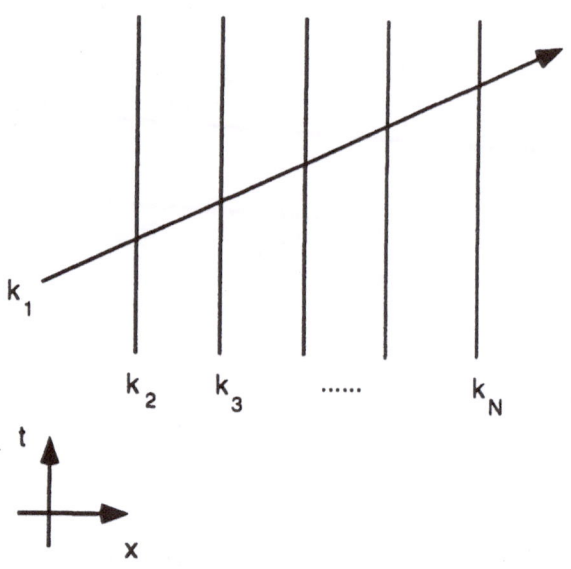

Fig. 10.11. Scattering of a momenta state through the rest of the momenta states.

brought to the vector amplitude corresponding to the permutation identity ξ_0 by successive application of the R matrix, as in (10.58), carelessly of the order chosen to achieve the initial state.

Now , in order for there to be a consistently defined theory on a circle we have to impose the periodic boundary conditions. For the momentum k_1 the figure 10.11 shows us that upon $N - 1$ scattering events the order of the momenta is $k_2, k_3, \ldots k_N, k_1$. Then, the initial vector amplitude satisfy the equation,

$$e^{ik_1 L} R_{N-1,N}(k_{1N}) \ldots R_{23}(k_{13}) \ R_{12}(k_{12}) \ \xi_0 = \tilde{P} \ \xi_0 \qquad (10.63)$$

where $\tilde{P}$ is the cyclic permutation, $\tilde{P}_{1,2,3,\ldots,N} (1, 2, 3, \ldots, N) = 2, 3, \ldots, N, 1$. Defining the S-matrix as usual,

$$S_{i,i+1} = P_{i,i+1} \ R_{i,i+1} \qquad (10.64)$$

the boundary condition translates into,

$$\lambda_1 \ S_{1,N}(k_{1N}) \ldots \ S_{1,3}(k_{13}) \ S_{1,2}(k_{12}) \ \xi_0 = \xi_0 \qquad (10.65)$$

where

$$\lambda_j \equiv e^{ik_j L} , \quad j = 1, 2, \ldots, N \qquad (10.66)$$

Notice that when there is not interaction $c = 0$, then the scattering matrix is the unit matrix, $S_{i,i+1} = id$

There are N equations like (10.63), one for each momenta. Thus the vector amplitude ξ_0 has to be simultaneously an eigenvector of N operators. These N operators can be seen to commute with each other, using

$$S_{ij}S_{ji} = 1 \tag{10.67}$$

$$S_{i+1,i+2}(k - k') \, S_{i,i+1}(k - k'') \, S_{i+1,i+2}(k' - k'') =$$
$$S_{i,i+1}(k' - k'')S_{i+1,i+2}(k - k'')S_{i,i+1}(k - k') \tag{10.68}$$

$$S_{ij}S_{kl} = S_{kl}S_{ij}, \quad \text{i,j,k and l all unequal} \tag{10.69}$$

Equation (10.68) is the famous Yang-Baxter equation which was firstly derived by Yang [2] in his solution of the δ-function many-body system. It was rediscovered later by Baxter [4] in his solution of the 8-vertex model in which it was called the star-triangle relation. As a matter of fact, the star-triangle relation goes back to the Onsager solution of the Ising model in two dimensions. The name Yang-Baxter is given to relation (10.68) after Faddeev and Takhatajan . The transposition operator P_{ij} acts as a representation of the permutation group S_N on the amplitude vectors ξ. This representation is reducible, thus it is useful to decompose it into irreducible ones so that the eigenvalue problem is reduced to one of smaller dimensions. Let us denote by R the irreducible representations of the permutation group S_N. To give an idea of the physical content of the eigenvalue equation (10.65), we consider to extreme cases. When $R = $ identity representation $= [N]$, then $P_{ij} = 1$, then equation (10.65) is an equation for numbers instead of matrices and the solution is the boson result of Lieb and Liniger [5]. When $R = $ antisymmetric representation $= [1^N]$, then $P_{ij} = -1$ and $S_{ij} = 1$ and the solution to (10.65) is $e^{ik_j L} = 1$, the non-interacting case corresponding to fermions (the trivial case).

Now, following Yang [2], to solve the eigenvalue problem (10.65) it is convenient to introduce an associated eigenvalue problem for the conjugate representation $\tilde{R}$ of R. The reason for this will be apparent soon.

Further notation is needed. The λ_j in (10.65) is a function of k, c and R that we denote by $\lambda_j(k, c, R)$. Then it is possible to show the following relation between the eigenvalue problems for the representation R and its conjugate:

$$\lambda_j(k, c, R) = \prod_{i \neq j} \frac{k_{ij} - ic}{k_{ij} + ic} \lambda_j(k, -c, \tilde{R}) \tag{10.70}$$

Another ingredient is the eigenvalue problem (10.65) for a problem in which the permutation P_{ij} is substituted by its opposite $-P_{ij}$, namely,

$$\lambda_1' \, S_{1,N}'(k_{1N}) \dots \, S_{1,3}'(k_{13}) \, S_{1,2}'(k_{12}) \, \xi_0' = \xi_0' \tag{10.71}$$

where the scattering matrix S_{ij}' is defined as usual in (10.47) with the change $P_{ij} \longrightarrow -P_{ij}$. Then it can also checked the relation,

$$\lambda_j'(k, c, \tilde{R}) = \lambda_j(k, c, R) \tag{10.72}$$

After introducing the previous two eigenvalues problems associated to the o-
riginal one, we are in position to evaluate λ_j for the representation $R =$
$[2^M 1^{N-2M}]$ of mixed symmetry that we considered in the begining of this model (neither bosons nor fermions). From (10.72) this is equivalent to evaluate
$\lambda'_j(k, c, [N - M, M])$. The key point to understand all the manipulations introduced so far is to notice that the symmetry properties of the eigenfunctions
ξ'_0 associated to $\lambda'_j(k, c, [N - M, M])$ can be realized by means of spin wave
functions for total z spin $= 1/2(N - 2M)$ transforming under S_N according to
a sum of irreducible representations,

$$[N] + [N - 1, 1] + [N - 2, 2] + \ldots [N - M, M] \tag{10.73}$$

This spin is a fictitious, spin nothing to do with any physical spin of the
system for the particles under consideration are spinless. It is related to the extra
degree of freedom present in the Yang model as compared to the Heisenberg
model. The eigenvalue equations for $\lambda'_j(k, c, [N - M, M])$ have then to be solved
for a ξ'_0 that belongs to the representation $[N - M]$. We may see this as a
Heisenberg model in the sector of M overturned spins already solved by Bethe
ansatz. Thus, here we have to perform an additional Bethe ansantz on top of
the initial one (10.52) to account for the two degrees of freedom of the system.
This is called the *nested Bethe ansatz*. Introducing "one-magnon" spin functions
$f(y; \Lambda)$ depending on the positions y and fictitious momemta Λ, the ansatz reads
as follows,

$$\xi'_0 = \sum_P A_P \prod_{i=1}^{M} f(y_i; \Lambda_i) \tag{10.74}$$

where $y_1 < y_2 < \ldots < y_M$ are the coordinates of the M "down spins". The
solution of this problem is achieved by standard methods and the result is,

$$f(y; \Lambda) = \prod_{j=1}^{y-1} \frac{ik_j - i\Lambda - c/2}{ik_{j+1} - i\Lambda + c/2} \tag{10.75}$$

$$\lambda'_j(k, c, [N - M, M]) = e^{ik_j L} = \prod_{\beta=1}^{M} \frac{ik_j - i\Lambda_\beta - c/2}{ik_j - i\Lambda_\beta - c/2} \tag{10.76}$$

$$-\prod_{j=1}^{N} \frac{ik_j - i\Lambda_\alpha - c/2}{ik_j - i\Lambda_\alpha - c/2} \prod_{\beta=1}^{M} \frac{-i\Lambda_\beta + i\Lambda_\beta + c}{-i\Lambda_\beta + i\Lambda_\beta - c} \tag{10.77}$$

Thus, the two equations (10.76) and (10.77) gives us the solution of the eigenvalue problem determining the allowed physical momenta and hence the energies
of the particles.
In passing let us mention that if we considered more general symmetry properties for the particles like that of the representation $[3^{M_3}, 2^{M_2}, 1^{M_1}]$ with
$3M_3 + 2M_2 + 1M_1 = N$, then to solve the eigenvalue problem we have to

introduce spins under the group $SU(3)$ and perform an extra Bethe ansatz, and so on and so forth.

In order to describe the excitation spectrum of the model, it is customary to present the eigenvalue equations (10.76) and (10.77) after taking logarithms. For N even and M odd we have,

$$k_j L = 2\pi I_j + \sum_{\alpha=1}^{M} \vartheta(2(\lambda_j - \lambda_\alpha)), \quad j = 1, \ldots, N. \tag{10.78}$$

$$-\sum_{j}^{N} \vartheta(2(\lambda_\alpha - k_j)) = 2\pi J_\alpha + \sum_{\beta \neq \alpha}^{M} \vartheta(\lambda_\beta - \lambda_\alpha), \quad \alpha = 1, \ldots, M. \tag{10.79}$$

where the $k's$ and $\Lambda's$ are a set of ascending real numbers,

$$\vartheta(k) = -2arctan(k/c), \quad (-\pi \leq \theta\pi) \tag{10.80}$$

and $\frac{1}{2} + I_j$ are successive integers from $1 - N/2$ to $N/2$, J_α are successive integers from $-\frac{1}{2}(M-1)$ to $+\frac{1}{2}(M-1)$.

As we are not interested in the physics of this model, we shall not pursue the study of these equations any further. However, they will show up in the solution of the Hubbard model in next section and then we will have the opportunity of analyzing its physical consequences.

10.2   Bethe Ansatz for the Hubbard Model

The basics ingredients to solve the Hubbard model have been introduced in the previous subsection, namely, the Bethe ansatz and its generalized version called nested Bethe ansatz. The reason why we shall be dealing with the nested version of the Bethe ansatz is because in the Hubbard model, unlike the Heisenberg model, there are two degrees of freedom: charge and spin. Here this spin is the real spin of the electrons and not a fictitious auxilary spin-like degree of freedom as in the Yang model.

As a result of the solution to be presented below, we shall be able to answer two crucial questions which are natural to pose when dealing with the Hubbard model, as we explained in chapter 7. These questions are,

- Is there a Mott transition for a finite value of the coupling constant U ?

- What does the ground state looks like ?

It is quite remarkable to be able to answer exactly these questions and that is why this one dimensional model is of great importance, for it serves as a sort of paradigm of what one envisages to answer in higher dimensions.

The procedure we are going to employ to solve the model is the nested Bethe ansatz which will lead us to an eigenvalue problem identical to the one appearing in the Yang model with the only difference that in the latter we have to substitute the momentum k by the rapidity $\sin k$ genuine of the Hubbard model.

10.2.1 Bethe ansatz and eigenstates.

Let us recall the Hubbard hamiltonian in one dimension,

$$H = -t \sum_{x\sigma}(c^\dagger_{x\sigma}c_{x+1\sigma} + h.c.) + U \sum_x n_{x\uparrow}n_{x\downarrow} \tag{10.81}$$

where periodic boundary conditions are assumed and the sum extends over $x = 1, 2, \ldots, N$, N being the total number of electrons. As this number is conserve, we can work in sectors with definite number of electrons. A generic state in the N-particle sector can be written as,

$$|\psi> = \sum_{x_1\sigma_1} \cdots \sum_{x_N\sigma_N} \psi(x_1\sigma_1, \ldots, x_N\sigma_N)c^\dagger_{x_1\sigma_1} \cdots c^\dagger_{x_N\sigma_N}|0> \tag{10.82}$$

Unlike in the Yang model, we shall impose from the very begining that the symmetry of this wave function is that of a fermionic system. Thus, this state satisfy antisymmetry under a permutation of any two particles,

$$\psi(x_{Q_1}\sigma_{Q_1}, \ldots, x_{Q_N}\sigma_{Q_N}) = (-1)^{\epsilon(Q)}\psi(x_1\sigma_1, \ldots, x_N\sigma_N) \tag{10.83}$$

where $\epsilon(Q)$ is the sign of the permutation,

$$\epsilon(Q) = \begin{cases} +1 & \text{for } Q \text{ even permutation} \\ -1 & \text{for } Q \text{ odd permutation} \end{cases}$$

Substitution of (10.82) in the Schrodinger equation

$$H|\psi> = E|\psi> \tag{10.84}$$

turns the eigenvalue problem into the following conditions among the wave functions,

$$-t \sum_{i=1}^N [\psi(x_1\sigma_1, \ldots, x_i + 1\sigma_i, \ldots, x_N\sigma_N) + \psi(x_1\sigma_1, \ldots, x_i - 1\sigma_i, \ldots, x_N\sigma_N)]$$

$$+U \sum_{i<j} \delta_{x_i,x_j}\psi(x_1\sigma_1, \ldots, x_N\sigma_N)$$

$$= E\psi(x_1\sigma_1, \ldots, x_N\sigma_N) \tag{10.85}$$

To obtain this expression we have used the antisymmetry property in (10.83).

To gain some insight into this relation let us proceed as in the previous section by investigating the structure of the solutions in the low number of particles sectors. Thus we shall see how similar the solutions are to Heisenberg and Yang models solutions.

In the $N = 1$ particle sector, the eigenvalue problem is simply

$$- t\left[\psi(x+1, \sigma) + \psi(x, \sigma)\right] = E\psi(x, \sigma) \tag{10.86}$$

and the solution is a plane wave function as usual,

$$\psi(x, \sigma) = A(p\sigma)e^{ipx} \tag{10.87}$$

$$E = -2t \cos p \tag{10.88}$$

In the $N = 2$ particle sector, now the eigenvalue problem incorporates the interaction between electrons for the first time,

$$-t[\psi(x_1 + 1\sigma_1, x_2\sigma_2) + \psi(x_1 - 1\sigma_1, x_2\sigma_2) + \psi(x_1\sigma_1, x_2 + 1\sigma_2)+$$

$$+ \psi(x_1\sigma_1, x_2 - 1\sigma_2)] + U\delta_{x_1 x_2}\psi(x_1\sigma_1, x_2\sigma_2) = E\psi(x_1\sigma_1, x_2\sigma_2) \tag{10.89}$$

To solve this relation, we devise a Bethe ansatz solution quite similar to the Yang model solution except for the fact that here we have to impose the antisymmetry of the wave functions,

$$\psi(x_1\sigma_1, x_2\sigma_2) = A(p_1\sigma_1, p_2\sigma_2)e^{i(p_1x_1 + p_2x_2)} - A(p_2\sigma_1, p_1\sigma_2)e^{i(p_2x_1 + p_1x_2)} \tag{10.90}$$

for $x_2 \geq x_1$ and

$$\psi(x_1\sigma_1, x_2\sigma_2) = A(p_2\sigma_2, p_1\sigma_1)e^{i(p_1x_1 + p_2x_2)} - A(p_1\sigma_2, p_2\sigma_1)e^{i(p_2x_1 + p_1x_2)} \tag{10.91}$$

for $x_2 \leq x_1$.

When the two particles do not coincide, the Bethe ansatz automatically satisfies the eigenvalue condition provided the energy of the state is,

$$E = -2t(\cos p_1 + \cos p_2) \tag{10.92}$$

When the two particles meet at a certain lattice site, say $x_1 = x_2 = x$, the eigenvalue problem provides a relation for the amplitudes A in the Bethe wave functions,

$$\left[-t(e^{-ip_1} + e^{ip_2}) + U - E\right] A(p_1\sigma_1, p_2\sigma_2)-$$

$$- \left[-t(e^{ip_1} + e^{-ip_2}) + U - E\right] A(p_2\sigma_1, p_1\sigma_2)$$

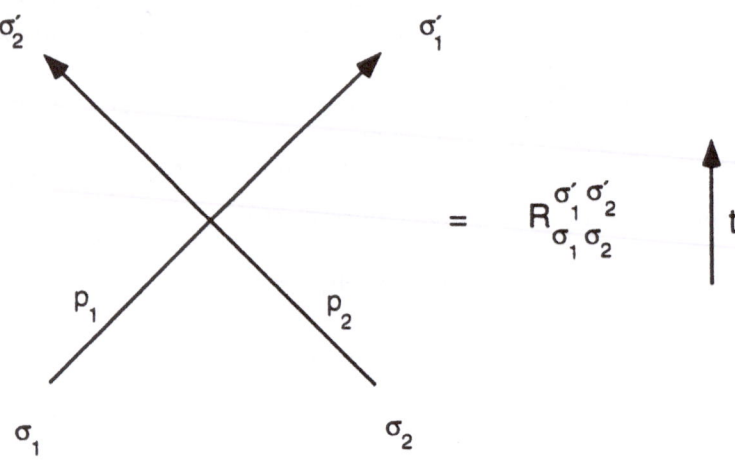

Fig. 10.12. Definition of the R-matrix for states carrying momentum and spin.

$$= -t(e^{-ip_1} + e^{ip_2})A(p_1\sigma_2, p_2\sigma_1) + +t(e^{ip_1} + e^{-ip_2})A(p_2\sigma_2, p_1\sigma_1) \qquad (10.93)$$

In addition to this condition we have to impose the continuity condition,

$$A(p_1\sigma_1, p_2\sigma_2) - A(p_2\sigma_1, p_1\sigma_2) = A(p_2\sigma_2, p_1\sigma_1) - A(p_1\sigma_2, p_2\sigma_1) \qquad (10.94)$$

With the aide of equations (10.93), (10.94) we can express two amplitudes in terms of the other two as,

$$A(p_2\sigma_2, p_1\sigma_1) = \sum_{\sigma_1'\sigma_2'} R^{\sigma_1\sigma_1'}_{\sigma_2\sigma_2'}(p_1, p_2)A(p_1\sigma_1', p_2\sigma_2') \qquad (10.95)$$

where the R-matrix is given by

$$R^{\sigma_1\sigma_1'}_{\sigma_2\sigma_2'}(p_1, p_2) = \frac{1}{\sin p_1 - \sin p_2 + i(U/2t)}$$

$$\left\{ (\sin p_1 - \sin p_2)\delta_{\sigma_1,\sigma_1'}\delta_{\sigma_2,\sigma_2'} + i(U/2t)\delta_{\sigma_2,\sigma_2'} \right\} \qquad (10.96)$$

and its physical meaning is depicted in figure 10.12.

Now that we got acquainted with the form of the solutions that emerge from the Bethe analysis, let us consider the general case of N-particle solutions. First of all, as the total spin is conserved, we shall be dealing with states with the total number of spins up fixed. The reason for this is that as in the Hubbard

model there are two degrees of freedom, charge and spin, we shall make a Bethe
ansatz for each degree of freedom one go at a time. This is the origin of the
nested version of the Bethe ansatz in this model.

To implement the fixing of the spin in the N-particle states, we introduce
a spatial ordering of the creation fermion operators $c_{x_i \sigma_i}^\dagger$ so that, given a total
number of spins up and an ordering of the particles, the N-particle state is
completely determined. Thus, we reexpress the many-particle state as,

$$|\psi> = \sum_Q \sum_{x_1 \leq \ldots \leq x_N} \phi(x_1 \sigma_{Q_1}, \ldots, x_N \sigma_{Q_N}) c_{x_1 \sigma_{Q_1}}^\dagger \cdots c_{x_N \sigma_{Q_N}}^\dagger |0> \qquad (10.97)$$

where the sum extends over all the permutations Q of the permutation group
S_N. In this fashion, once the total spin and the ordering of the particles is given,
the permutations Q label the wave functions. Thus, we introduce the notation,

$$\psi(x_1, \ldots, x_N|Q) \equiv \phi(x_1 \sigma_{Q_1}, \ldots, x_N \sigma_{Q_N}) \qquad (10.98)$$

With this notation the generalization of the Bethe ansatz to the many-particle
case is straightfoward,

$$\psi(x_1, \ldots, x_N|Q) = \sum_P (-1)^{\epsilon(P)} A_{Q,P} e^{i \sum_j x_j P_{P_j}} \qquad (10.99)$$

where the summation runs over all the permutations of the permutation group
S_N.

With the problem thus posed, there are $N! \times N!$ amplitude coefficients to
be determined. We shall prove that these coefficients are completely calculated
when the following conditions are imposed,

- The eigenvalue condition (Schrödinger equation).

- The continuity conditions.

- The boundary conditions.

10.2.2 The eigenvalue condition.

To translate the eigenvalue problem (10.84) into a condition on the coefficient
amplitudes, we proceed as usual. We have to distinguish between two cases,

1. No particles meet at a lattice point. In this case (10.84) gives us a relation
   where there is no interacting term,

$$\sum_P (-1)^{\epsilon(P)} \left\{ -t \sum_{j=1}^N (e^{i P P_j} + e^{-i P P_j}) - E \right\} A_{Q,P} e^{i \sum_i x_i P P_i} = 0 \qquad (10.100)$$

This condition determines the energy E of the many-particle state, for imposing all the momenta to be different makes the bracket to vanish identically,

$$E = -2t \sum_{j=1}^{N} \cos p_i \qquad (10.101)$$

2. The case when any two particles coincide at a lattice site, say $x_i = x_{i+1}$. Then, inserting the Bethe ansatz in the Schrodinger equation (10.4) yields the equation,

$$\sum_{P} (-1)^{\epsilon(P)} \{ \left[t(e^{ip} + e^{-ip'}) + U \right] A_{Q,P} +$$

$$+ t(e^{-ip} + e^{ip'}) A_{Q',P} \} e^{i \sum_j x_j p_{P_j}} = 0 \qquad (10.102)$$

where we have used equation (10.101) along with the notation $p = p_{P_i}, p' = p_{P_{i+1}}$. As $x_i = x_{i+1}$, equation (10.102) can be further simplified,

$$\left[t(e^{ip} + e^{-ip'}) + U \right] A_{Q,P} + t(e^{-ip} + e^{ip'}) A_{Q',P}$$

$$- \left[t(e^{-ip} + e^{ip'}) + U \right] A_{Q,P'} - t(e^{ip} + e^{-ip'}) A_{Q',P'} = 0 \qquad (10.103)$$

where the permutations Q, Q' are defined by

$$Q = (Q_1, \ldots, Q_i, Q_{i+1}, \ldots, Q_N), \ Q' = (Q_1, \ldots, Q_{i+1}, Q_i, \ldots, Q_N) \qquad (10.104)$$

and similarly for the permutations P, P'.

10.2.3 Continuity conditions.

By consistency with the antisymmetric character of the many-particle wave functions, we must require

$$\psi(x_1, \ldots, x_N | Q) = -\psi(x_1, \ldots, x_N | Q') \qquad (10.105)$$

for two permutations related by (10.104). These conditions in turn yield the relations,

$$A_{Q,P} - A_{Q,P'} = A_{Q',P} - A_{Q',P'} \qquad (10.106)$$

10.2.4 Compatibility of eigenvalue and continuity conditions: the Yang-Baxter equation.

Now, the eigenvalue conditions (10.103) and continuity conditions (10.106) can expressed in the following form,

$$A_{Q,P} = \frac{(\sin p - \sin P')A_{Q',P} + i(U/2t)A_{Q,P}}{\sin p - \sin P') + i(U/2t)} \tag{10.107}$$

$$A_{Q',P'} = \frac{(\sin p - \sin P')A_{Q,P} + i(U/2t)A_{Q',P}}{\sin p - \sin P') + i(U/2t)} \tag{10.108}$$

These relations are the generalizations of (10.95) and (10.96). They can be casted in a more compact form if we adopt the convention of introducing $N!$ dimensional column vectors A_P, analogously to the ξ_P amplitude vectors in the Yang model, the components of which are the $N!$ amplitude coefficients $A_{Q,P}$ with P fixed and Q running over all permutations Q. Under this convention, equations (10.107) and (10.108) are simplified as,

$$A_{P'} = \frac{(\lambda - \lambda')P_{i,i+1} + i(U/2t)}{\lambda - \lambda' + i(U/2t)}A_P \tag{10.109}$$

where we have introduced the rapidities $\lambda = \sin p$, $\lambda' = \sin p'$ and the permutation operator $P_{i,i+1}$ acts on the permutations by simply interchanging Q_i and Q_{i+1}.

Equation (10.109) sintetizes the eigenvalue and continuity conditions and it calls for the introduction of the R-matrix for the Hubbard model,

$$R_{i,i+1}(\lambda - \lambda') = \frac{(\lambda - \lambda')P_{i,i+1} + i(U/2t)}{\lambda - \lambda' + i(U/2t)} \tag{10.110}$$

$$A_{P'} = R_{i,i+1}(\lambda - \lambda')A_P \tag{10.111}$$

It is clear that equation (10.109) represent $N! \times (N-1)$ conditions for $N!$ unknown amplitude coefficients A_P. So the system of equations is highly redundant. Recall that in the analysis of the Yang model we alrady met with this same problem and the solution to it was found through the Yang-Baxter equation that the S-matrix is shown to satisfy.

$$R_{23}(\lambda - \lambda')R_{12}(\lambda - \lambda'')R_{23}(\lambda' - \lambda'') = R_{12}(\lambda' - \lambda'')R_{23}(\lambda - \lambda'')R_{12}(\lambda - \lambda') \tag{10.112}$$

Given any amplitude vector A_P we can obtain it from the vector A_0 corresponding to the identity permutation by successive application of the scattering S-matrix. The Yang-Baxter equation enforces that this connection between amplitudes for different permuations does not depend on the particular sequence of scattering matrices used to do the job (see figure 10.9).

10.2.5 Periodic boundary conditions.

As must be quite well-known by now, in order to have a consistently well-defined theory on a lattice ring we need to impose periodic boundary conditions on the many-body solutions.

In the $N = 1$ particle sector the boundary condition turns out to be,

$$\psi(x + L, \sigma) = \psi(x, \sigma) \tag{10.113}$$

where ψ is the plane wave solution (10.7). Then, the condition that determines the momenta is

$$e^{ipL} = 1 \tag{10.114}$$

which yields

$$p = \frac{2\pi J}{L} \quad \text{with} \quad J = 0, 1, \ldots, L - 1 \tag{10.115}$$

The ground state corresponds to $J = 0$.

Using periodic boundary conditions

$$c_{x_N \sigma} = c_{x_0 \sigma} \quad \text{with} \quad x_N = x_0 + L \tag{10.116}$$

we can set up the usual boundary conditions on the many-paticle wave functions, namely

$$\psi(x_1, \ldots, x_{N-1}, x_0 + L | Q) = (-1)^{N-1} \psi(x_0, x_1, \ldots, x_{N-1} | \tilde{Q}) \tag{10.117}$$

where $\tilde{Q}$ is the cyclic permutation $\tilde{Q} = (Q_N, Q_1, \ldots, Q_{N-1})$.

In order for this equation to hold, the amplitude coefficients must satisfy the equations,

$$A_{Q,P} e^{ip_{P_N} L} = A_{\tilde{Q}, \tilde{P}} \tag{10.118}$$

Using equations (10.109) and (10.112) the boundary conditions can be expressed as,

$$A_P e^{ip_{P_N} L} = P_{N,N-1} \ldots P_{32} P_{21} A_{\tilde{P}}$$

$$= P_{N,N-1} \ldots P_{32} P_{21} R_{12}(\lambda_{p_1} - \lambda_{p_N}) \cdots R_{N-1,N}(\lambda_{p_{N-1}} - \lambda_{p_N}) A_P \tag{10.119}$$

The meaning of these relations is that the same $N!$ amplitude vector must be an eigenvector of N different operators. These relations are familiar to us, for they appeared in the solution of the Yang model withe the exception that the momenta variables p there must be substituted by the rapidities $\sin p$ here. This is the key observation made by Lieb and Wu in solving [3] the Hubbard model.

It is of no importance that the Yang model is defined on the continuum while the Hubbard model is defined on the lattice, for the algebraic structure of both models is the same.

To simplify the analysis of the eigenvalue problem (10.119), we shall choose a particular permutation P defined by $(P_1, \ldots, P_N) = (1, \ldots, j-1, j+1, N, j)$. We prefer to express equation (10.119) in terms of the amplitude vector corresponding to the permutation identity A_0. This can be achieved by applying a string of scattering matrices, the result being indepent of the particular choice of the string. Let us use the string,

$$A_0 = R_{j,j+1}(\lambda_{j+1} - \lambda_j) \cdots R_{N-1,N}(\lambda_N - \lambda_j)A_0 \qquad (10.120)$$

The periodic boundary conditions (10.39) are then expressed as,

$$e^{ip_j L} A_0 = R_{j,j+1}(\lambda_{j+1} - \lambda_j) \cdots R_{N-1,N}(\lambda_N - \lambda_j)$$

$$P_{N,N-1} \ldots P_{32} P_{21} R_{12}(\lambda_1 - \lambda_j) \cdots R_{j-1,j}(\lambda_{j-1} - \lambda_j) A_0 \qquad (10.121)$$

Then, as in the case of the Yang model, it is convenient to introduce scattering S-matrices by,

$$S'_{i,j}(\lambda_i - \lambda_j) = P_{i,j} R_{ij}(\lambda_i - \lambda_j) \qquad (10.122)$$

In terms of these S-matrix operators, the boundary conditions read as follows,

$$e^{ip_j L} A_0 = S_{j,j+1}(\lambda_{j+1} - \lambda_j) S_{j,j+2}(\lambda_{j+2} - \lambda_j) \cdots$$

$$S_{j,N}(\lambda_N - \lambda_j) S_{1,j}(\lambda_1 - \lambda_j) \cdots S_{j-1,j}(\lambda_{j-1} - \lambda_j) A_0 \qquad (10.123)$$

These N eigenvalue equations are sufficient conditions to construct eigenstates of the Hubbard hamiltonian satisfying periodic boundary conditions.

10.2.6 Nested Bethe ansatz.

Recall that, so far, we have not made any use of the spin degree of freedom. As a matter of fact, in all the previous analysis we fixed from the very begining the conserved total spin of the many-particle states. Now, time has come to make use the symmetries of the wave function amplitudes with respect to the exchange of spins.

In general, we shall have M particles with spin down and $N - M$ with spin up.

Remark. Any permutation of two like spins does not change the state, and the wave functions with the same sequence of spins are equal.

We could proceed as usual starting with the sector of no down spins, then we would turn down one, then two and so on and so forth until achieve the general solution in the sector of M overturned spins. However, we shall make a shortcut to solve the model by mapping it onto the Yang solution. This mapping is by now clear from the discussion thus far, the key point being that both models have the same algebraic structure (Yang-Baxter equation) and the fictitious spin degree of freedom introduced in the solution of the Yang model is the actual spin of the electrons in the Hubbard model.

More explicitly, the R matrix of the Hubbard model has the same structure as that of the Yang model if the following substitutions are made,

$$
\begin{aligned}
k &\longrightarrow \sin p \\
c &\longrightarrow U/2t \\
\Lambda &\longrightarrow \Lambda
\end{aligned}
$$

With this identification at hand we can read off inmediately the solution to the eigenvalue problem. From equations (10.78), (10.79) this is given by,

$$
p_j L = 2\pi J_j + \sum_{\alpha=1}^{M} \vartheta(2(\sin p_j - \Lambda_\alpha)), \quad j = 1,\ldots,N. \tag{10.124}
$$

$$
\sum_{j}^{N} \vartheta(2(\sin p_j - \Lambda_\alpha)) = 2\pi I_\alpha + \sum_{\beta \neq \alpha}^{M} \vartheta(\Lambda_\beta - \Lambda_\alpha), \quad \alpha = 1,\ldots,M. \tag{10.125}
$$

where

$$
e^{i\vartheta(x)} \equiv \frac{(U/2t) - ix}{-(U/2t) - ix}, \quad -\pi < \vartheta(x) \leq \pi \tag{10.126}
$$

and the quantum numbers $\{J_j\}$ are all different from each other. They are integers if M is even and half-odd integers (HOI) if M is odd. Similarly, the quantum numbers $\{I_\alpha\}$ are all distinct from each other. They are integers if $N - M$ is odd and HOI if $N - M$ is even. Altogether, they both are defined modulo L. In addition, there is the restriction

$$
|I_\alpha| < (N - M + 1)/2 \tag{10.127}
$$

These are the two set of equations, first obtained by Lieb and Wu [3], characterizing the eigenstates of the one dimensional Hubbard model for the case of M spin down electrons and $N - M$ spin up electrons on a lattice ring of L sites. Notice that the coupling dependence in U is encoded in the $\vartheta(x)$ function introduced in (10.126).

As an example of how we can retrieve basic physical features of the Hubbard model from these sets of equations, let us see how to determine the ground state of the model, one of the paradigms one envisages to characterize in higher dimensions.

10.2.7 Ground state of the Hubbard model.

An immeditate consequence of equations (10.124), (10.125) is,

$$\sum_{j=1}^{N} k_j = \frac{1}{L}\left(\sum_{j=1}^{N} I_j + \sum_{\alpha=1}^{M} J_\alpha\right) \tag{10.128}$$

For the ground state, J_α and I_j are consecutive integers (or half-odd integers) centered aroun the origin and satisfying

$$\sum_{j=1}^{N} k_j = 0 \tag{10.129}$$

10.3 Physical Consequences of Lieb-Wu's Equations

The study of the Lieb-Wu's equations yields the answer to the central questions of interest posed at the begining of the previous section. As we shall see below, the outcome of that study is,

Mott transition There is NO Mott transition for nonzero coupling constant U. Actually, the ground state for a half-filled band is insulating for any nonzero U, and conducting for $U = 0$.

Ground state The magnetic nature of the ground state is *antiferromagnetic*.

To address these physical questions we need to perform the thermodynamic limit of Lieb-Wu's equations. This amounts to take the limits $N \longrightarrow \infty$, $M \longrightarrow \infty$ and $L \longrightarrow \infty$ keeping the ratios N/L and M/L finite. These ratios represent the total density of charge and spin, respectively.
As it happens, the real numbers p_j and λ_α are uniformly distributed between say, $-p_0$ and $p_0 \leq \pi$ and $-\lambda_0$ and $\lambda_0 \leq \infty$. These cutoffs are related to the largest quantum numbers J_j and I_α, i.e, to the total number of particles N and down spins M respectively.
Introducing density distributions $\rho(p)$ and $\sigma(\lambda)$ for the charge and spin degrees of freedom, respectively, equations (10.124), (10.125) then lead to the following coupled integral equations for the unknown densitiy distributions,

$$1 = 2\pi\rho(p) - 2\cos p \int_{-\lambda_0}^{\lambda_0} d\lambda \sigma(\lambda) \frac{U/t}{(U/2t)^2 + 4(\sin p - \lambda)^2} \tag{10.130}$$

$$2\int_{-p_0}^{p_0} dp \rho(p) \frac{U/t}{(U/2t)^2 + 4(\sin p - \lambda)^2} =$$

$$2\pi\sigma(\lambda) + \int_{-\lambda_0}^{\lambda_0} d\lambda'\sigma(\lambda')\frac{U/t}{(U/2t)^2 + 4(\sin p - \lambda)^2} \tag{10.131}$$

The cutoffs p_0 and λ_0 are determined by the conditions,

$$\int_{-p_0}^{p_0} dp\rho(p) = \frac{N}{L} \tag{10.132}$$

$$\int_{-\lambda_0}^{\lambda_0} d\lambda\sigma(\lambda) = \frac{M}{L} \tag{10.133}$$

In the continuum representation the ground state energy (10.104) now becomes,

$$E_0 = -2tL \int_{-p_0}^{p_0} dp\rho(p)\cos p \tag{10.134}$$

This expression is formally the same as that for a system of non-interacting fermions, the difference being that the interactions produce phase shifts leading to a modified density of states.

From the analysis of the above equations, Lieb and Wu established the following,

1. The coupled integral equations have a unique solution which is positive for all the allowed values of the cutoffs p_0 and λ_0.

2. The ratio M/N of the total density of spin to the total density of charge is a monotonically increasing function of p_0 reaching a maximum of $\frac{1}{2}$ at $B = \infty$. This is the antiferromagnetic case, $S_z = 0$, and corresponds to the absolute ground state.

3. The total density of charge is a monotonically increasing function of λ_0, reaching a maximum at $p_0 = \pi$.

Lieb and Wu [3] succeded in solving analitically the coupled integral equations for the case of a half-filled band $N = L$. For other band fillings those equations have been solved both numerically for general values of U [6] and analitically for large U [7].

To determine whether the Hubbard model is metal or an insulator in its ground state, we have to compute the chemical potentials μ_+ and μ_- defined as

$$\mu_+ \equiv E(M+1, M; U) - E(M, M; U)$$

$$\mu_- \equiv E(M, M; U) - E(M-1, M; U) \tag{10.135}$$

where the groung state energy is denoted by $E(M, N-M; U)$.

If μ_+ and μ_- are equal, the system behaves as a conductor, while if $\mu_+ > \mu_-$ it behaves as an insulator. From the calculation of these chemical potentials using the Lieb-Wu's one finds that, indeed,

$$\mu_+ > \mu_- \quad \text{for} \quad U > 0. \tag{10.136}$$

and

$$\lim_{U \to 0} \mu_{\pm} = 0 \tag{10.137}$$

10.3.1 Excitation Spectrum.

On top of the ground state described above we may have four types of excited states which are obtained by varying the quantum numbers $\{J_j\}$ and $\{I_\alpha$ in the Lieb-Wu's equations (10.124), (10.125) [8]. These are spinons, holons, particle states and a hole states. The sequence of quantum numbers defining the ground state is, for a system with $N = 4n + 2$ electrons,

$$\{J_j\} = \{-(N-1)/2, \ldots, (N-1)/2\} \tag{10.138}$$

$$\{I_\alpha\} = \{-(N/2-1)/2, \ldots, (N/2-1)/2\} \tag{10.139}$$

Holons The holons are charged states, but spinless, which are obtained varying the quantum numbers $\{J_j\}$ in (10.138) associated to charge degrees of freedom. There are many possibilities the simplest of which is to remove one value of J_j from inside the ground-state string (10.138) and place anther value, say J_0, outside the string. That is to say,

$$\{J_j\} = \{-(N-1)/2, \ldots, -(N-1)/2 + j_0 - 1,$$
$$-(N-1)/2 + j_0 - 1, \ldots, (N-1)/2, J_0\} \quad |J_0| > (N-1)/2 \tag{10.140}$$

$$\{I_\alpha\} = \{-(N/2-1)/2, \ldots, (N/2-1)/2\} \tag{10.141}$$

This is a two-parameter family of excited states (j_0, J_0).

Spinons The spinons are states carrying spin but not charge. They are obtained varying the quantum numbers $\{I_\alpha\}$ which are associated to the magnons of the Bethe solution. The simplest excitations may have total spin $S = 1$ (triplets) or $S = 0$ (singlets).

The triplets occur when all the momenta $p's$ and $\Lambda's$ real. This is only possible for $M < N/2$. Moreover, the condition $S = 1$ demands $M = N/2 - 1$. Then, the simplest case corresponds to making one hole in the ground-state string (10.139) and creating another quantum number outside the string as in (10.140) for the holons. This is a two-parameter family.

The singlets occur when there exists a pair of complex congugated $\Lambda's$ among the solutions to the original equations (10.124), (10.125). The condition $S = 1$ demands $M = N/2$. Then their construction goes step by step as in the triplet case. These triplet and singlet states can be thought of as formed out of non-interacting spin-1/2 particles called spinons.

Particle States One particle is added to the ground state leaving the number of overturned spins M unchanged. This forces all the quantum numbers to be half-odd integers. It can be shown that once the particle is created it evolves into separated charge and spin degrees of freedom [8]. This corroborates the results obtained from bosonization techniques which are appropiate for the description of the low-energy lying states. What is particular of the Hubbard model is that is spin-charge separation also occurs for higher excited states.

Hole States These are created by removing one particle from the ground state. Then $N \rightarrow N-1$, $M = (N-1)/2$. This forces all the quantum numbers to be integers now. The subsequent evolution of the hole states is similar to the particles states, namely, they split into spin-charge degrees of freedom (spinons and holons).

References

[1]Bethe, H.A., *1931, Z. Phys.* **71**, *205*

[2]Yang, C.N., *1967, Phys. Rev. Lett.* **19**, *1312*

[3]Lieb, E.H., Wu, F. Y., *1968, Phys. Rev. Lett.* **20**, *1445*

[4]Baxter, R.J., *Exactly Solved Models in Statistical Mechanics* Academic Press, London, 1982.

[5]Lieb, E.H., Liniger, W., *1963, Phys. Rev.* **130**, *1605*

[6]Shiba, H., *1972, Phys. Rev. B* **6**, *930*

[7]Carmelo, J., Baeriswyl, D., *1988, Phys. Rev. B* **37**, *7541*

[8]Schulz, H.J., *Interacting Fermions in One Dimension: From Weak to Strong Correlation* Proceedings of the Jerusalem Winter School for Theoretical Physics on "Correlated Electron Systems". Ed. V.J. Emery. World Scientific 1992.

11. New and Old Real-Space Renormalization Group Methods for Quantum Lattice Hamiltonians

11.1 Introduction

The study of correlated system such as the Hubbard model requires a combination of different techniques in order to derive the relevant physics. Mean field methods, although not recommended in correlated systems, may describe nevertheless some particular situations.

Other situations may however require more powerful techniques already existing in the literature, and quite likely new methods need to be invented to handle complicated cases. This is probably the situation one is facing in connection with high-T_C models. One may think that the renormalization group methods (RG) will play an important role in the study of these problems. In parts I and II of this book (Chaps. 2, 3) a momentum-space RG-approach in the framework of the solid state physics has been addressed. For the sake of completeness, we would like to review briefly the real space RG-approach. Moreover, it turns out that momentum-space RG-formulations readly become perturbative thereby giving only information in a small neighbourhood of the coupling constant around a critical point. On the contrary, we shall see that real-space RG may lead to non-perturbative information in the whole range of the coupling constant either by using analytical schemes or numerical schemes. From the point of view of Chap. 8 of Part III it is also consistent to pursue the real-space RG since it is a natural continuation of the strong coupling analysis of Chap. 8. As a matter of fact, we have proved there that the t-J model can be obtained as the strong coupling limit (i.e. $U/t \longrightarrow \infty$) of the Hubbard model. This map (Hubbard $\longrightarrow$ t-J) is nothing but an elimination of the high energy mode (double occupied states) and the construction of a low energy effective Hamiltonian (the t-J Hamiltonian) in the sheer RG-philosophy. One would like to continue this mode elimination procedure down to lower and lower energies. For that purpose the real-space RG-method seems in principle quite suitable and it is surprising that it has not recieved more attention in the past. Only recently there has been new developments along this direction.

11.2 Foundations of the Real-Space Renormalization Group

Let us summarize the main features of the real-space RG. The problem that one faces generically is that of diagonalizing a quantum lattice Hamiltonian H, i.e.,

$$H|\psi> = E|\psi> \tag{11.1}$$

where $|\psi>$ is a state in the Hilbert space $\mathcal{H}$. If the lattice has N sites and there are k possible states per site then the dimension of $\mathcal{H}$ is simply

$$dim\mathcal{H} = k^N \tag{11.2}$$

The examples we have considered in the previous chapters correspond to $k = 4$ (Hubbard model), $k = 3$ (t-J model), $k = 2$ (Heisenberg model) etc.
When N is large enough the eigenvalue problem (11.1) is out of the capability of any human or computer means unless the model turns out to be integrable which only happens in $d = 1$.

These facts open the door to a variety of approximate methods among which the RG-approach, specially when combined with other techniques (e.g. numerical, variational etc.), is one of the most relevant. The main idea of the RG-method is the mode elimination or thinning of the degrees of freedom followed by an iteration which reduces the number of variables step by step until a more managable situation is reached. These intuitive ideas give rise to a well defined mathematical description of the RG-approach to the low lying spectrum of quantum lattice hamiltonians.

To carry out the RG-program we shall introduce the following objects:

- $\mathcal{H}$: Hilbert space of the original problem.

- $\mathcal{H}'$: Hilbert space of the effective degrees of freedom.

- H: Hamiltonian acting in $\mathcal{H}$.

- H': Hamiltonian acting in $\mathcal{H}'$ (effective Hamiltonian).

- T : embedding operator : $\mathcal{H}' \longrightarrow \mathcal{H}$

- $T^\dagger$:truncation operator : $\mathcal{H} \longrightarrow \mathcal{H}'$

The problem now is to relate H, H' and T. The criterium to accomplish this task is that H and H have in common the low lying spectrum. An exact implementation of this is given by the following equation:

$$HT = TH' \tag{11.3}$$

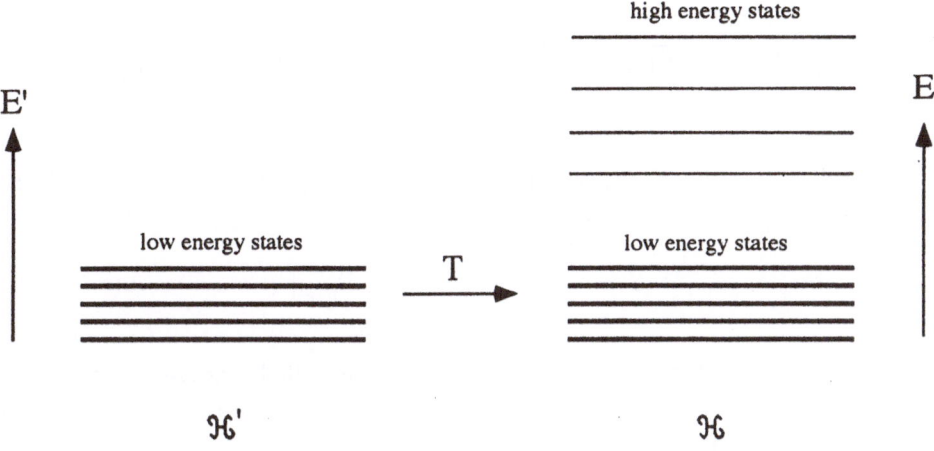

Fig. 11.1. Action of the intertwiner operator T as an embedding operator for the low liying states of the spectrum.

which imply that if $\Psi'_{E'}$ is an eigenstate of H' then $T\Psi'_{E'}$ is an eigenstate of H with the same eigenvalue (unless it belongs to the kernel of T: $T\Psi'_{E'} = 0$), indeed,

$$HT\Psi'_{E'} = TH'\Psi'_{E'} = E'T\Psi'_{E'} \tag{11.4}$$

To avoid the possibility that $T\Psi' = 0$ with $\Psi' \neq 0$, we shall impose on T the condition,

$$T^{\dagger}T = \mathbb{1}_{\mathcal{H}'} \tag{11.5}$$

such that

$$\Psi = T\Psi' \Rightarrow \Psi' = T^{\dagger}\Psi \tag{11.6}$$

Condition (11.5) thus stablishes a one to one relation between $\mathcal{H}'$ and $\mathrm{Im}(T)$ in $\mathcal{H}$ (see Fig.11.1)
Observe that Eq. (11.3) expresses the commutativity of the following diagram:

$$
\begin{array}{ccc}
\mathcal{H}' & \xrightarrow{T} & \mathcal{H} \\
H' \downarrow & & \downarrow H \\
\mathcal{H}' & \xrightarrow{T} & \mathcal{H}
\end{array}
$$

In fact, T is nothing but an *intertwiner operator* for the action of the Hamiltonians H and H' in their corresponding Hilbert spaces.

A formal solution of Eqs. (11.3) and (11.5) which explains the philosophy underlying the RG-approach is given by:

$$H = \sum_{E_{low}} |E_{low} > E_{low} < E_{low}| + \sum_{E_{high}} |E_{high} > E_{high} < E_{high}| \quad (11.7)$$

$$H' = \sum_{E_{low}} |E_{low} >' E_{low}' < E_{low}| \quad (11.8)$$

$$T = \sum_{E_{low}} |E_{low} > \,' < E_{low}| \quad (11.9)$$

Eqs. (11.3) and (11.5) characterize what may be called exact renormalization group method (ERG) in the sense that the whole spectrum of H' is mapped onto a part (usually the bottom part) of the spectrum of H. In practical cases though the exact solution of Eqs. (11.3) and (11.5) is not possible so that one has to resort to approximations (see later on).

At this point it is worthwhile to point out that Eqs. (11.7)- (11.9) represent in the framework of hte canonical quantization the equivalent implementation of RG ideas in the framework of the path integral formalism in which the RG program is carried out by integrating out the high energy classical fields Φ_{high} and retaining the low energy ones Φ_{low} in the path integral measure (recall Chap. 2):

$$\mathcal{Z} = \int [d\Phi] e^{-S[\Phi]} = \int [d\Phi_{low}][d\Phi_{high}] e^{-S[\Phi_{low} + \Phi_{high}]}$$

$$= \int [d\Phi_{low}] e^{-S_{eff}[\Phi_{low}]} \quad (11.10)$$

Considering Eqs. (11.3) and (11.5) we can set up the effective Hamiltonian H' as:

$$H' = T^\dagger H T \quad (11.11)$$

This equation does not imply that the eigenvectors of H' are mapped onto eigenvectors of H. Notice that Eq.(11.8) together with (11.5) does not imply Eq. (11.3) (recall that $TT^\dagger \neq \mathbb{1}_{\mathcal{H}}$).

What Eq.(11.8) really implies is that the mean energy of H' for the states Ψ' of $\mathcal{H}'$ coincides with the mean energy of H for those states of $\mathcal{H}$ obtained through the embedding T, namely,

$$< \Psi'|H'|\Psi' > = < \Psi'T|H'|T\Psi' > \quad (11.12)$$

In other words $T\Psi'$ is used as a variational state for the eigenstates of the Hamiltonian H. In particular T should be chosen in such a way that the states truncated in $\mathcal{H}$, which go down to $\mathcal{H}'$, are the ones expected to contribute the most to the ground state of H. Thus Eq. (11.8) is the basis of the so called variational renormalization group method (VRG). As a matter of fact, the VRG method was the first one to be proposed. The ERG came afterwards as a perturbative extension of the former (see later on).

More generally, any operator $\mathcal{O}$ acting in $\mathcal{H}$ can be "pushed down" or renormalized to a new operator $\mathcal{O}'$ which acts in $\mathcal{H}'$ defined by the formula,

$$\mathcal{O}' = T^\dagger \mathcal{O} T \tag{11.13}$$

The practical implementation of the VRG and ERG methods requires the introduction of the concept of the block. In this sense both of them are block methods.

11.3 Block Methods (BRG)

Once we have established the main features of the RG-program, there is quite freedom to implement specifically this fundamentals. We may classify this freedom in two aspects:

- The choice of how to reduce the size of the lattice.

- The choice of how many states to be retained in the truncation procedure.

We shall address the first aspect now. There are mainly two procedures to reduce the size of the lattice:

- by dividing the lattice into blocks with n_s sites each. This is the blocking method introduced by Kadanoff to treat spin lattice systems.

- by retrieving site by site of the lattice at each step of the RG-program. This is the procedure used by Wilson in his RG-treatment of the Kondo problem. This method is clearly more suitable when the lattice is one-dimensional.

We shall be dealing with the block methods mainly because they are well suited to perform analytical computations and because they are conceptually easy to be extended to higher dimensions.

To exemplify this method we shall study a 1d-lattice Hamiltonian. The main ideas are also valid in higher dimensions although computations are more involved. We shall also follow to the chinese addaggio saying that one picture is worth more than thousand words.

Hence we shall be dealing with a one-dimensional lattice, usually a periodic chain (see Fig. 11.2).

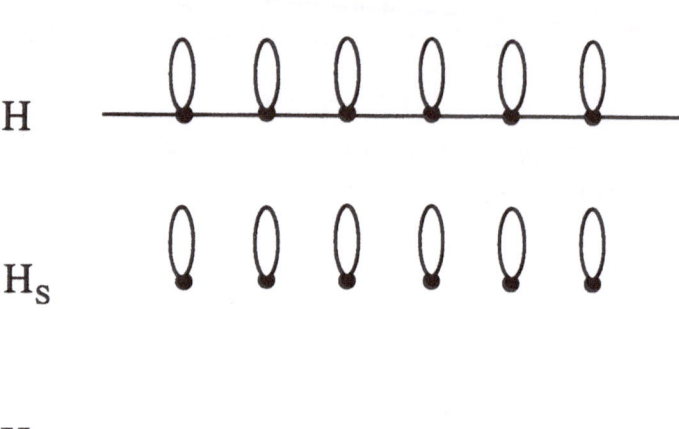

Fig. 11.2. One dimensional open chain.

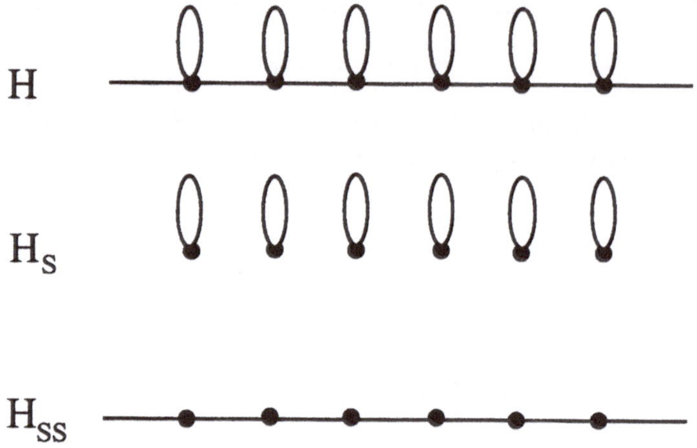

Fig. 11.3. Pictorical decomposition of the Hamiltonian H into single-site part H_S and two-nearest-neighbour-site part H_{SS}.

In every site of the chain there are k degrees of freedom, hence:

$$\mathcal{H} = \mathbb{C}^k \otimes \overset{N}{\ldots} \otimes \mathbb{C}^k := \otimes^N \mathbb{C}^k \qquad (11.14)$$

We shall consider Hamiltonians H containing operators which involve only a single-site part H_{s} or two-nearest-neighbour-site part H_{ss} and will be simbolically depicted as follows (see Fig. 11.3),
in such a way that

$$H = H_{\mathrm{s}} + H_{\mathrm{ss}} \qquad (11.15)$$

As a matter of illustration, let us give one example of this decomposition which will turn out to be very useful in putting many key ideas to the test.

11.3.1 Ising Model in a Transverse Field (ITF)

The Ising Model in a Transverse Field is originally a one-dimensional quantum lattice system with quantum critical properties equal to the well-known thermal

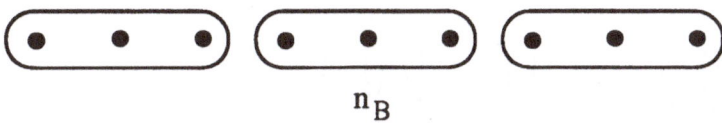

Fig. 11.4. Block decomposition of the open chain into blocks with $n_B = 3$ sites.

critical properties of the classical 2D-Ising Model. The lattice Hamiltonian of the ITF model is:

$$H_N(\Gamma, J) = -\Gamma \sum_{j=1}^{N} \sigma_j^x - J \sum_{j=1}^{N} \sigma_j^z \sigma_{j+1}^z \qquad (11.16)$$

The Hilbert space of states and the intrablock and interblock Hamiltonians for this model are, respectively:

$$\mathcal{H} = \otimes_1^N \mathbb{C}^2 \qquad (11.17)$$

$$H_S = -\Gamma \sum_{j=1}^{N} \sigma_j^x \qquad (11.18)$$

$$H_{SS} = -J \sum_{j=1}^{N} \sigma_j^z \sigma_{j+1}^z \qquad (11.19)$$

The first step of the BRG method consists in asembling the set of lattice points into diconnected blocks of n_B sites each, as in Fig. 11.4,

In this fashion there are a total of $N' = N/n_B$ blocks in the whole chain. This partition of the lattice into blocks induces a decomposition of the Hamiltonian (11.15) into an intrablock Hamiltonian H_B and a interblock Hamiltonian H_{BB} as illustrated in Fig. 11.5,

Observe that the block Hamiltonian H_B is a sum of commuting Hamiltonians each acting on every block. The diagonalization of H_B can thus be achieved for small n_B either analytically or numerically. The content of Fig. 11.5 can be written as

$$H = H_B + \lambda H_{BB} \qquad (11.20)$$

where λ is a coupling constant which is already present in H or else it can be introduced as a parameter characterizing the interblock coupling and in this latter case one can set it to one at the end of the discussion.

Eq. (11.20) suggests that we should search for solutions of the intertwiner equation (11.3) in the form of a perturbative expansion in the interblock coupling constant parameter λ, namely,

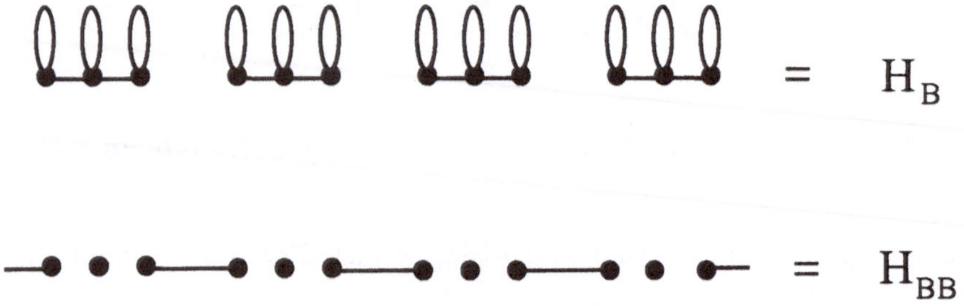

Fig. 11.5. Pictorical representation of the block Hamiltonian H_B and the interblock Hamiltonian H_{BB} for the ITF model.

$$T = T_0 + \lambda T_1 + \lambda^2 T_2 + \dots \tag{11.21}$$

$$H' = H'_0 + \lambda H'_1 + \lambda^2 H'_2 + \dots \tag{11.22}$$

To zeroth order in λ Eq. (11.3) becomes

$$H_B T_0 = T_0 H'_0 \tag{11.23}$$

Since H_B is a sum of disconnected block Hamiltonians $h_{j'}^{(B)}$, $j' = 1, \dots, N'$ implicitly defined through the relation

$$H_B = \sum_{j'=1}^{N'} h_{j'}^{(B)} \tag{11.24}$$

one can search for a solution of T_0 in a factorized form

$$T_0 = \prod_{j'=1}^{N'} T_{0,j'} \tag{11.25}$$

and an effective Hamiltonian H'_0 which acts only at the site j' of the new chain,

$$H'_0 = \sum_{j'=1}^{N'} h_{j'}^{(s')} = H'_{s'} \tag{11.26}$$

Observe that $H'_{s'}$ is nothing but a site-Hamiltonian for the new chain. Eq. (11.23) becomes for each block:

$$h_{j'}^{(B)} T_{0,j'} = T_{0,j'} h_{j'}^{(s')} \tag{11.27}$$

The diagonalization of $h_{j'}^{(B)}$ for $j' = 1, \ldots, N'$ will allow us to write

$$h_{j'}^{(B)} = \sum_{i=1}^{k'} |i>_{j'} \; \epsilon_i \; _{j'} <i| + \sum_{\alpha=1}^{k^{n_s}-k'} \alpha >_{j'} \epsilon_\alpha \; _{j'} < \alpha| \qquad (11.28)$$

where $|i>_{j'}$ for $j = 1, \ldots, k'$ are the k'-lowest energy states of $h_{j'}^{(B)}$. Moreover, we suppose that $h_{j'}^{(B)}$ is the same Hamiltonian for each block so that ϵ_i does not depend on the block.

The truncated Hamiltonian $h_{j'}^{(s)}$ and the intertwiner operator $T_{0,j'}$ are then given by:

$$h_{j'}^{(s')} = \sum_{i=1}^{k'} |i>_{j'}' \; \epsilon_i \; _{j'}^{\,'} <i| \qquad (11.29)$$

$$T_{0,j'} = \sum_{i=1}^{k'} |i>_{j'} \; _{j'}^{\,'} <i| \qquad (11.30)$$

The reader is urged to compare Eqs. (11.28)-(11.30) with Eqs. (11.7)-(11.9). Later on we shall show examples of these relations.

To obtain the first order correction to the Hamiltonian H_1' we must consider Eq. (11.3) to first order in λ:

$$H_{BB}T_0 + H_B T_1 = T_0 H_1' + T_1 H_0' \qquad (11.31)$$

Multiplying the left hand side by $T_0^\dagger$ and using $T_0^\dagger T_0 = 1$ along with $H_B T_0 = T_0 H_0'$ we readly obtain:

$$T_0^\dagger H_{BB}T_0 + H_0' T_0^\dagger T_1 = H_1' + T_0^\dagger T_1 H_0' \qquad (11.32)$$

We would like to kill the term proportional to $T_0^\dagger T_1$. For this purpose Eq. (11.5) which implies $T_0^\dagger T_1 + T_1 T_0^\dagger = 1$ is not very useful. A resolution of this problem can be accomplished if instead of the operator T one uses another operator $\tilde{T}$ satisfying the defining equations:

$$H\tilde{T} = \tilde{T} H' \qquad (11.33)$$

$$T_0^\dagger \tilde{T} = 1\!\!1_{\mathcal{H}'} \qquad (11.34)$$

Then $\tilde{T}_0 = T_0$ and $T_0^\dagger \tilde{T}_1 = 0$ in which case Eq. (11.32) simply becomes:

$$H_1' = T_0^\dagger H_{BB}T_0 = H_{s's'}' \qquad (11.35)$$

We can summarize these results saying that up to first order in λ, the effective Hamiltonian H' can be obtained using simply the zeroth order intertwiner operator T_0:

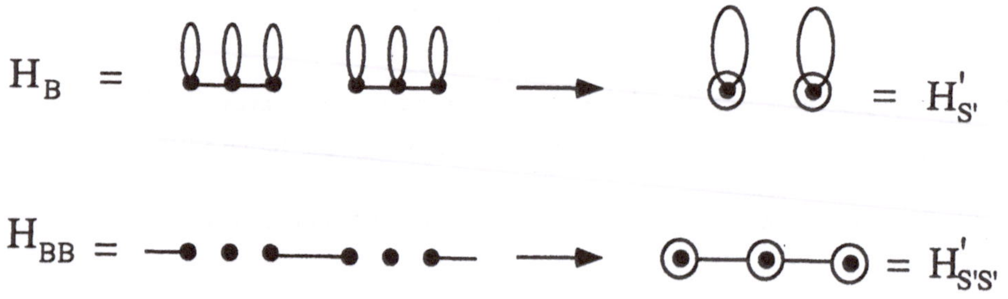

Fig. 11.6. Pictorical representation of the truncation procedure for the block and interblock Hamiltonians in the ITF model.

$$H'_{(\text{up to order } \lambda)} = H'_{s'} + H'_{s's'} = T_0^\dagger (H_B + \lambda H_{BB}) T_0 \qquad (11.36)$$

See Fig. 11.6,

This is precisely the prescription of Drell et al.

The second order correction to H' can be obtained again from Eqs. (11.33)-(11.34) and is given by

$$H'_2 = T_0^\dagger H_{BB} \tilde{T}_1 \qquad (11.37)$$

There is a close parallelism between the perturbative solution of Eqs. (11.3) or (11.33)- (11.34) and the pertubation theory of the Schrodinger equation for a Hamiltonian of the form $H_0 + \lambda H_1$. As a matter of fact, the normalization condition (11.34) for operators is equivalent to the standard normalization for wavefunctions $< \Psi_0|\Psi(\lambda) >=< \Psi_0|\Psi_0 >= 1$ that is adopted to avoid normalization complications. In what follows we shall mainly concentrate on the first order solution Eq. (11.36).

The final outcome of this analysis is that the effective Hamiltonian H' has a similar structure to the one we started with, H. The operators involved in $H'_{s'}$ and $H'_{s's'}$ may by all means differ from those of H_s and H_{ss}, but in some cases the only difference shows up as a change in the coupling constants. This is known as the renormalization of the bare coupling constants. When this is the case, one may easily iterate the RG-transformation and study the RG-flows.

Let us summarize the RG-prescription we have introduced so far in a table 11.1.

We have denoted this prescription by BRG1(n_s,k') where n_s and k' have been defined earlier and 1 denotes that we are working to first order in perturbation theory in solving Eqs. (11.7) -(11.9).

Table 11.1. Block Renormalization Group Method BRG1 (n_B, k')

Steps of the BRG1 Method
1) Blocking Transformation: $H = H_B + H_{BB}$
2) Diagonalization of H_B
3) Truncation within each block: T_0
4) Renormalization of H_B and H_{BB}
$H'_{s'} = T_0^\dagger H_B T_0$
$H'_{s's'} = T_0^\dagger H_{BB} T_0$
5) Iteration: Repeat 1) $\rightarrow$ 4) for $H' = H'_{s'} + H'_{s's'}$

We shall next carry out this BRG1 formalism to the models displayed in Table 11.1.

11.4 Ising Model in a Transverse Field (ITF)

We shall consider blocks of two sites ($n_s = 2$) and truncation to two states ($k' = 2$).

The block Hamiltonian has two sites and has the form,

$$h^{(B)} = -\Gamma(\sigma_1^x + \sigma_2^x) - J\sigma_1^z\sigma_2^z \qquad (\Gamma, J > 0) \qquad (11.38)$$

The eigenstates of this block Hamiltonian (11.38) are given in increasing order of energies by,

$$|G> = \frac{1}{\sqrt{1+a^2}}(|00> +a|11>) \quad E = -\sqrt{J^2 + 4\Gamma^2} \qquad (11.39)$$

$$|E> = \frac{1}{\sqrt{2}}(|00> +|11>) \qquad E = -J \qquad (11.40)$$

$$|E'> = \frac{1}{\sqrt{2}}(|00> -|11>) \qquad E = J \qquad (11.41)$$

$$|E''> = \frac{1}{\sqrt{1+a^2}}(-a|00> +|11>) \quad E = \sqrt{J^2 + 4\Gamma^2} \qquad (11.42)$$

$|0>$ and $|1>$ are the eigenstates of σ^x,

$$\sigma^x|0> = |0>, \quad \sigma^x|1> = -|1> \qquad (11.43)$$

and $a = a(g)$ is the following function of the ratio $J/2\Gamma := g$,

$$a(g) = \frac{-1 + \sqrt{1+g^2}}{g} \qquad (11.44)$$

which in turn satisfies

$$a(0) = 0, \quad a(\infty) = 1 \tag{11.45}$$

The intertwiner operator whithin each block has the form

$$T_0(a) = |G>'<0| + |E>'<1| \tag{11.46}$$

where $|0>'$ and $|1>'$ form a basis of states at each point of the new chain. The effective Hamiltonian H' up to order J can be computed from Eq. (11.36).

Thus to get $\mathcal{H}'$ we have to study the renormalization of the various operators entering in its definition. Using (11.46) and (11.39)-(11.42) one obtains after some elementary algebra:

$$T^\dagger \sigma_j^x T = \frac{1-a^2}{2(1+a^2)}(\mathbb{1}+\sigma_{j'}^x) \tag{11.47}$$

$$T^\dagger \sigma_j^z T = \frac{1+a}{\sqrt{2(1+a^2)}}\sigma_{j'}^z \tag{11.48}$$

$$T^\dagger \sigma_{2j-1}^z \sigma_{2j}^z T = \frac{(1+a)^2}{2(1+a^2)}\mathbb{1} + \frac{(1-a)^2}{2(1+a^2)}\sigma_{j'}^x \tag{11.49}$$

$$T^\dagger \sigma_{2j}^z \sigma_{2j+1}^z T = \frac{(1+a)^2}{2(1+a^2)}\sigma_{j'}^z \sigma_{j'+1}^z \tag{11.50}$$

The range of the indexes run as follows:

$$j = 2j' - 1 + p \tag{11.51}$$

$$j = 1, \ldots, N \tag{11.52}$$

$$j' = 1, \ldots, N/2 \quad\quad p = 0, 1 \tag{11.53}$$

Applying Eqs. (11.47)- (11.50) to the ITF Hamiltonian, one gets

$$T^\dagger H_N(\Gamma, J)T = \Delta E + H_{N/2}(\Gamma', J') \tag{11.54}$$

where

$$\Delta E = -\frac{N}{2}\left[\Gamma\frac{1-a^2}{(1+a^2)} + \frac{J}{2}\frac{(1+a)^2}{(1+a^2)}\right] \tag{11.55}$$

$$\Gamma' = \Gamma\frac{1-a^2}{(1+a^2)} - J\frac{(1+a)^2}{2(1+a^2)} \tag{11.56}$$

$$J' = J\frac{(1+a)^2}{2(1+a^2)} \tag{11.57}$$

The derivation of Eqs. (11.47)- (11.50) and (11.54)-(11.57) does not make use of Eq. (11.44) and hence have a more general validity. In other words, we can use the function $a(g)$ as a variational function in order to construct better ground states in the spirit of the VRG.

Eq. (11.44) is one of the numerous choices we can make. We shall consider later on other examples. Schematically we can set up the following relationship,

$$\text{RG-Prescription with } n_s = 2, k' = 2 \iff \begin{cases} a(g) \geq 0 \\ a(0) = 0, \quad a(\infty) = 1 \end{cases} \quad (11.58)$$

The physical properties of $H_N(\Gamma, J)$ depend only upon the ratio $g = J/2\Gamma$. If $0 \leq g \leq 1/2$ one in a disordered region characterized by a unique ground state with unbroken symmetry $(< \sigma^z >= 0)$. If $g > 1/2$ the $\mathbb{Z}_2$ symmetry associated to the operator $Q = \prod_j \sigma_j^z$, which commutes with H_N for N even, is broken. This is the ordered phase which has two degenerate ground states corresponding to $< \sigma^z >= \pm m \neq 0$.

At $g = g_c = 1/2$ the system is critical and belongs to the same universality class as the 2D-classical Ising model. The critical exponents can be defined in terms of the behaviour of the "quantum observables" as functions of $g_c - g$.

As in the theory of critical phenomena in statistical mechanics, most of the exponents can be computed from the properties of the RG-transformation. In the case of the ITF model the RG-transformation can be obtained from Eq.(11.55)-(11.57):

$$g' := \frac{J'}{2\Gamma'} := R(g) = \frac{1}{2} \frac{g(1 + a(g))^2}{1 - a(g)^2 - g(1 - a(g))^2} \quad (11.59)$$

For any function $a(g)$ satisfying Eq. (11.58), this transformation has 3 fixed points $g_* = 0, g_c, \infty$. The fixed points $g_* = 0$ and ∞ are attractive and correspond to the disordered and ordered phases respectively. The fixed point at g_c is repulsive and correspond to the critical point of the ITF Hamiltonian. The value of the function $a(g)$ at $g = g_c$ is a function only of g_c and does not depend on the particular prescription chosen, that is, it is a *universal function* :

$$g_c = R(g_c) \Rightarrow a(g_c) = \frac{2\sqrt{1 - 2g_c} + 2g_c - 1}{3 + 2g_c} \quad (11.60)$$

This equation implies in particular that whatever prescription is chosen, the critical value obtained from Eqs.(11.55)-(11.57) will always be less than the exact value $1/2$.

$$g_c \leq g_c^{\text{exact}} = 1/2 \quad (11.61)$$

To get the value of g_c for a given prescription one has simply to find the intersection of the function $a(g)$ and the function,

$$f(g) = \begin{cases} \frac{2\sqrt{1-2g}+2g-1}{3+2g} & 0 \leq g \leq 1/2 \\ 0 & g \geq 1/2 \end{cases} \quad (11.62)$$

as is shown in Fig. 11.7,

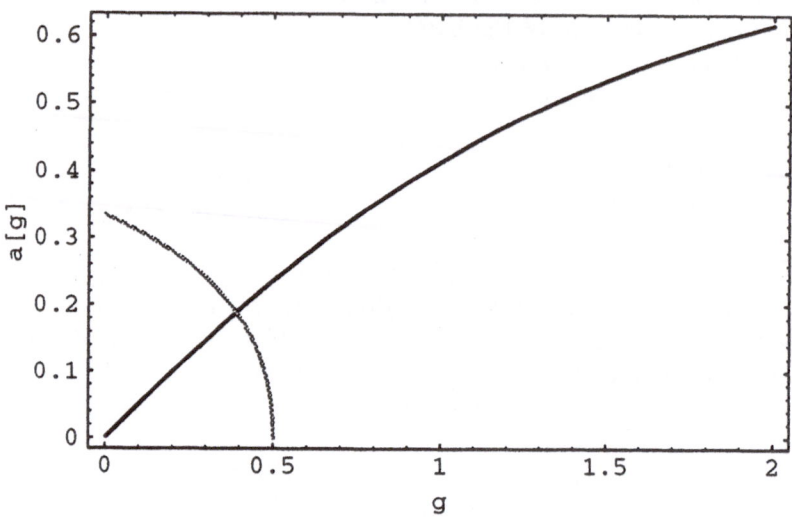

Fig. 11.7. Plot of the $a(g)$ function (solid line) for the ITF model and its cut with the universal function $f(g)$ (grey line).

The analysis of the RG-equations usually has to be done numerically. However, there is a great deal of information that can be retrieved without completely solving the RG-equations if we know wheather the successive coupling constants $g_n \longrightarrow g_{n+1} = R(g_n)$ increase or decrease during the iteration procedure. To this end it is convenient to introduce the the familiar beta function $\beta(g)$ of quantum field theory which in this context is,

$$\beta(g) := R(g) - g \tag{11.63}$$

In Fig. 11.8 we have plotted the beta function for the ITF model we are analyzing. A fixed point of the transformation occurs at values of g which reproduce themselves under the RG-iteration, i.e., they are the zeroes of the beta funciton:

$$\beta(g_c) = 0 \tag{11.64}$$

Thus Fig. 11.9 shows that there are 3 fixed points for besides the two zeroes at $g = 0$ and 0.39, $g = \infty$ is also a fixed point for it cannot be reduced by further iterations.

There is additional qualitative information which can be extracted from the shape of the beta function $\beta(g)$. In particular, the sign of $\beta(g)$ is responsible for the stability character of the fixed point. When $\beta(g) < 0$ (> 0) this means that g decreases (increases) after one iteration and the resulting g' lies to the left (right) of the g we started with. For the ITF model, Fig. 11.9 shows that $\beta(g)$ is negative for all $g < g_c$, thus if we start at a point $g < g_c$ successive iterations of the BRG1-method will drive us towards an effective Hamiltonian

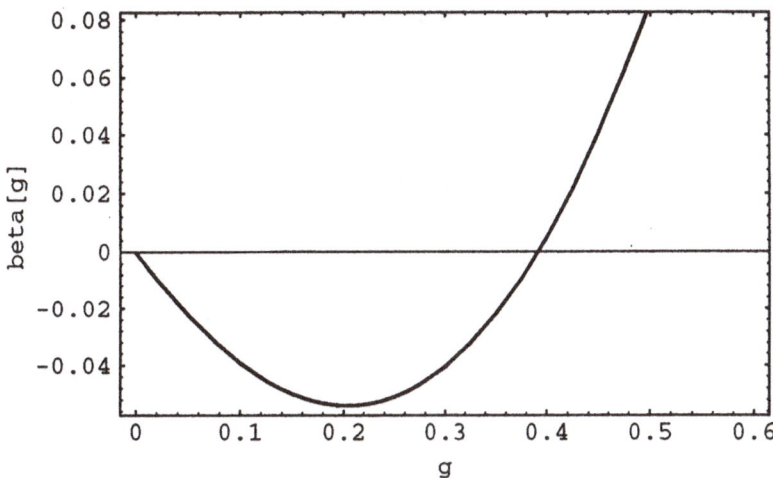

Fig. 11.8. Plot of the beta function $\beta(g)$ according to the BRG method for the ITF model.

Fig. 11.9. The RG-flow diagram of the ITF model.

corresponding to weak coupling perturbation theory. On the other hand, $\beta(g)$ is positive for $g > g_c$ and then if we start this region we will be driven towards $g = \infty$ which corresponds to the strong coupling limit of the ITF model.

The outcome of this RG-analysis can be summarized by saying that the fixed points $g_c = 0, \infty$ are *stable fixed points* while $g_c = 0.39$ is an *unstable fixed point*. Graphically this information in depicted in Fig. 11.9 in a one-dimensional RG-flow diagram.

Given the RG-transformation Eq. (11.59) we can compute several critical exponents and compare them with the exact results in order to check the accuracy of the method.

11.4.1 Correlation Length Exponent ν.

It gives the behaviour of the correlation length in the vecinity of g_c

$$\xi \sim (g - g_c)^{-\nu} \tag{11.65}$$

Under the RG-transformation (11.59) $\xi \to \xi' = \xi/2$ which leads inmediately to an expression for ν in terms of the derivative of $R(g)$ evaluated at the critical point,

$$\frac{1}{\nu} = \frac{\ln R'(g_C)}{\ln 2} \tag{11.66}$$

From Eq. (11.59) we can evaluate $R'(g_C)$ as a function of $a_C = a(g_C)$ and $a'_C = \frac{da}{dg}|_C$,

$$\lambda_T := R'(g_C) = 1 + 2g_C(\frac{1 - a_C}{1 + a_C})^2 + 4\frac{a'_C g_C \sqrt{1 - 2g_C}}{(1 + a_C)^2} \tag{11.67}$$

In Table 11.7 we show the value of ν obtained for different choices of the function $a(g)$.

11.4.2 Dynamical Exponent z.

At the critical point where $g' = g$ holds, the Hamiltonian changes by an overall factor which in turn defines the dynamical exponent z

$$H_C \longrightarrow H'_C = \frac{1}{2^z} H_C \tag{11.68}$$

In order to get z we notice that

$$\frac{1}{2^z} = (\frac{J'}{J})_C = (\frac{\Gamma'}{\Gamma})_C \tag{11.69}$$

Hence

$$z = \frac{\ln(\frac{J'}{J})}{\ln 2} = 1 + \frac{\ln\left[(1 + a_C^2)/(1 + a_C)^2\right]}{\ln 2} \tag{11.70}$$

It follows from Eq. (11.70) and the positivity property of a_C that z is always less than the exact value,

$$z \le z^{exact} = 1 \tag{11.71}$$

11.4.3 Magnetic Exponent β.

This critical exponent is defined through the spontaneous magnetization M_z in the ordered phase,

$$M = <\sigma_j^z> \tag{11.72}$$

Above a_C but close to the critical point we will have

$$M \sim (g - g_C)^\beta \tag{11.73}$$

11.4.4 Gap Exponent *s*.

Using scaling arguments satisfied by the BRG method, the gap G behaves like

$$G \sim [g - g_c]^s \tag{11.74}$$

with

$$s = \nu z \tag{11.75}$$

Equation (11.48) relates the magnetization M and the one obtained after the RG-transformation

$$M = \frac{1 + a}{\sqrt{2(1 + a^2)}} M' \tag{11.76}$$

Combining Eqs.(11.76), (11.73) and $g' = R(g)$ we arrive at,

$$\frac{M'}{M} = [R'(g_c)]^\beta = 2^{\beta/\nu} = \frac{1 + a_c}{\sqrt{2(1 + a_c^2)}} \tag{11.77}$$

Using Eq.(11.70) we are able to relate the critical exponents β, ν and z through the following scaling relation,

$$\beta = \frac{1}{2} z\nu \tag{11.78}$$

This relation is valid for any choice of the function $a(g)$ and therefore it is characteristic of using a block containing two sites. Observe that the exact exponents of the ITF model never satisfy this scaling relation (11.78). This shows the limitation of the block method when using a two-site block.

11.5 Antiferromagnetic Heisenberg Model

Prior to considering in detail the AF Heisenberg model we shall make some general considerations concerning Hamiltonians which commute with a symmetry group $\mathcal{G}$. Notice that for the AF Heisenberg model $\mathcal{G}$ is nothing but the rotation group $SU(2)$. Let us call g an element of the group $\mathcal{G}$ and $\Pi_{\mathcal{H}}(g)$ a representation of g acting on the Hilbert space $\mathcal{H}$. We say that $\mathcal{G}$ is a symmetry group of the Hamiltonian H if

$$[H, \Pi_{\mathcal{H}}(g)] = 0 \quad \forall g \in \mathcal{G} \tag{11.79}$$

Similarly we want the effective Hamiltonian H' to be invariant under the action of $\mathcal{G}$ acting now on the Hilbert space $\mathcal{H}'$,

$$[H', \Pi_{\mathcal{H}'}(g)] = 0 \quad \forall g \in \mathcal{G} \tag{11.80}$$

For this to be the case, the RG-transformation must preserve the symmetry of the original hamiltonian. This can be simply achieved if one choses T as the intertwiner or Clebchs-Gordan operator. Indeed we may recall that if a representation say Π_{V_3} is contained in the tensor product $\Pi_{V_1} \otimes \Pi_{V_2}$ then one can define the CG-operator as,

$$C_3^{12} : \quad V_3 \longrightarrow V_1 \otimes V_2 \tag{11.81}$$

which in fact satisfies the intertwiner condition:

$$(\Pi_{V_1} \otimes \Pi_{V_2})(g) \, C_3^{12} = C_3^{12} \Pi_{V_3}(g) \tag{11.82}$$

This equation expresses the commutativity of the following diagram,

$$
\begin{array}{ccc}
V_3 & \xrightarrow{C_3^{12}} & V_1 \otimes V_2 \\
\Pi_3 \downarrow & & \downarrow \quad \Pi_1 \otimes \Pi_2 \\
V_3 & \xrightarrow{C_3^{12}} & V_1 \otimes V_2
\end{array} \tag{11.83}
$$

Thus from a theoretical point of view, the truncation process pertaining to the real-space RG is nothing but a tensor product decomposition of representations of the group $\mathcal{G}$. To make this point more explicit let us suppose that $\mathcal{H}$ and $\mathcal{H}'$ are given as tensor products as:

$$\mathcal{H} = \otimes_1^N V \tag{11.84}$$

$$\mathcal{H}' = \otimes_1^{N'} V' \tag{11.85}$$

where V and V' are irreducible representation spaces of $\mathcal{G}$. Then the block method of Sect. 11.3 applied to this case is equivalent to the tensor product decomposition:

$$V \otimes \overset{n_B}{\ldots} \otimes V \longrightarrow V' \tag{11.86}$$

In Eq. (11.86) one is establishing that the irrep $\Pi_{V'}$ is contained in the tensor product of N copies of the irrep Π_V.

Obviously, the tensor product decomposition contains many irreps. The criterion to choose a particular irrep, or a collection of irreps is the one of *minimum energy*. All the states of a given irrep V' will have the same energy.

The summary of the discussion so far is that the intertwiner operator T_0 can be identified with the Clebsch-Gordan operator,

$$T_0 = C_{V'}^{V \otimes \overset{n_B}{\ldots} \otimes V} : V' \longrightarrow V \otimes \overset{n_B}{\ldots} \otimes V \tag{11.87}$$

Let us illustrate this ideas with the AF Heisenberg-Ising model whose Hamiltonian is given by:

$$H_N = J \sum_{j=1}^{N-1} (S_j^x S_{j+1}^x + S_j^y S_{j+1}^y + \Delta S_j^z S_{j+1}^z) \tag{11.88}$$

where $\Delta \geq 0$ is the anisotropic parameter and $J > 0$ for the antiferromagnetic case. If $\Delta = 1$ one has the AF-Heisenberg model which was solved by Bethe in 1931. If $\Delta = 0$ one has the XX-model which can be trivially solved using a Jordan-Wigner transformation which maps it onto a free fermion model. For the remaing values of Δ the model is also solvable by Bethe ansatz and it is the 1D relative of the 2D statistical mechanical model known as the 6-vertex or XXZ-model.

The region $\Delta > 1$ is massive with a doubly degenerate ground state in the thermodynamic limit $N \to \infty$ characterized by the non-zero value of the staggered magnetization,

$$m_{st} = \langle \frac{1}{N} \sum_j S_j^z (-1)^j \rangle \tag{11.89}$$

The region $0 \leq \Delta \leq 1$ is massless and the ground state is non-degenerate with a zero staggered magnetization. The phase transition between the two phases has an essential singularity.

We would like next to show which of these features are captured by a real-space RG-analysis. The rule of thumb for the RG-approach to half-integer spin model or fermion model is to consider blocks with an *odd number of sites*. This allows in principle, although not necessarilly, to obtain effective Hamiltonians with the same form as the original ones. Choosing for (11.88) blocks of 3 sites we obtain the block Hamiltonian:

$$\frac{1}{J}H = \mathbf{S}_1 \cdot \mathbf{S}_2 + \mathbf{S}_2 \cdot \mathbf{S}_3 + \epsilon(S_1^z S_2^z + S_2^z S_3^z)$$

$$= \frac{1}{2}\{[\mathbf{S}_1 + \mathbf{S}_2 + \mathbf{S}_3]^2 - (\mathbf{S}_1 + \mathbf{S}_3)^2 - 3/4\} + \epsilon(S_1^z S_2^z + S_2^z S_3^z) \tag{11.90}$$

$\epsilon := \Delta - 1$.

If $\epsilon = 0$ the block Hamiltonian H_B is invariant under the $SU(2)$ group and according to the introduction to this section, we should consider the tensor product decomposition:

$$\frac{1}{2} \otimes \frac{1}{2} \otimes \frac{1}{2} = \frac{1}{2} \oplus \frac{1}{2} \oplus \frac{3}{2} \tag{11.91}$$

The particular way of writing H_B given in Eq. (11.90) suggests to compose first $\mathbf{S}_1$ and $\mathbf{S}_3$ and then, the resulting spin with $\mathbf{S}_2$. The result of this of this compositions is given as follows:

$$|\frac{3}{2}, \frac{3}{2}\rangle = |\uparrow\uparrow\uparrow\rangle \quad E_B = J/2 \tag{11.92}$$

$$|\frac{3}{2}, \frac{1}{2}\rangle = \frac{1}{\sqrt{3}}(|\uparrow\downarrow\uparrow\rangle + |\downarrow\uparrow\uparrow\rangle + |\uparrow\uparrow\downarrow\rangle) \quad E_B = J/2 \tag{11.93}$$

$$|\frac{1}{2}, \frac{1}{2}\rangle_1 = \frac{1}{\sqrt{2}}(|\uparrow\uparrow\uparrow\rangle - |\downarrow\uparrow\uparrow\rangle) \quad E_B = 0 \tag{11.94}$$

$$|\frac{1}{2},\frac{1}{2}\rangle_0 = \frac{1}{\sqrt{6}}(2|\uparrow\downarrow\uparrow\rangle - |\downarrow\uparrow\uparrow\rangle - |\uparrow\uparrow\downarrow\rangle) \quad E_B = -J \qquad (11.95)$$

Hence for $\Delta = 0$ we could choose the spin 1/2 irrep. with basis vectors $|\frac{1}{2},\frac{1}{2}\rangle_0$ and $|\frac{1}{2},-\frac{1}{2}\rangle_0$ in order to define the intertwiner operator T_0.

However, if $\Delta \neq 0$ the states (11.92) -(11.95) are not eigenstates of (11.90). The full rotation group is broken down to the rotation around the z-axis. The states $|\frac{3}{2},\frac{1}{2}\rangle$ and $|\frac{1}{2},\frac{1}{2}\rangle_1$ are mixed in the new ground state which is given by:

$$|+\frac{1}{2}\rangle = \frac{1}{\sqrt{1+2x^2}}(2|\frac{1}{2},\frac{1}{2}\rangle_1 + \sqrt{2}x|\frac{3}{2},\frac{1}{2}\rangle) \qquad (11.96)$$

where

$$x = \frac{2(\Delta - 1)}{8 + \Delta + 3\sqrt{\Delta^2 + 8}} \qquad (11.97)$$

and its energy is,

$$E_B = -\frac{J}{4}[\Delta + \sqrt{\Delta^2 + 8}] \qquad (11.98)$$

along with its $|-\frac{1}{2}\rangle$ partner. This are now the two states retained in the RG method. To be more explicit, we have

$$|+\frac{1}{2}\rangle = \frac{1}{\sqrt{6(1+2x^2)}}[(2x+2)|\uparrow\downarrow\uparrow\rangle + (2x-1)|\uparrow\uparrow\downarrow\rangle + (2x-1)|\downarrow\uparrow\uparrow\rangle] \quad (11.99)$$

$$|-\frac{1}{2}\rangle = -\frac{1}{\sqrt{6(1+2x^2)}}[(2x+2)|\downarrow\uparrow\downarrow\rangle + (2x-1)|\downarrow\downarrow\uparrow\rangle + (2x-1)|\uparrow\downarrow\downarrow\rangle] \quad (11.100)$$

The intertwiner operator T_0 reads then,

$$T_0 = |+\frac{1}{2}\rangle\langle'\uparrow| + |-\frac{1}{2}\rangle\langle'\downarrow| \qquad (11.101)$$

where $|\uparrow\rangle'$ and $|\downarrow\rangle'$ form a basis for the space $V' = \mathbb{C}^2$. The RG-equations for the spin operators S_i ($i = 1, 3$) are then given by

$$T_0^\dagger S_i^x T_0 = \xi^x S_i'^x \quad i = 1, 3. \qquad (11.102)$$

$$T_0^\dagger S_i^y T_0 = \xi^y S_i'^y \quad i = 1, 3. \qquad (11.103)$$

$$T_0^\dagger S_i^z T_0 = \xi^z S_i'^z \quad i = 1, 3. \qquad (11.104)$$

where ξ^x, etc are the renormalization factors which depend upon the anisotropy parameter by,

Table 11.2. Fixed Points of the Anisotropic AF-Heisenberg Model

Δ	0	1	∞
R^{BRG}_{∞}	-0.2828	-0.3913	$-\frac{1}{4}\Delta$
e^{exact}_{∞}	-0.3183	-0.4431	$-\frac{1}{4}\Delta$
$\frac{e^B - e^{exact}}{e^{exact}} \times 100$	11%	12%	0

$$\xi^x = \xi^y := \frac{2(1+x)(1-2x)}{3(1+2x^2)} \tag{11.105}$$

$$\xi^z := \frac{2(1+x)^2}{3(1+2x^2)} \tag{11.106}$$

Observe the symmetry between the sites $i = 1$ and 3 which is a consequence of the even parity of the states (11.99) -(11.100).

The renormalized Hamiltonian can be easily obtained using Eqs.(11.102)-(11.106) and (11.88), and apart from and additive constant it has the same form as H, namely,

$$T_0^\dagger H_N(J,\Delta)T_0 = \frac{N}{3}e_B(J,\Delta) + H_{N/3}(J',\Delta') \tag{11.107}$$

where

$$J' = (\xi^x)^2 J \tag{11.108}$$

$$\Delta' = \left(\frac{\xi^z}{\xi^x}\right)^2 \Delta \tag{11.109}$$

Iterating these equations we generate a family of Hamiltonians $H^{(m)}_{N/3^m}(J^{(m)}, \Delta^{(m)})$. The energy density of the ground state of H_N in the limit $N \to \infty$ is then given by,

$$\lim_{N \to \infty} \frac{E_0}{N} = e^{BRG}_{\infty} = \sum_{m=0}^{\infty} \frac{1}{3^{m+1}} e_B(J^{(m)}, \Delta^{(m)}) \tag{11.110}$$

where initially $J^{(0)} = J$, $\Delta^{(0)} = \Delta$ and Eqs.(11.108) -(11.109) provide the flow of the coupling constants.

The analysis of Eq.(11.109) shows that there are 3 fixed points corresponding to the values $\Delta = 0$ (isotropic XX-model), $\Delta = 1$ (isotropic Heisenberg model) and $\Delta = \infty$ (Ising model). The properties of these fixed points are given in Table 11.2 and the RG-flow is depicted in Fig. 11.10,
The computation of e^{BRG}_{∞} in this case is facilitated by the fact that (11.110) becomes a geometric series at the fixed point. The exact results concerning the models $\Delta = 0$ and $\Delta = 1$ are extracted from references [1] and [2]. The case with $\Delta \to \infty$ is exact because the states $|\pm \frac{1}{2}\rangle$ given in (11.99) - (11.100) tend

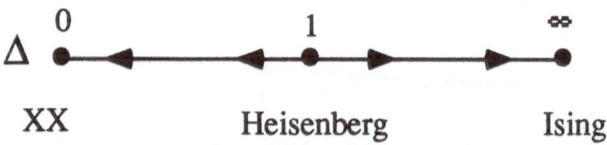

Fig. 11.10. RG-flow diagram for the XXZ model according to the BRG method.

in that limit to the exact ground state $|\uparrow\downarrow\uparrow\rangle$ and $|\downarrow\uparrow\downarrow\rangle$ of the Ising model. As a matter of fact,

$$|+\tfrac{1}{2}\rangle \simeq_{\Delta\to\infty} |\uparrow\downarrow\uparrow\rangle - \tfrac{1}{\Delta}|\uparrow\uparrow\downarrow\rangle - \tfrac{1}{\Delta}|\downarrow\uparrow\uparrow\rangle$$

$$|-\tfrac{1}{2}\rangle \simeq_{\Delta\to\infty} -|\downarrow\uparrow\downarrow\rangle - \tfrac{1}{\Delta}|\downarrow\downarrow\uparrow\rangle - \tfrac{1}{\Delta}|\uparrow\downarrow\downarrow\rangle$$

The region $0 < \Delta < 1$ which flows under the RG-transformation to the XX-model is massless since both $J^{(m)}$ and $\Delta^{(m)}$ go to zero. We showed at the begining of this section that all this region is critical (a line of fixed points) and therefore massless. The RG-equations (11.108) -(11.109) are not able to detect this criticality except at the point $\Delta = 0$. Only the masslessness property is detected.

The region $\Delta > 1$ which flows to the Ising model is massive and this follows from the fact that the product $J^{(m)}\Delta^{(m)}$ goes in the limit $m \to \infty$ to a constant quantity $J^{(\infty)}\Delta^{(\infty)}$ which can be computed from Eqs. (11.108) -(11.109) and (11.101),

$$J^{(\infty)}\Delta^{(\infty)} = \prod_{m=0}^{\infty} \frac{4}{9} \frac{(1+x_m)^4}{(1+2x_m^2)^2} \qquad (11.111)$$

where x_m is given by (11.97) with Δ replaced $\Delta^{(m)}$. This quantity gives essentially the mass gap above the ground state and also the end-to-end or LRO order (Long Range Order) given by the expectation value $|\langle \mathbf{S}(1)\cdot\mathbf{S}(N)\rangle|$ in the limit $N \to \infty$.

In summary, the properties of the Heisenberg-Ising model are qualitatively and quantitatively well described in the massive region $\Delta > 1$ while in the massless region $0 < \Delta < 1$ one predicts the massless spectrum *but no criticality at each value of* Δ. This latter fact is rather subtle and elusive. One would like to construct a RG-formalism such that the Hamiltonian $H_N(\Delta)$ would be a fixed point Hamiltonian for every value of Δ in the range from -1 to 1. In a sense this can be achieved in momentum-space renormalization group using the RG-techniques of Chapter 2 if one considers the Wigner-Jordan transformation of the Heisenberg-Ising model which goes to a model with interacting spinless fermions. Hence we postpone this discussion to next section.

The phase transition between the two regimes is correctly predicted to happen at the value $\Delta = 1$. This is a consequence of the rotational symmetry, namely at $\Delta = 1$ the system is $SU(2)$ invariant and the RG transformation has been defined as to preserve this symmetry. When $\Delta \neq 1$ the $SU(2)$ symmetry is broken and this is reflected later on in the RG-flow of the coupling constant Δ. We may wonder whether the criticality of the region $|\Delta| \leq 1$ is due to some non-trivial symmetry underlying the anisotropic Hamiltonian H. We shall show in what follows that this is indeed the case.

11.6 Quantum Groups and the Block Renormalization Group Method

We present in this section a novel treatment of the Block Renormalization Group method for one dimensional quantum Hamiltonians based on the introduction of a mathematical structure called *quantum group*. Our aim is to address the important questions left open in the previous section and to clarify the peculiar role played by the Renormalization group in the one dimensional physics [3].

Let us consider the following open spin chain Hamiltonian,

$$H_N(q, J) = \frac{J}{4} \left\{ \sum_{j=1}^{N-1} \sigma_j^x \sigma_{j+1}^x + \sigma_j^y \sigma_{j+1}^y + \frac{q + q^{-1}}{2} \sigma_j^z \sigma_{j+1}^z - \frac{q - q^{-1}}{2} (\sigma_1^z - \sigma_N^z) \right\}$$

(11.112)

where q is an arbitrary quantum parameter[1]. This Hamiltonian is known to be integrable [5], [6], [7], [8]. As a matter of fact, it is invariant under the following quantum group generators [9]: $S^+, S^- and S^z$,

$$S^z = \frac{1}{2} \sum_{j=1}^{N} \sigma_j^z$$

(11.113)

$$S^{\pm} = \sum_{j=1}^{N} q^{-\frac{1}{2}(\sigma_1^z + \cdots \sigma_{j-1}^z)} \sigma_j^{\pm} q^{\frac{1}{2}(\sigma_{j+1}^z + \cdots \sigma_N^z)}$$

(11.114)

which satisfies the quantum group algebra:

$$[S^+, S^-] = \frac{q^{2S^z} - q^{-2S^z}}{q - q^{-1}}$$

(11.115)

$$[S^z, S^{\pm}] = \pm S^{\pm}$$

(11.116)

---

[1]The reader not familiar with the mathematical aspects and physical applicantions of Quantum Groups may find convenient to consult a manual on this topics like that of Gómez et al. [4] and references therein.

In the limit $q \to 1$ one recovers from Eqs.(11.113) -(11.116) the usual algebra and addition rules of $su(2)$. For $q \neq 1$ this algebra is called the *quantum universal enveloping algebra* $\mathcal{U}_q(su(2))$, or simply the quantum $su(2)$ group denoted by $SU_q(2)$. The important property is that the generators (11.113)-(11.114) commute with the Hamiltonian (11.112):

$$[H_N(q), S^{\pm}] = [H_N(q), S^z] = 0 \tag{11.117}$$

If we set

$$\Delta = \frac{q + q^{-1}}{2} \tag{11.118}$$

we observe that the bulk terms of Eq.(11.112) and the one in Eq.(11.88) coincide. The only difference appears in the boundary term of Eq.(11.112) which is essential for the existence of the quantum group symmetry.
There are two important cases which we can consider:

$$q : \text{real and positive} \quad \Rightarrow \quad \Delta \geq 1 \tag{11.119}$$

$$|q| = 1 \quad \Rightarrow \quad |\Delta| \leq 1 \tag{11.120}$$

If q is real then $H_N(q)$ is Hermitean while if q is a phase then $H_N(q)$ is not Hermitean but nevertheless its spectrum is real for $H_N(q)$ and $H_N(q^{-1})$ are related by a similarity transformation. This case is the most interesting one. In particular if we write q as a root of unity $q = e^{\frac{i\pi}{\mu+1}}$, then the Hamiltonian (11.112) is a critical Hamiltonian with a Virasoro central algebra given by,

$$q = e^{\frac{i\pi}{\mu+1}} \quad \Rightarrow \quad c = 1 - \frac{6}{\mu(\mu+1)} \tag{11.121}$$

However for the time being let us keep q as an arbitrary parameter characterizing the anisotropy of the model (11.118).

Let us try to apply the Block Renormalization Group Method to the analysis of the Hamiltonian (11.112). To this end we have to write (11.112) in the following form:

$$H_N(q) = \sum_{j=1}^{N-1} h_{j,j+1}(q, J) \tag{11.122}$$

$$h_{j,j+1}(q, J) = \frac{J}{4}[\sum_{j=1}^{N-1} \sigma_j^x \sigma_{j+1}^x + \sigma_j^y \sigma_{j+1}^y + \frac{q+q^{-1}}{2}\sigma_j^z \sigma_{j+1}^z - \frac{q-q^{-1}}{2}(\sigma_j^z - \sigma_{j+1}^z)] \tag{11.123}$$

Observe that each boundary term in $h_{j,j+1}$ cancels one another leaving only those at the end of open chain (i.e. $j = 1$ and N). The nice feature of the site-site Hamiltonians (11.123) is that all of them commute independently with the q-group generators $S^{\pm}$ and S^z;

$$[h_{j,j+1}, S^{\pm}] = [h_{j,j+1}, S^z] = 0 \quad \forall j = 1, \ldots, N-1 \tag{11.124}$$

Using $h_{j,j+1}$ we can construct q-group invariant block Hamiltonians. If the block has for example 3 sites then we will have,

$$H_B = h_{1,2} + h_{2,3} \tag{11.125}$$

In the isotropic case (i.e. $q = 1$) we employ the Clebsch-Gordan decomposition (11.91) in order to find the eigenstates of the block Hamiltonian H_B ($\epsilon = 0$) given in (11.90). For quantum groups, we can also perform q-CG decompositions. The new feature is that now the q-CG coefficients depend on the value of q. For generic values of q the analoge of Eqs.(11.92) -(11.95) are given by:

$$|\frac{3}{2}, \frac{3}{2}\rangle = |\uparrow\uparrow\uparrow\rangle \tag{11.126}$$

$$e_B = J\frac{q+q^{-1}}{4}$$

$$|\frac{3}{2}, \frac{1}{2}\rangle = \frac{1}{\sqrt{[3]_q}}(|\uparrow\downarrow\uparrow\rangle + q|\downarrow\uparrow\uparrow\rangle + q^{-1}|\uparrow\uparrow\downarrow\rangle) \tag{11.127}$$

$$e_B = J\frac{q+q^{-1}}{4}$$

$$|\frac{1}{2}, \frac{1}{2}\rangle_1 = \frac{1}{\sqrt{2(q+q^{-1}-1)}}(q^{1/2}|\uparrow\uparrow\downarrow\rangle + (q^{1/2} - q^{-1/2})|\uparrow\downarrow\uparrow\rangle - q^{-1/2}|\downarrow\uparrow\uparrow\rangle) \tag{11.128}$$

$$e_B = J\frac{2-q-q^{-1}}{4}$$

$$|\frac{1}{2}, \frac{1}{2}\rangle_0 = \frac{1}{\sqrt{2(q+q^{-1}+1)}}((q^{1/2} + q^{-1/2})|\uparrow\downarrow\uparrow\rangle - q^{-1/2}|\downarrow\uparrow\uparrow\rangle - q^{1/2}|\uparrow\uparrow\downarrow\rangle) \tag{11.129}$$

$$e_B = -J\frac{2+q+q^{-1}}{4}$$

$$|\frac{1}{2}, -\frac{1}{2}\rangle_0 = \frac{1}{\sqrt{2(q+q^{-1}+1)}}(-(q^{1/2} + q^{-1/2})|\downarrow\uparrow\downarrow\rangle - q^{1/2}|\downarrow\downarrow\uparrow\rangle - q^{1/2}|\uparrow\downarrow\downarrow\rangle) \tag{11.130}$$

$$e_B = -J\frac{2 + q + q^{-1}}{4}$$

where $[3]_q$ denotes the q-number with the usual definition

$$[n]_q = \frac{q^n - q^{-n}}{q - q^{-1}} \quad n \in \mathbb{N} \tag{11.131}$$

The normalization of the states in Eqs. (11.126) - (11.130) is as if q were always a real number, a bra vector $\langle|$ means transposing a ket vector.

If q goes to 1 the vectors in Eqs. (11.126) - (11.130) go over the vectors in Eqs. (11.92) - (11.95). On the other hand, for $q \neq 1$ there is a certain similarity between the states $|\frac{1}{2}, \pm\frac{1}{2}\rangle_0$ and the states $|\pm\frac{1}{2}\rangle$ given in Eqs. (11.99) - (11.100). However the main difference is that $|\frac{1}{2}, \pm\frac{1}{2}\rangle_0$ are not parity invariant $(1 \leftrightarrow 3), (2 \leftrightarrow 2)$, while states (11.99) - (11.100) are.

If we let q go to zero then Δ goes to $+\infty$. Let us recall that in this limit the states $|\pm\frac{1}{2}\rangle$ (11.99) - (11.100) go to the exact ground state of the Hamiltoninan (11.90). However in the case of the states $|\frac{1}{2}, \pm\frac{1}{2}\rangle_0$ one does not recover the exact eigenstates in this limit. This shows that the states (11.126) - (11.130) are not appropiate for a discussion of the AF region $\Delta > 1$ (or q real). Hence we shall confine ourselves to the critical region $|\Delta| < 1$ where q is a pure phase. As we did already for the isotropic case we shall truncate the basis (11.126) - (11.130) to the states $|\frac{1}{2}, \pm\frac{1}{2}\rangle_0$ and therefore the intertwiner operator T_0 is given by,

$$T_0 = |\frac{1}{2}, \frac{1}{2}\rangle_0\langle'\uparrow| + |\frac{1}{2}, -\frac{1}{2}\rangle_0\langle'\downarrow| \tag{11.132}$$

which satisfies the normalization condition,

$$T_0^t T_0 = \mathbb{1} \tag{11.133}$$

where the superscript t stands for the transpose of the operator (instead of the adjoint).

This means that we can get the effective Hamiltonian H' through the formula:

$$H' = T_0^t H T_0 \tag{11.134}$$

The renormalized spin operators can be computed similarly and one gets, where $|\uparrow\rangle'$ and $|\downarrow\rangle'$ form a basis for the space $V' = \mathbb{C}^2$. The RG-equations for the spin operators $\mathbf{S}_i$ $(i = 1, 3)$ is then given by

$$T_0^t \mathbf{S}_i^x T_0 = \xi \mathbf{S}'^x \quad i = 1, 3. \tag{11.135}$$

$$T_0^t \mathbf{S}_i^y T_0 = \xi \mathbf{S}'^y \quad i = 1, 3. \tag{11.136}$$

$$T_0^t \mathbf{S}_i^z T_0 = \xi \mathbf{S}'^z + \eta_i \mathbb{1}' \quad i = 1, 3. \tag{11.137}$$

where ξ is the renormalization factor which depends upon the anisotropy parameter through the q-parameter in the following fashion,

$$\xi = \frac{q + q^{-1} + 2}{2(q + q^{-1} + 1)} \tag{11.138}$$

and

$$\eta_1 = -\eta_3 := \eta = \frac{q - q^{-1}}{4(q + q^{-1} + 1)} \tag{11.139}$$

Observe that there are quite a few remarkable differences between this "quantum" renormalization prescription Eqs. (11.135) - (11.138) with respect to the ordinary renormalization expressed in Eqs. (11.102) - (11.106). To begin with, the renormalization constant ξ is common to all the spin operators regardless of its spatial component. This is a reflection of the $SU_q(2)$ preservation of the RG method adopted. And last but not least, observe the presence in Eq.(11.137) of an extra term proportional to the identity operator. We may call this term a *quantum group anomaly*.

Using these equations we can compute the renormalization of the block-block Hamiltonian h_{BB} which turns out to be of the same form as the original site-site Hamiltonian (11.123) with *the same value of q*, namely

$$T_0^t h_{3k,3k+1}(q, J) T_0 = \frac{J}{4}\xi^2[\sigma_k'^x \sigma_{k+1}'^x + \sigma_k'^y \sigma_{k+1}'^y + \frac{q + q^{-1}}{2}\sigma_k'^z \sigma_{k+1}'^z \tag{11.140}$$

$$-\frac{q - q^{-1}}{2}(\sigma_k'^z - \sigma_{k+1}'^z) + e_{BB}(q, J)]$$

with

$$e_{BB}(q, J) = J\frac{(q - q^{-1})^2(3q + 3q^{-1} + 4)}{32(q + q^{-1} + 1)^2} \tag{11.141}$$

Combining Eqs.(11.6), (11.141) and (11.126) - (11.130) we finally arrive at quantum group RG-equations,

$$T_0^t H_N(q, J) T_0 = H_{N/3}(q', J') + \frac{N}{3}e_B(q, J) + (\frac{N}{3} - 1)e_{BB}(q, J) \tag{11.142}$$

with

$$e_B(q, J) = -J\frac{2 + q + q^{-1}}{4} \tag{11.143}$$

$$q' = q \tag{11.144}$$

$$J' = \xi^2 J \tag{11.145}$$

Hence we obtain a quite remarkable result we were searching of, namely, that the coupling constant Δ or alternatively q *does not flow under the RG-transformation*, while $J^{(m)}$ goes to zero in the limit when $m \to \infty$, which in

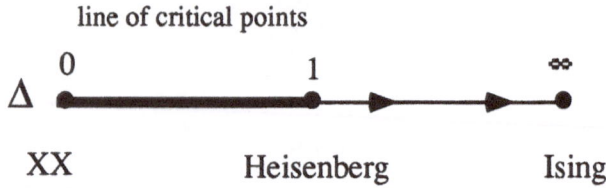

Fig. 11.11. RG-flow diagram for the XXZ model according to the q-RG method. It coincides with the exact flow diagram.

turn implies that theory is massless. In Fig. 11.11 we show the the RG-flow diagram obtained with our q-RG method in which we get the correct line of critical points that the exact model has. Comparing with Fig. 11.10 we see that the q-RG method amounts to a sensible improvement with respect to the standard BRG treatment.

The computation of the ground state energy of H_N for any value of N is very simple since one has only to compute a geometrical series. The result is

$$E_0(N) = N\frac{1 - (\xi^2/3)^M}{3 - \xi^2}(e_B + e_{BB}) - \frac{1 - \xi^{2M}}{1 - \xi^2}e_{BB} \qquad (11.146)$$

where M is the number of RG-steps we have to make in order to resolve the chain of $N = 3^M$ sites.

As a check of the validity of this expression and because of its own interest as well, we shall consider the case $q = e^{i\pi/3}$ in (11.146) which yields,

$$E_0(N, q = e^{i\pi/3}) = -\frac{3}{8}N + \frac{3}{8} \qquad (11.147)$$

This expression coincides with the *exact* result obtained through Bethe ansatz in reference [5]. Since the term proportional to $1/N$ is absent in (11.147) one observes that this value of q corresponds to a central extension $c = 0$ of the Virasoro algebra, in agreement with equation (11.121) ($\mu = 2$).

At first sight it looks surprising that an approximation method such as the Block Renormalization Group yields the exact result at least in the case $q = e^{i\pi/3}$. The peculiarity of this value of q, and in general when q is a root of unity, has been noticed in various contexts [4]: conformal field theory, quantum groups and in fact they are intimately related.

The first thing to be noticed is that at $q = e^{i\pi/3}$ the two denominators of the states $|\frac{3}{2}, \frac{1}{2}\rangle$ and $|\frac{1}{2}, \frac{1}{2}\rangle_1$ *vanishes* reflecting the fact that the "norm" of these states is zero, i.e., they are null states and therefore they must be droped out in a consistent theory. On the other hand, the energy of this states become the same (i.e. $e_B = J/4$). It has been shown in [9] that because of $(S^\pm)^3 = 0$ the 6 states $|\frac{3}{2}, m\rangle$ and $|\frac{1}{2}, m\rangle_1$ do not form two irreps of dimensions 4 and 2 but rather a simple indecomposable but not irreducible representation of $SU_q(2)$.

All these facts motivates that the tensor product decomposition (11.91) for the case $q = e^{i\pi/3}$ should be really be taken as

$$\left(\frac{1}{2} \otimes \frac{1}{2} \otimes \frac{1}{2}\right)_{q=e^{i\pi/3}} = \frac{1}{2} \qquad (11.148)$$

where the irrep $1/2$ on the right hand side denotes the one generated by $|\frac{1}{2}, m\rangle_0$.

This truncation is mathematically consistent and concides precisely with the truncation we have adopted in our Block Renormalization Group approach to the q-group invariant Hamiltonian (11.112).

From a physical point of view, the truncation (11.113)-(11.114) means that the states $|\frac{3}{2}, \frac{1}{2}\rangle$ and $|\frac{1}{2}, \frac{1}{2}\rangle_1$ are not "good excited states" above the "local ground state" given by $|\frac{1}{2}, m\rangle_0$. In other words, above the ground state there are not well behaved excited states. This is why the central extension is $c = 0$ which means that the unique state in the theory is actually the ground state. What the BRG method does is to pick up that piece of the ground state which projects onto a given block! In the case of $q = e^{i\pi/3}$ we have therefore construct for chains with $N = 3^M$ sites the exact ground state of the model through the BRG method. It is worthwhile to point out that this derivation is independent of the Bethe ansatz construction and relies completely on the quantum group symmetry.

11.7 Density Matrix Renormalization Group Methods: Introduction

In the previous section we have considered an significant improvement of the Block Renormalization Group ideas based on the quantum group symmetry. In this section we shall address another line of reasoning to improve the BRG method based upon the work by S.R. White and collaborators starting as of 1992.

Let us recall that Wilson developed his numerical renormalization group procedure to solve the Kondo problem. It was clear from the beginning that one could not hope to achieve the accuracy Wilson obtained for the Kondo problem when dealing with more complicated many-body quantum Hamiltonians as the ones we are addressing in this book. The *key difference* is that in the Kondo model there exists a *recursion relation* for Hamiltonians at each step of the RG-elimination of degrees of freedom. Squematically,

$$H_{N+1} = H_N + \text{hopping boundary term} \qquad (11.149)$$

$$H_{N+1} = R(H_N) \qquad (11.150)$$

The existence of such recursion relation facilitates enormously the work, but as it happens it is specific of *impurity problems*.

From the numerical point of view, the Block Renormalization Group procedure proved to be not fully reliable in the past particularly in comparison with

Table 11.3. BRG results for the Critical Exponents of the 1D Hubbard Model

n_s	s	z	ν	β
3	1.588	0.738	2.151	1.15
5	1.760	0.833	2.114	1.09
7	1.878	0.891	2.108	1.07
∞	2	1	2	1
Exact	∞	1	∞	∞

other numerical approaches, such as the Quantum MonteCarlo method which were being developed at the same time. This was one of the reasons why the BRG methods remained undeveloped during the '80's until the begining of the '90's when they are making a comeback as one of the most powerful numerical tools when dealing with zero temperature properties of many-body systems, a situation where the Quantum MonteCarlo methods happen to be particularly badly behaved as far as fermionic systems is concenrned [10].

It must be apparent by now that the BRG gives a good qualitative picture of many properties exhibited by quantum lattice Hamiltonians: Fixed points, RG-flow, phases of the system etc. as well as good quantitative for some properties such as ground state energy and others. However in some important instances the BRG method is off the correct values of critical exponents by a sensible amount.

As a matter of illustration, we are given in Table 11.3 [11] an account of the critical exponents s, z, ν and β for the case of Spinless Fermions in 1D, along with its exact values for the is exactly solvable. The number of sites n_s kept at each BRG truncation is also shown. The lesson that we can retrieve from this example is twofold. On one hand, only the z exponent is tends to the correct value as n_s is increased. On the other hand, no matter how much n_s is increased the s, ν and β exponents tend to values that are incorrect. (The ∞ value reflects the essential singularity of the fixed point).

From the above discussion one is tempted to think that the BRG method always misses essential singular fixed points but this is not quite the case. For the Hubbard model in 1D, Hirchs [12] was able to detect the essential singularity with a BRG procedure keeping 3 sites per block. It turns out is that the energy gap E_g as a funciton of the coupling constant is approximaltely given by,

$$E_g^{\mathrm{BRG}} \sim e^{-c(t/U)^2}, \quad c = 7.4 \tag{11.151}$$

while the exact result is

$$E_g^{\mathrm{exact}} \sim e^{-2\pi(t/U)} \tag{11.152}$$

Thus the BRG shows up the essential singularity although it misses the correct power.

The first reason that was given in order to explain these failures was that the states kept during the truncation procedure were very few, tipically 2 or 3 depending on the characteristics of the model. However, Bray and Chui [13] undertook the numerical task of keeping a high number of states at each step of the BRG. Their results were deceiving. For the 1D-Hubbard model they kept ~ 1000 states and for chains of $L = 16$ sites they got energies for the lowest few levels which were off by $5 - 10\%$. For $L = 32$ sites, the results were not considered reliable and were not presented.

Therefore, even if more states are retained, the method does not improved sensitivily and it must be due to other more intrinsic causes to the BRG method which were not understood at that time. We shall dwell upon them in the forthcoming sections.

11.8 The Role of Boundary Conditions: the CBC Method

The first advance in trying to understand the sometimes bad numerical performance of the BRG methods came in the understanding of the effect of *boundary conditions* (BC) on the standard RG procedure [14].

White and Noak [14] pointed out that the standard BRG approach of neglecting all connections to the neighbouring blocks during the diagonalization of the block Hamiltonian H_B introduces large errors which cannot be corrected by any reasonable increase in the number of states kept. Moreover, in order to isolate the origin of this problem they study an extremely simple model: a free particle in a 1D lattice. As a matter of fact, it was Wilson [15] who pointed out the importance of understanding real-space RG in the context of this simple tight-binding model where the standard BRG clearly fails as we are going to show.

The reason for this failure can be traced back to the importance of the boundary conditions in diagonalizing the states of a given block Hamiltonian H_B in which the lattice is decomposed into. Notice that in this fashion we are isolating a given block from the rest of the lattice and this applies a *particular BC* to the block. However, the block is not truly isolated! A statement which is the more relevant the more strongly correlated is the system under consideration. Thus, if the rest of the lattice were there it would apply different BC's to the boundaries of the block. This in turn makes the standard block-diagonalization conceptually not faithfully suited to account for the interaction with the rest of the lattice.

Once the origin of the problem is brought about the solution is also apparent: devise a method to change the boundary conditions in the block in order to mimic the interaction with the rest of the lattice. This is called the Combination of Boundary Conditions (CBC) method.

After this conceptual preamble let us see these ideas at work. We shall not dwell upon all the details (many of them quite technical) for this CBC method

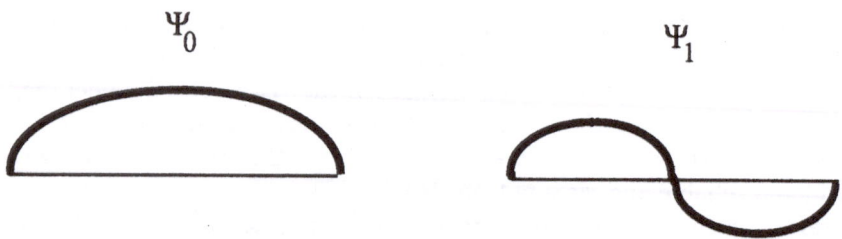

Fig. 11.12. Ground state and first excited state for the 1D free fermion with fixed boundary conditions.

is just one step towards more developed procedures to be discussed in the next section.

To begin with the case of a single-particle problem in a box, let us recall the continuum version of this Hamiltonian with fixed boundary conditions:

$$H = -\frac{\partial^2}{\partial x^2}, \quad \psi(0) = \psi(L) = 0 \tag{11.153}$$

where L is the size of the box. The problem is trivially solvable and the first two free states are depicted in Fig. 11.12,

The lattice version of this Hamiltonian is given by the tight-binding model,

$$H = \begin{pmatrix} 2 & -1 & & & \\ -1 & 2 & -1 & & \\ & -1 & 2 & -1 & \\ & & -1 & 2 & -1 \\ & & & & \ddots \end{pmatrix} \tag{11.154}$$

For this free problem the number of states of the system grows linearly with the system size L instead of the exponential growth 2^L or 4^L tipical of the interacting systems. Therefore, the lattice is built up by using direct sums $\oplus$ instead of tensor products $\otimes$. Thus, the lattice Hamiltonian exhibiting the blocking structures of the BRG method introduced in the previous sections reads as follows:

$$H_{\text{particle}} = \begin{pmatrix} H_B & H_{BB} & & & \\ H_{BB}^t & H_B & H_{BB} & & \\ & H_{BB}^t & H_B & H_{BB} & \\ & & H_{BB}^t & H_B & H_{BB} \\ & & & & \ddots \end{pmatrix} \tag{11.155}$$

This is a block tridiagonal matrix in terms of the diagonal blocks H_B and the off-diagonal inter-block Hamiltonians H_{BB}. Initially they both are 1×1 equal to

Table 11.4. Exact and Standard RG Values of the Low Lying States for the 1D Tight-Binding Model

	Exact	RG
E_0	2.351×10^{-6}	1.9207×10^{-2}
E_1	9.403×10^{-6}	1.9209×10^{-2}
E_2	2.116×10^{-5}	1.9214×10^{-2}
E_3	3.761×10^{-5}	1.9217×10^{-2}

2 and -1 respectively. Then the standard procedure starts combining two blocks resulting in new Hamiltonians:

$$\tilde{H}_B = \begin{pmatrix} H_B & H_{BB} \\ H_{BB}^t & H_B \end{pmatrix} \tag{11.156}$$

$$\tilde{H}_{BB} = \begin{pmatrix} 0 & 0 \\ H_{BB} & 0 \end{pmatrix} \tag{11.157}$$

Next step is to diagonalize $\tilde{H}_B$ to get its eigenvectors V_l, $l = 1, \ldots, n_B$. With these eigenvectors we form the intertwiner operator T_0 which in matricial form looks like:

$$T_0 = \begin{pmatrix} \vdots & \vdots & & \vdots \\ V_1 & V_2 & \cdots & V_{n_B} \\ \vdots & \vdots & & \vdots \end{pmatrix} \tag{11.158}$$

With the help of T_0 we change basis and truncate as usual:

$$H_B' = T_0^t H_B T_0 \tag{11.159}$$

$$H_{BB}' = T_0^t H_{BB} T_0 \tag{11.160}$$

These are the effective Hamiltonians that come out of the first iteration in the BRG method. The rest of the task is devoted to perform as many iterations as one needs.

Despite the simplicity of this free model the performance of the BRG algorithm is bad. To give an idea we give in Table 11.4 a comparison with the exact results for 10 blockings and $n_B = 8$ states. The origin of this big errors is intrinsic to the BRG procedure. In Fig. 11.13 it is shown the shape of the Ψ_0 wave function for the whole chain with fixed boundary conditions at the ends, as well as the corresponding wave functions ψ_0^L and ψ_0^R of the ground states for each block assumming for simplicity that 2 blocks make up the lattice. It is apparent from the figure 11.13 that in using ψ_0^L and ψ_0^R we are introducing spurious anomalous behaviour in the middle of the chain which has nothing to do with the exact wave function Ψ_0.

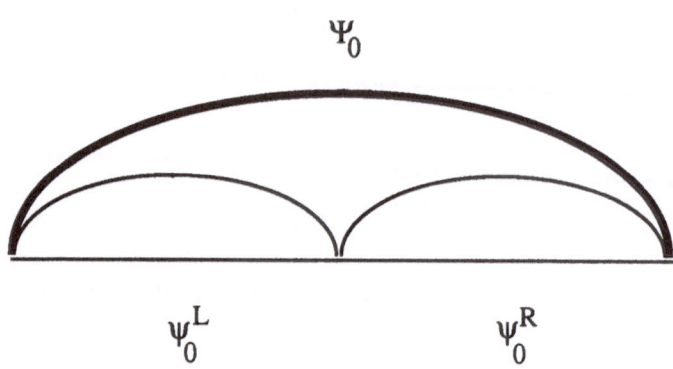

Fig. 11.13. Block decomposition of the ground state Ψ_0 in terms of states ψ_0^L, ψ_0^R with fixed boundary conditions.

What we are doing wrong is to use the same BC's that Ψ_0 fulfills as appropiate BC's for the wave functions ψ_0^L and ψ_0^R in the blocks that make up the lattice. This is a clear way to see that the rest of the lattice "changes" the BC's of the block. As a matter of fact, to remove the spurious "kink" in Fig. 11.13 one would need to keep almost all the states thereby producing a useless algorithm.

A natural way to fix the problem is to incorporate in the BRG method as many different boundary conditions for the block as possible. Otherwise stated, devise an intertwiner operator $T)^{CBC}$ made up of eigenvectors corresponding to different BC's [14]. For the case of a simple free particle there are not too many different BC's. White and Noack combined 4 possible cases: Fixed-Fixed (ff), Fixed-Free (fo), Free-Fixed (of) andFree-Free (oo) BC's. Then, at each iteration they perform 4 diagonalizations of the intrablock Hamiltonian H_B, one for each election of BC's and retain $n_B/4$ states for different diagonalizations. This amounts to n_B states again to construct the intertwiner T_0^{CBC} which in matricial form looks like,

$$
T_0^{CBC} = \begin{pmatrix} \vdots & & \vdots & \vdots & & \vdots & \vdots & & \vdots & \vdots & & \vdots \\ V_1^{ff} & \cdots & V_{n_B/4}^{ff} & V_1^{fo} & \cdots & V_{n_B/4}^{fo} & V_1^{of} & \cdots & V_{n_B/4}^{of} & V_1^{oo} & \cdots & V_{n_B/4}^{oo} \\ \vdots & & \vdots & \vdots & & \vdots & \vdots & & \vdots & \vdots & & \vdots \end{pmatrix}
$$

$$(11.161)$$

Except for this change in the embedding operator, otherwise the procedure is identical to the standard BRG, but the performance improves dramatically. In Table 11.5 it is shown the results for the same situation as in Table 11.4 ($n_B = 8$, 10 blockings).

So far the good news. Bad news is that the CBC method does not work well for interacting systems for which we neeed too many states to represent the response

Table 11.5. Exact, Standard RG and CBC Values of the Low Lying States for the 1D Tight-Binding Model

	Exact	Standard RG	CBC
E_0	2.351×10^{-6}	1.9207×10^{-2}	2.351×10^{-6}
E_1	9.403×10^{-6}	1.9209×10^{-2}	9.403×10^{-6}
E_2	2.116×10^{-5}	1.9214×10^{-2}	2.116×10^{-5}
E_3	3.761×10^{-5}	1.9217×10^{-2}	3.761×10^{-5}

of the block to lots of possible BC's. Moreover, it is not clear whatsoever how to choose to vary the BC's in interacting systems. White tried several methods for Heisenberg chains but none of them worked.

11.9 Density Matrix Renormalization Group Foundations

The Density Matrix RG-method (DMRG) is an improved version of the real-space renormalization group methods introduced by White [16] as a futher e-laboration of the ideas concerning the Combination of Boundary Conditions method.

The fundamental difficulty of the BRG method lies in choosing the eigenstates of the block Hamiltonian H_B to be the states kept. Since H_B contains no connections to the rest of the lattice its eigenstates have inappropiate features at the block ends. The CBC method of the previous section is a first attempt to solving this intrinsic problem. The rationale of this method was that quantum fluctuations in the rest of the system effectively apply a variety of boundary conditions to the block. States from any single boundary condition cannot respond properly to these fluctuations. The CBC method proved to be very effective for the simple single-particle problem studied by White and Noak, but it happens to be ill-suited to interacting systems. The origin of the difficulties is that in order to properly represent the block, the states kept not only need to allow for different end behaviour for different particles but also must represent a complete range of boundary behaviour. Therefore, it is difficult to arrive at a systematic way of finding a complete set of BC's.

The importance of the CBC method relies more on the lessons we can learn from it rather than the specific technicalities pertaining the simple case where it is applied successfully. There are two main ideas in order to proceed further towards a more feasible method, namely:

- The block is not isolated.

- The eigenstates retained are not eiegenstates of a unique block Hamiltonian H_B.

Based upon these points White and Noack suggested a second alternative to the standard BRG method called the Superblock Method [14]. It shares with the CBC method a common feature: the states that are kept are not eigenstates of the block Hamiltonian H_B. The Superblock method is a precursor of the Density Matrix RG method. Briefly explained, in this method one diagonalizes a larger system: the "superblock", composed of 3 or more blocks which includes the original two blocks, say BB which are used to form later the new effective block, say B'. The wave functions for the Superblock are projected onto the two blocks BB, and these projected states of BB are kept to construct the intertwiner operator T_0.

The performance of this method when applied to the simple 1D tight-binding model is also very good [14]. However, for interacting systems the method is again not workable for the diagonalization of the Superblock amounts to a huge computational task which usually goes out the reach of computers power. In addition, not all the information contained in the states of the Superblock is necessary for intuitively not all the states of the Superblock are equally important. Therefore, the problem is to find the most relevant states and the answer is that the Density Matrix tells us which states are the most important.

The DRMG method is in a sense an evolution of the CBC method in which we "let the system" to choose the best boundary conditions. From a pedagogical point of view, it is instructive to introduce the Density Matrix RG by analogy with a situation taken from thermodynamics and statistical mechanics. For an isolated block at finite temperature $T > 0$, the probability that the block is in an eigenstate say α of the block Hamiltonian H_B is proportional to its Boltzmann weight $e^{-\beta E_\alpha}$. As it happens, the Boltzmann weight is an eigenvalue of the following (density) matrix:

$$\rho = e^{-\beta H_B} \tag{11.162}$$

and an eigenstate of the block Hamilotian is also an eigenstate of the density matrix. Since lowest energy corresponds to highest probability in the Boltzmann weight, thus we can view the standard BRG approach as choosing the n_B most probable eigenstates to represent the block given the *assumption* that the block is *isolated*. However, we must remark the following facts:

- The block is *not isolated*, for there exists the rest of the lattice in interaction.

- The density matrix is *not* $e^{-\beta H_B}$.

- Eigenstates of H_B are *not* eigenstates of the block's density matrix.

The lesson that we draw from this comparison is that for a system which is strongly coupled to the outside "universe" (the rest of the lattice), it is much more appropiate to use *eigenstates of the block density matrix* to describe the system (block), see Fig. 11.14, rather than the eigenstates of the system's Hamiltonian. Thus we arrive to a natural generalization of the BRG method:

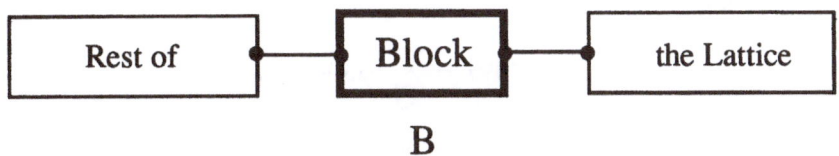

Fig. 11.14. Graphical representation of the block B and the rest of the lattice.

Lattice Decomposition

Fig. 11.15. Lattice decomposition into "system"- and "universe"-parts.

- Choose to keep the n_B most probable eigenstates of the block density matrix.

It is possible to show that keeping the most probable eigenstates of the density matrix gives the most accurate representation of the state of the system as a whole [16].

Let us suppose that we have diagonalize a superblock and thereby obtained one particular state $|\psi\rangle$ which is called the *target state* and probably will be the ground state. Let $\{|i\rangle, i = 1, \ldots, I\}$ be a complete set of states of the the block B which we call "the system". Let also $\{|j\rangle, j = 1, \ldots, J\}$ be a complete set of states of the superblock which we call "the universe" (see Fig. 11.15). Now we proceed to decompose the target state $|\psi\rangle$ into its system- and universe-parts according to the following equation,

$$|\psi\rangle = \sum_{i,j} \psi_{ij} |i\rangle |j\rangle \tag{11.163}$$

Next we want to devise a procedure for producing a set of states of the system denoted by

$$|u^\alpha\rangle, \quad \alpha = 1, \ldots, n_B \quad \text{with} \quad |u^\alpha\rangle = \sum_i u_i^\alpha |i\rangle \tag{11.164}$$

which are optimal for representing the target state $|\psi\rangle$ in a sense to be specified
below. The number of states kept is such that $n_B < l$ so that $|\psi\rangle$ is represented
approximately, that is,

$$|\psi\rangle \approx |\tilde{\psi}\rangle = \sum_{\alpha,j} a_{\alpha,j}|u^\alpha\rangle\, |j\rangle \qquad (11.165)$$

where $a_{\alpha,j}$ are components to be determined by demanding that the following
distance to be a minimun:

$$\mathcal{D} := \||\psi\rangle - |\tilde{\psi}\rangle|^2 \qquad (11.166)$$

This minimization problem requires to vary over all $a_{\alpha,j}$ and u^α subject to the
condition,

$$\langle u^\alpha|u^{\alpha'}\rangle = \delta_{\alpha\alpha'} \qquad (11.167)$$

In order to solve for this minimization problem it is better put it in a matricial
form for then we shall make contact with known problems in Linear Algebra.
To this end let us write without loss of generality,

$$|\tilde{\psi}\rangle = \sum_{alpha} a_\alpha|u^\alpha\rangle\, |v^\alpha\rangle \qquad (11.168)$$

where

$$v_j^\alpha = \langle j|v^\alpha\rangle = N_\alpha a_{\alpha,k} \qquad (11.169)$$

N_α : normalization constants such that $\sum_j |v_m^\alpha|^2 = 1$ $\qquad (11.170)$

Now the minimization problem takes the following matrix form,

$$\mathcal{D} = \sum_{ij}(\psi_{ij} - \sum_{\alpha=1}^{n_B} a_\alpha u_i^\alpha v_j^\alpha)^2 \qquad (11.171)$$

and the variation is taken over all values of u^α, v^α and a^α once the values of
states to be retanined n_B is specified. The solution comes from Linear Algebra
through what is called the *singular value decomposition* (SVD) of a matrix ψ
[16], [17]. This is a theorem in Linear Algebra which states that any $M \times N$
matrix $\mathbf{A}$ whose number of rows M is greater than or equal to its number of
columns N, can be written as the product of an $M \times N$ column-orthogonal
matrix $\mathbf{U}$, an $N \times N$ diagonal matrix $\mathbf{D}$ with positive or zero elements (the
singular values), and the transpose of an $N \times N$ orthogonal matrix $\mathbf{V}$, that is
to say,

$$\mathbf{A} = \mathbf{U} \cdot \mathbf{D} \cdot \mathbf{V}^t \qquad (11.172)$$

The uselfulness of this decomposition is related to the linear problem of finding a solution $\mathbf{x}$ to the following set of simultaneous equations:

$$\mathbf{A} \cdot \mathbf{x} = \mathbf{b} \tag{11.173}$$

where $\mathbf{A}$ is a square matrix for simplicity and $\mathbf{b}$ and $\mathbf{x}$ are vectors. If $\mathbf{A}$ is singular, then there is some subspace of $\mathbf{x}$, called the *null space*, that is mapped to zero, $\mathbf{A} \cdot \mathbf{x} = 0$. When the vector $\mathbf{b}$ on the RHS is not zero, the important question is whether it lies in the range of $\mathbf{A}$ or not. If it does, the the singular set of equations does have a solution $\mathbf{x}$; as a matter of fact, it has more than one solution since any vector in the null space (i.e., any column of $\mathbf{V}$ with a corresponding zero d_j) can be added to $\mathbf{x}$ in any linear combination. If we want to single out one particular member of this solution-set of vectors as a representative we might want to pick one with the smallest length $|\mathbf{x}|^2$. The prescription that the SVD theorem gives to find such a vector is simply *replace $1/d_j$ by zero if $d_j = 0$* and compute the following expresion [17]:

$$\mathbf{x} = \mathbf{V} \cdot [diag(1/d_j)] \cdot (\mathbf{U}^t \cdot \mathbf{b}) \tag{11.174}$$

This will be the solution vector of smallest length; the columns of $\mathbf{V}$ that are in the null space complete the specification of the solution set.

However it may also happen that $\mathbf{b}$ is not in the range of the singular matrix $\mathbf{A}$, then the set of equations (11.173) has no solution. Nevertheless, if $\mathbf{b}$ is not in the range of $\mathbf{A}$, then equation (11.174) can still be used to construct a "solution" vector $\mathbf{x}$. This vector $\mathbf{x}$ will not exactly solve (11.173), but among all possible vectors $\mathbf{x}$ it will still do the closest possible job in the least square sense. In other words, Eq. (11.174) finds the vector $\mathbf{x}$ which minimizes the distance

$$d := |\mathbf{A} \cdot \mathbf{x} - \mathbf{b}| \tag{11.175}$$

Once we have arrived to this point of the SVD decomposition, here comes the connection with our original problem in Eq. (11.171). It should be apparent by now that what we need is the SVD decomposition of the matrix ψ,

$$\psi = \mathbf{U} \cdot \mathbf{D}\mathbf{V}^t \tag{11.176}$$

where $\mathbf{U}, \mathbf{D}$ are $l \times l$ matrices, $\mathbf{U}$ is orthogonal while $\mathbf{D}$ is diagonal. $\mathbf{V}$ is a $l \times J$ matrix which is column-orthogonal. Then the u^α, v^α and a_α which minimizes S in (11.171) are given as follows:

- The n_B largest in magnitude diagonal elements of $\mathbf{D}$ are the a_α.

- The corresponding columns of $\mathbf{U}$ and $\mathbf{V}$ are the u^α and v^α.

The *optimal states* u^α are also eigenvectors of the reduced density matrix of the system as part of the universe.

The reduced for the system depends on the state of the universe which in this case is a pure state $|\psi\rangle$.

Therefore, the density matrix for the system is given by,

$$\rho_{ii'} = \sum_j \psi_{ij}\psi_{i'j} \tag{11.177}$$

Inserting the SVD decomposition of ψ in Eq. (11.177) we may express ρ in matricial form as,

$$\rho = UD^2U^t \tag{11.178}$$

that is to say, U diagonalizes the density matrix. The eigenvalues of ρ are $w_\alpha = a_\alpha^2$ and the optimal states are the eigenstates of ρ with the largest eigenvalues. The w_α represents the probability of the system being in the state u^α, subject to the condition

$$\sum_{\alpha=1}^{n_B} w_\alpha = 1 \tag{11.179}$$

A natural way to measure the accuracy of the truncation to n_B states is to measure the deviation from unity of the quantity,

$$p_{n_B} = \sum_{\alpha=1}^{n_B} w_\alpha \tag{11.180}$$

To summarize, *when the entire lattice is assumed to be in a pure state, the optimal states to be kept are them n_B most significant states of the reduced density matrix of the block, say B.*

11.9.1 Density Matrix Algorithm.

A density matrix algorithm is defined mainly by two features the same as the standard BRG algorithm is, namely, according to the form of the superblock and the manner in which the block are enlarged, e.g., doubling the block B' = B B (Kadanoff) or adding a single site B' = B + site (Wilson in Kondo).

As far as computer power is concerned, generally is more efficient to enlarge the block by adding a *single site* rather than doubling the block. The reason for this is that the diagonalization of a superblock composed of say p identical blocks is difficult for a many-particle interacting system for the dimension of the Hilbert space of states goes like n_B^p, assuming that n_B states are kept per block.

A target state is an eigenstate of the superblock used to form the *block density matrix.* The most efficient algorithms [16] use only a single target state, usually the ground state, to construct the block density matrix.

White has found [16] that a better approach than the standard blocking is to enlarge the lattice according to the superblock configuration depicted in Fig. 11.16 which follows Wilson's strategy. The notation for this configuration is $B_l \bullet \bullet B_{l'}^R$ where l, l' are the number of sites per block (usually equal) and

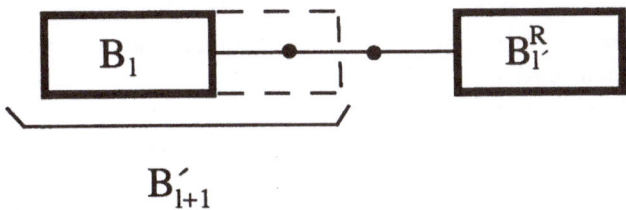

Fig. 11.16. Pattern of lattice growth according to the infinite chain DMRG method.

Table 11.6. DMRG Algorithm for Infinite Chains.

Steps of the DMRG Method
1) Diagonalize a big system (Lanczos or Davidson) to get the ground state ψ
2) Extract ρ for B' from ψ, $(B' = B\bullet)$
3) Diagonalize ρ. Form the intertwiner T_0 from the n_B most probable eigenvectors of ρ
4) Change basis and truncate B'
5) Iteration: Go to step 1).

$B_{l'}^R$ denotes the reflected block constructed out of the states of B_l with right interchanged by left.

This configuration $B_l \bullet \bullet B_{l'}^R$ can be used in two different ways:

- Infinite Chain Method.

- Finite Chain Method.

For the sake of illustration we shall consider here only the infinite chain algorithm (the reader is referred to [16] for more algorithms). It is ussually employed when one is interested in ground state properties of the infinite system. The steps of this DMRG algorithm for interacting systems is 1D is squematically written in Table 11.6.

Some explanations are in order regarding this infinite chain DMRG algorithm. In the first step one starts with a, say, 4 site chain and the superblock is simply $B_1 \bullet \bullet B_1^R$ and then diagonalize the corresponding Hamiltonian for this first superblock. Then choose a target state and calculate the *block density matrix*, and finally form the effective Hamiltonian for $B_2 = B_1\bullet$. This procedure constitutes the first iteration of the algorithm. Next iteration starts with the diagonalization of the new superblock $B_2 \bullet \bullet B_2^R$, and so on and so forth, continuing in this manner diagonalizing the configuration $B_l \bullet \bullet B_l^R$ and setting $B_{l+1} = B_l\bullet$. Notice that at each step both blocks increase in length by one site [2].

---

[2]There are technical reasons for not to use a simpler superblock configuration like $B_l \bullet B_l^R$, see [16].

White has applied this DMRG method to the $S = 1/2$ and $S = 1$ Heisenberg antiferromagnetic spin chains [16]. The results are substantially better than the best available from Monte Carlo calculations, similar in accuracy and the variety of properties measured to what one expects from exact diagonalization, but one can treat lattices hundreds of sites long. For example, the DMRG method accurately recovers the exact energy of the $S = 1/2$ chain to almost six digits. The $S = 1$ case is not solvable by Bethe ansatz and has been the subject of considerable numerical effort since Haldane argued that the infinite system has a finite gap between the ground state and the first excited state (see Chapter 9). Remarkably enough, with the help of the DMRG method it has been able to calculate several of the most interesting properties of the $S = 1$ Heisenberg chains with an accuracy not achieved with other methods. For example, White has computed the Haldane gap Δ_L (L lattice size) finding the following finite-size scaling values:

$$\Delta_L = \Delta + a/L^2 \tag{11.181}$$

with $\Delta = 0.4107(3)$ and $a = 67.9$, while extrapolations based on Monte Carlo results give $\Delta \simeq 0.41$.

Results for the correlation length and other properties can be found in [16].

It can be asserted that the iterative DMRG method devised by White is arguably the leading numerical method for 1D systems. The problem currently is how to make it work efficiently in 2 or more dimensions. Notice that the algorithm depicted Fig. 11.16 is intrinsically one-dimensional. Some attempts at this challenge have recently appeared for two coupled-chains of spinless fermions [18].

11.10 DMRG Study of the ITF Model

The DMRG algorithms studied in the previous section are intrinsically numerical for they are based upon Wilson's procedure of enlarging the system size. On the contrary, the BRG study of the ITF model carried out in this chapter was done in an analytical fashion mainly because it is a blocking procedure. We shall present hereby a new treatment [21] of the ITF model along the lines of the density matrix RG method which is *analytical.*

To this purpose, we have incorporated the block method in the DMRG algorithm. In addition, we shall incorporate another ingredient for the algorithm to become analytical: we shall choose the target state (ground state) using a variational method or a Fokker-Planck method, and then compare the results obtained in these two fashions with the standard BRG results of the previous sections.

The way to combine the above tools is as follows. We first set up a variational ground state for the *whole chain* whose energy is determined by solving the corresponding minimization equations. Next we use this state to construct a block density matrix ρ for a block having two sites in the phylosophy of the

BRG method. This ρ turns out to be (see below) a 4×4 matrix whose two largests eigenvalues denoted by $|A\rangle$ and $|B\rangle$ are kept to construct the intertwiner operator T_0 as in Section 11.2. Once we make contact with the blocking method, the iteration procedure goes over and over. The important point now is that we can keep the DMRG study at an analytical level.

It is worth noticing that we are introducing two new features in this fashion. In the original DMRG method of last section the target state selected comprises just a few sites of the chain while now we are using the whole chain. On the contrary, in doing so we have to resort to variational (or Fokker-Planck later) methods to handle the problem, while in the numerical DMRG the target state selected is exact for the particular size of the superblock chosen.

11.10.1 Variational DMRG.

Let us start searching for the block density matrix using the variational method. Recall that the density matrix in Eq. (11.177) can be rewritten as a scalar product of two block states $|\psi\rangle_B$ and $|\psi'\rangle_{B'}$ given by their defining system-universe decompositon of the previous section:

$$|\psi\rangle_B = \sum_{i \in B} \sum_{j \in B^c} \psi_{ij} |i\rangle |j\rangle \tag{11.182}$$

$$|\psi'\rangle_{B'} = \sum_{i' \in B'} \sum_{j' \in B'^c} \psi_{i'j'} |i'\rangle |j'\rangle \tag{11.183}$$

where i, i' are system-indexes, j, j' are universe-indexes, B, B' are system-blocks and B^c, B'^c are universe-blocks. The density matrix ρ is given by the scalar product of these two block states:

$$\rho =_B \langle \psi | \psi' \rangle_{B'} \tag{11.184}$$

for upon substitution of Eqs. (11.182), (11.183) we arrive at

$$\rho = \sum_{ij} \sum_{i'j'} \langle i | \langle j | \psi_{ij} \psi_{i'j'} | i' \rangle | j' \rangle$$

$$= \sum_{i'i} | i' \rangle \langle i | \rho_{ii'} \tag{11.185}$$

with $\rho_{ii'}$ as in Eq.(11.177).

Now let us consider the following variational state ansatz for the ground state $|\psi_0\rangle_{1,2,3,...,N}$ of the ITF model in a lattice with N sites:

$$|\psi_0\rangle_{1,2,3,...,N} = \exp^{\left(\frac{\alpha}{2} \sum_{j=1}^{N} \sigma_j^z \sigma_{j+1}^z\right)} |0\rangle_{1,2,3,...,N} \tag{11.186}$$

where α is a variational parameter which is determined by minimizing the vacuum expectation value of the ITF Hamiltonian in this state, thereby making the parameter to become a function $\alpha = \alpha(J/2\Gamma)$ of the coupling constant of the model; and $|0\rangle_{1,2,3,...,N} = |0\rangle_1 \otimes ... \otimes |0\rangle_N$ with $|0\rangle$ being the ground state of the σ^x matrix. This exponential-variational ansatz constitutes part of a method called Perturbative-Variational method (PV) developed in [19] for spin systems and in [20] for fermionic systems. The explanation of the origin and uselfulness of this method is beyond the scope of this book and the reader is referred to the bibliography above (and referenced therein) for a comprenhensive account on this issue. It suffices to say here that the PV method is essentially a cluster method which combines perturbative and variational techniques. Using Eq. (11.184) we construct the block density matrix out of this target state for a block containing 2 sites as:

$$\rho^{PV} = \frac{1}{Z_N(\alpha)} \,_{1,2}\langle\psi_0|\psi_0\rangle_{1',2'} \tag{11.187}$$

where $Z_N(\alpha)$ is a normalization factor being the norm of the state which can be interpreted as the partition function of a certain associated statistical model [19]. As a matter of fact, in the large N limit, it turns out to be simply:

$$Z_N(\alpha) = \langle\psi_0|\psi_0\rangle = (\cosh\alpha)^N, \quad N \gg 1 \tag{11.188}$$

Inserting the ansatz (11.186) into (11.187) we can express the density matrix in a more appealing form, namely,

$$\rho^{PV} = \frac{1}{Z_N(\alpha)} \,_{1,2}\langle 0| \exp^{\frac{\alpha}{2}(\sigma_1^z\sigma_2^z+\sigma'^z_1\sigma'^z_2)} \, Z^{(0)}_{N-2}(\alpha,h,h')|0\rangle_{1',2'} \tag{11.189}$$

where we have defined the following quantities,

$$Z^{(0)}_{N-2}(\alpha,h,h') :=_{3,...,N} \langle 0| \exp^{(\alpha\sum_{j=3}^{N-3}\sigma_j^z\sigma_{j+1}^z+h\sigma_3^z+h'\sigma_N^z)} |0\rangle_{3,...,N} \tag{11.190}$$

$$h := \frac{\alpha}{2}(\sigma_2^z + \sigma'^z_2) \tag{11.191}$$

$$h' := \frac{\alpha}{2}(\sigma_1^z + \sigma'^z_1) \tag{11.192}$$

For the time being, it is convenient to shift $N - 2 \rightarrow N$ in order to make expressions easier (at the end we shall come back to the correct value). It is possible to recast $Z^{(0)}_{N-2}(\alpha,h,h')$ into the following form,

$$Z^{(0)}_N(\alpha,h,h') =_{1,...,N} \langle 0| \exp^{(\alpha\sum_{j=1}^{N-1}\sigma_j^z\sigma_{j+1}^z+h\sigma_1^z+h'\sigma_N^z)} |0\rangle_{1,...,N}$$

$$= \frac{1}{2^N} \sum_{\{\sigma_1,...,\sigma_N\}} \exp^{(\alpha\sum_{j=1}^{N-1}\sigma_j\sigma_{j+1}+h\sigma_1+h'\sigma_N)} \tag{11.193}$$

Now we can recognize this equation as the partition function for the Ising model on an open chain of N sites subject to an external magnetic field applied only at the ends of the chain. This partition function can be worked out exactly using standard transfer matrix calculations [21] yielding the result:

$$Z_N^{(0)} = \frac{1}{2^N}([\cosh(h+h') + \cosh(h-h')](2\cosh\alpha)^{N-1}$$

$$+ [\cosh(h+h') - \cosh(h-h')](2\sinh\alpha)^{N-1}) \tag{11.194}$$

In the $N \to \infty$ limit in which we are interested in, it further simplifies to:

$$Z_{N\to\infty}^{(0)} = \frac{1}{2}(\cosh\alpha)^{N-1}[\cosh(h+h') + \cosh(h-h')] \tag{11.195}$$

Inserting now Eq.(11.195) in Eq.(11.189) we arrive at the following expression for the PV block density matrix:

$$\rho^{PV} = \frac{1}{2(\cosh\alpha)^3} \, {}_{1,2}\langle 0| \exp^{\frac{\alpha}{2}(\sigma_1^z\sigma_2 + \sigma_1'^z\sigma_2'^z)}$$

$$\times [\cosh\frac{\alpha}{2}(\sigma_1^z + \sigma_2^z + \sigma_1'^z + \sigma_2'^z) + \cosh\frac{\alpha}{2}(\sigma_1^z - \sigma_2^z + \sigma_1'^z - \sigma_2'^z)]|0\rangle_{1',2'} \tag{11.196}$$

This is a nice result. Observe that the piece $_{1,2}\langle 0| \exp^{\frac{\alpha}{2}(\sigma_1^z\sigma_2)} \exp^{(\sigma_1'^z\sigma_2'^z)} |0\rangle_{1',2'}$ corresponds to a density matrix of a pure state $\rho = |\phi\rangle_{1',2'} \, {}_{1,2}\langle\phi|$ is the projection of the target state ψ_0 onto the block $(1,2)$. The extra terms in Eq.(11.196) are the novel features that the DMRG(PV) method brings about.

To proceed further and give ρ^{PV} a simple matricial form, it is convenient to change basis from eigenstates $|0\rangle$, $|1\rangle$ of σ^x to eigenstates $|+\rangle$, $|-\rangle$ of σ^z. The notation is,

$$|+\rangle = \begin{pmatrix} 1 \\ 0 \end{pmatrix}, \quad |-\rangle = \begin{pmatrix} 0 \\ 1 \end{pmatrix} \tag{11.197}$$

$$|0\rangle = \frac{1}{\sqrt{2}}\begin{pmatrix} 1 \\ 1 \end{pmatrix}, \quad |1\rangle = \frac{1}{\sqrt{2}}\begin{pmatrix} 1 \\ -1 \end{pmatrix} \tag{11.198}$$

In the new basis $\{|+\rangle, |-\rangle\}$ the components of ρ^{PV} are:

$$\rho^{PV}_{\sigma_1'\sigma_2'\sigma_1\sigma_2} = \frac{1}{8(\cosh\alpha)^3} \exp\frac{\alpha}{2}(\sigma_1\sigma_2 + \sigma_1'\sigma_2')$$

$$\times [\cosh\frac{\alpha}{2}(\sigma_1 + \sigma_2 + \sigma_1' + \sigma_2') + \cosh\frac{\alpha}{2}(\sigma_1 - \sigma_2 + \sigma_1' - \sigma_2')] \tag{11.199}$$

Now ρ^{PV} takes the following matricial form in the basis $\{++, --, +-, -+\}$:

$$\rho^{PV} = \frac{1}{4(\cosh\alpha)^3} \begin{pmatrix} e^\alpha(\cosh\alpha)^2 & e^\alpha & \cosh\alpha & \cosh\alpha \\ e^\alpha & e^\alpha(\cosh\alpha)^2 & \cosh\alpha & \cosh\alpha \\ \cosh\alpha & \cosh\alpha & e^{-\alpha}(\cosh\alpha)^2 & e^{-\alpha} \\ \cosh\alpha & \cosh\alpha & e^{-\alpha} & e^{-\alpha}(\cosh\alpha)^2 \end{pmatrix} \tag{11.200}$$

We can readily check that ρ^{PV} is normalized:

$$tr\rho^{PV} = 1 \tag{11.201}$$

Next step in the DMRG algorithm is to diagonalize the density matrix (11.200) and truncate to the largest ones; in this case the truncation is to two states to be denoted by $|A\rangle$ and $|B\rangle$ and will play the parallel role of the states $|G\rangle$ and $|E\rangle$ for the block Hamiltonian H_B of Section 11.4.

We do not need to make a "blind" diagonalization of this 4×4 matrix for we may take advantage of what we have learnt in Section 11.4 about the eigenstates of the H_B in the ITF model. Then, the largest eigenvector say $|A\rangle$ will be in the even sector $\{|00\rangle, |11\rangle\}$, while the next to the largest one, say $|B\rangle$, will be in the odd sector $\{|01\rangle, |10\rangle\}$. According to this analysis, we may write those states as,

$$|A\rangle = x_{00}|00\rangle + x_{11}|11\rangle \tag{11.202}$$

$$|B\rangle = x_{01}|01\rangle + x_{10}|10\rangle \tag{11.203}$$

where $x_{00}, x_{11}, x_{01}, x_{10}$ are the components to be determined. Expressing the states $|00\rangle, |11\rangle \ldots$ in the basis of $\{|+\rangle, |-\rangle\}$, the diagonalization of the density matrix (11.200) yields the following eigenvalues:

$$w_0 = \frac{1}{4(\cosh \alpha)^3}[\cosh \alpha(1 + \cosh^2 \alpha) + \sqrt{\cosh^2 \alpha(1 + \cosh^2 \alpha)^2 - \sinh^4 \alpha}] \tag{11.204}$$

$$w_1 = \frac{1}{4(\cosh \alpha)^3}[\cosh \alpha(1 + \cosh^2 \alpha) - \sqrt{\cosh^2 \alpha(1 + \cosh^2 \alpha)^2 - \sinh^4 \alpha}] \tag{11.205}$$

$$w_3 = \frac{1}{4(\cosh \alpha)^3}e^{\alpha} \sinh^2 \alpha \tag{11.206}$$

$$w_4 = \frac{1}{4(\cosh \alpha)^3}e^{-\alpha} \sinh^2 \alpha \tag{11.207}$$

For α small it is easy to see that the eigenvalues are sorted according to

$$w_0 > w_3 > w_1 > w_4 \tag{11.208}$$

For arbitrary values of α, which in turn amounts to arbitrary values of the coupling constant $g = J/2\Gamma$ due to the variational equations to be given bellow, we have plotted these 4 eigenvalues in Fig. 11.17 observing that there are not level crossings in the whole range of variation of α and that the sorting in Eq.(11.208) holds all over, not just for small α. It is very important for our DMRG(PV) method to work properly that this property holds up, for when we

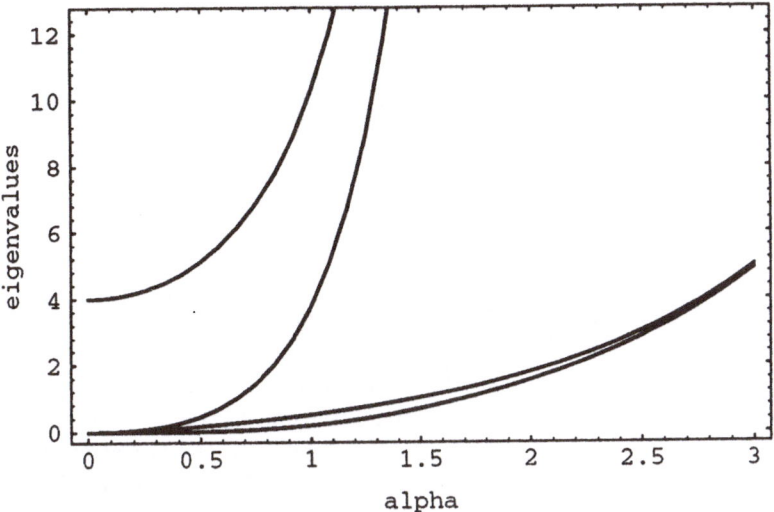

Fig. 11.17. The 4 eigenvalues w_0, w_1, w_3 and w_4 of the block density matrix corresponding to the variational DMRG method.

truncate to the eigenstates of the largest eigenvalues, w_0 and w_3, the physics will not change qualitatively when varying the coupling constant g.

Moreover, it is also posible to show that the eigenvectors $|A\rangle$ (of w_1) and $|B\rangle$ (of w_3) are given by:

$$|A\rangle = \frac{|00\rangle + a^{PV}|11\rangle}{\sqrt{1 + (a^{PV})^2}} \qquad (11.209)$$

$$|B\rangle = \frac{|01\rangle + |10\rangle}{\sqrt{2}} \qquad (11.210)$$

with the function a^{PV} now the following function of the parameter α (after some tedious algebra [21]),

$$a^{PV} = \frac{\sqrt{\cosh^2\alpha(1 + \cosh^2\alpha)^2 - \sinh^4\alpha} - 2\cosh\alpha}{\sinh\alpha(1 + \cosh^2\alpha)} \qquad (11.211)$$

Notice that these states which now come from a DMRG(PV) analysis have the same form as the states $|G\rangle$, $|E\rangle$ of the block Hamiltonian H_B, the difference being in the dependence of the function a upon the coupling constant. This means that we have again the same structure as in the BRG analysis where the intertwiner operator T_0 was fully determined, and consequently the whole RG procedure, by a single function $a = a(g)$ of the coupling constant. To obtain $a^{PV} = a^{PV}(g)$ we need the variational equation relating g with α. This can be found in [19] and is given by,

$$g = \frac{J}{2\Gamma} = \tanh \alpha \qquad (11.212)$$

Observe that when $J/\Gamma \ll 1$ then $\alpha = J/2\Gamma$ and thus $a^{PV} \sim \alpha/2 = J/4\Gamma$. Now it is possible to eliminate the intermediate parameter α between Eqs. (11.211) and (11.212) yielding the desired formula,

$$a^{PV}(g) = \frac{\sqrt{1 - g^2 + \frac{g^6}{4}} - (1 - g^2)}{g(1 - \frac{g^2}{2})} \qquad (11.213)$$

Therefore we may appreciate that this function shares the same qualitatives properties as a_{BRG} does Eqs. ,namely it goes to zero linearly when $g \to 0$ and it is bounded below 1, i.e., $0 < a^{PV} < 1$ for $0 < g < 1$. It addition, it has a singularity at $g = \sqrt{2}$ which prevents the extension of this method beyond that singular point. The origin of this singularity is due to the variational nature of the method (see [19]). Nevertheless the critical region lies within the region of applicability of the PV method.

Furthermore, it is also possible to define a variational DMRG method valid for the whole range of variation of the coupling constant. To do this, we simply recall that for small α the coupling constant depends linearly on the variational parameter,

$$g \sim \alpha, \quad \text{for } \alpha \text{ small} \qquad (11.214)$$

Thus, we may define another function say $a^{PV'}(g)$ by simply substituting α by g in $a(\alpha)$ (11.211),

$$a^{PV'} = \frac{\sqrt{\cosh^2 g(1 + \cosh^2 g)^2 - \sinh^4 g} - 2\cosh g}{\sinh g(1 + \cosh^2 g)} \qquad (11.215)$$

In this fashion, $a^{PV'}(g)$ shares the same properties with a_{BRG}, without any singularity in the range of g. In fact, $a^{PV'}(g)$ has a horizontal asintota at 1 as a_{BRG} does (see Fig 11.18).

11.10.2 Fokker-Planck DMRG.

This is another approximated version for preparing the target state to be projected onto the block-system in order to construct another analytical DMRG method based upon a blocking procedure. The details of how the Fokker-Planck (FP) method is applied to construct an approximate version of the real ground state of the ITF model are beyond the scope of this book and the reader is referred to the original paper [22]. Briefly stated, what the FP method does is to start with an exponential ansatz as in the variational method Eq. (11.186), but instead the parameter α is fixed by demanding that a certain Fokker-Planck

Hamiltonian H^{FP} (to be determined along the way) satisfies the eigenvalue e-
quation to a certain order ν in the perturbative expansion of the parameter α,
that is,

$$H^{FP}(\alpha)\Psi^{FP}(\alpha) = E^{FP}(\alpha)\Psi^{FP}(\alpha) \qquad (11.216)$$

such that the exact ITF Hamiltonian H_{ITF} and its Fokker-Planck approximation
version $H^{FP}(\alpha)$ differ in operators say V_I which involve interactions between
lattice sites a $\nu + 1$ distant apart or larger. Squematically,

$$H_{ITF} - H^{FP}(\alpha) = \sum_{I>\nu} C_I(\alpha)V_I \qquad (11.217)$$

Correspondingly, the FP-energy $E^{FP}(\alpha)$, although incorporating non-perturbative
effects, should agree with the exact ground state energy up to order $\nu + 1$ in α.
In a sense, this gives the best "exact" aproximation to the Hamiltonian H_{ITF}
to order ν in perturbation theory [22].
It is possible to show that these conditions fix the relationship between α and
g. To lowest order this is given by [22],

$$g = \frac{1}{2}\sinh(2\alpha) \qquad (11.218)$$

Again we see that for small coupling constant α and g are equal as in the
variational method.

In order to obtain the Fokker-Planck function, say $a^{FP}(g)$, we first notice
that as in this FP method we start with the same exponential ansatz (11.186) as
in the variational method, the same function $a(\alpha)$ in Eq. (11.211) is valid here.
The new feature is that we have to use the relation (11.218) now to express the
function $a^{FP}(g)$ in terms of the coupling constant. After some tedious algebra
we arrive at the following expression [21]:

$$a^{FP}(g) = \frac{\sqrt{(8 + 26g^2 + 4g^4) + (8 + 6g^2)\sqrt{1 + 4g^2} - 2(1 + \sqrt{1 + 4g^2})}}{g(3 + \sqrt{1 + 4g^2})}$$

$$(11.219)$$

This function contains all the information which upon inserted in the intertwiner
operator T_0 gives rise to what we denote by a DMRG(FP) method. By looking
at Fig. 11.18 we notice again that $a^{FP}(g)$ has the same qualitative properties
as a_{BRG} does.

In this section we have introduced 3 functions $a(g)$ namely, $a^{PV}(g)$, $a^{PV'}(g)$
and $a^{FP}(g)$, related to different analytical realizations of the density matrix
RG ideas. It is our purpose now to check the goodness of those methods by
comparing their predictions for the critical exponents with the exact values
already found by the standard BRG method in Sect. 11.4.

In Fig. 11.18 we have plotted the 4 functions $a(g)$ along with the universal
function $a(g_c)$ introduced in Eq. (11.60) whose cuts with the functions $a(g)$
gives the predictions on the location of the critical point g_c for each method.

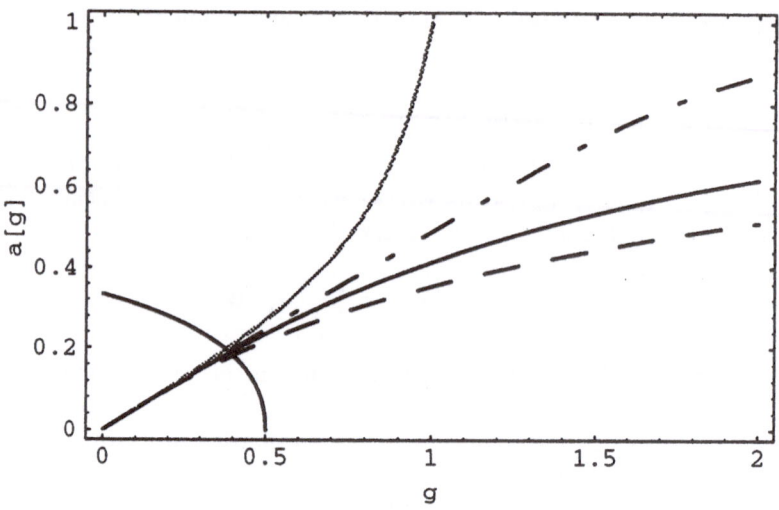

Fig. 11.18. Plot of the functions $a(g)$ according to the methods: BRG (solid line), DMRG(PV) (grey line), DMRG(FP) (dashed line), DMRG(PV').

Recalling that the exact value is $g_c^{exact} = 1/2$, we see that the closest value to this one is produced by the Fokker-Planck version of the DMRG method, even better than the standard BRG. Nevertheless, we may appreciate form Fig. 11.18 that the 4 methods lie rather close to one another within the critical region, the major differences being present off criticality when entering the strong coupling region. The particular values of g_c are gathered in Table .

Another interesting function to be plotted is the beta function $\beta(g)$ obtained for each method according to the analysis of Sect. 11.4, Eq. (11.63). We show these results in Fig. 11.19 where we observe that the 3 new beta functions introduced in this section by means of variational and Fokker-Planck DMRG methods have the same qualitative behaviour as the standard BRG beta function of Sect. 11.4. We know that in particular this means that the unstable character of the fixed point g_c is preserved by these new methods. Moreover, we notice again that in the critical region the differences are small, namely, the cut with the g-axis and the slope at the corresponding g_c. These two latter properties are related to critical exponents.

As far as the critical exponents is concerned, we have collected in Table the exponents computed previously for the standard BRG method: correlation length exponent ν, dynamical exponent z, magnetic exponent β and the gap exponent s. Notice first that the 3 new DMRG methods also satisfy the scaling relation,

$$\beta = \frac{1}{2}z\nu$$

which we know does not hold for the real ITF model.

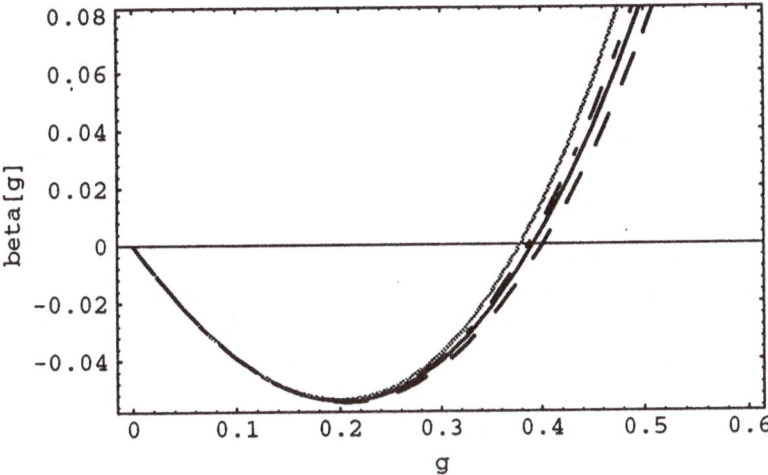

Fig. 11.19. Plot of the beta functions $\beta(g)$ according to the methods: BRG (solid line), DMRG(PV) (grey line), DMRG(FP) (dashed line), DMRG(PV').

Table 11.7. Critical exponents for the ITF model according to different RG methods and exact solution.

Method	g_c	ν	z	β	s
BRG	0.3916	1.4820	0.5515	0.4086	0.8173
DMRG(PV)	0.3790	1.4073	0.5353	0.3767	0.7534
DMRG(FP)	0.4011	1.5177	0.5647	0.4285	0.8570
DMRG(PV')	0.3874	1.4579	0.5459	0.3980	0.7958
Exact Solution	0.5	1	1	0.125	1

From Table 11.7 we arrive at the following conclusions,

- The best correlation length exponent ν is provided by the DMRG Perturbative-Variational method.

- The best dynamical exponent z is provided by the DMRG Fokker-Planck method.

- The best magnetic exponent β is provided by the DMRG Perturbative-Variational method.

- The best gap exponent s is provided by the DMRG Fokker-Planck method.

From these results we may draw the conclusion that the RG methods based on block density matrix, either variational or Fokker-Planck, provide an improvement respect to the standard BRG methods, though it is not a major improvement. One of the reasons why this improvement is not as good as the numerical results obtained by White [16] relies on the fact that we have just kept 2 states in our analytical DMRG method while in the numerical treatment quite a lot states are kept. In addition, there is another line of work in order to improve our analytical DMRG methods via the exact diagonalization of the block density matrix corresponding to 4 lattice sites (starting from the exact ground state for 4 lattice sites) and then to proceed in the usual fashion to construct the corresponding new $a(g)$ function which would allow us to carry the blocking method. We leave this posibility and others open for future work to be published elsewhere.

References

[1]Lieb, E., Schultz, T. and Mattis, D. *1961, Ann. Phys.* **16**, *407*

[2]Orbach, R., *1958, Phys. Rev.* **112**, *309*

[3]Martín-Delgado, M.A., Sierra, G., *1995, UCM-CSIC preprint, to be published.*

[4]Gómez, C., Ruiz-Altaba, M. and Sierra, G., "Quantum Groups in Two-Dimensional Physics". Cambridge University Press (to be published).

[5]Alcaraz, F.C., Barber, M.N., Batchelor, M.T., Baxter, R.J. and Quispel, G.R.W. *1987, J. Phys. A***20**, *6397*

[6]Gaudin, M., *1971, Phys. Rev. A* **4**, *386*

[7]Sklyanin, E.K., *1988, J. Phys. A***21**, *2375*

[8]Cherednik, I.V., *1984, Theor. Math. Phys.* **61**, *977*

[9]Pasquier, V. and Saleur, H. *1990, Nucl. Phys. B***330**, *523*

[10]Hirsch, J. *1983, Phys. Rev. B* **28**, *4059, 1985, Phys. Rev. B* **31**, *4403*; Hirsch, J. and Lin, H.Q., *1988, Phys. Rev. B* **37**, *5070.*

[11]Pfeuty, P.Jullien, R. and Penson, K.A., in "Real-Space Renormalization", editors Burkhardt, T.W. and van Leeuwen, J.M.J., series topics in Current Physics **30**, Springer-Verlag 1982.

[12]Hirsch, J. *1980, Phys. Rev. B* **22**, *5259.*

[13]Bray, J.W. and Chui, S.T., *1979, Phys. Rev. B* **19**, *4876*

[14]S.R. White, R.M. Noack, *Phys. Rev. Lett.* **68**, 3487 (1992).

[15]Wilson, K.G., *1986, unpublished informal talk.*

[16]S.R. White, *Phys. Rev. Lett.* **69**, 2863 (1992); *Phys. Rev. B* **48**, 10345 (1993).

[17]Press, W.H., Flannery, B.P., Teukolsky, S.A. and Vetterling, W.T., "Numerical Recipes: The Art of Scientific Computing". Cambridge, New York, NY, 1986.

[18]R.M. Noack, Scalapino, D.J. and S.R. White, *1994, preprint to be published.*

[19]J. Esteve and G. Sierra, *1995, Phys. Rev. B* **51**, *8928.*

[20]M.A. Martín-Delgado and G. Sierra, *1995, UCM-CSIC preprint, to be published*

[21]M.A. Martín-Delgado and G. Sierra, unpublished work.

[22]F. Jimenez and G. Sierra, *1995, CSIC preprint, to be published*

Lecture Notes in Physics

For information about Vols. 1–425
please contact your bookseller or Springer-Verlag

New Series m: Monographs